Scientific Advances in STEM: Synergies to Achieve Success, 3rd Volume

Scientific Advances in STEM: Synergies to Achieve Success, 3rd Volume

Topic Editors

Yadir Torres Hernández
Ana María Beltrán Custodio
Manuel Félix Ángel

MDPI

Basel • Beijing • Wuhan • Barcelona • Belgrade • Novi Sad • Cluj • Manchester

Topic Editors
Yadir Torres Hernández
Materials and Transport Science and Engineering
Universidad de Sevilla
Sevilla
Spain

Ana María Beltrán Custodio
Materials and Transport Science and Engineering
Universidad de Sevilla
Sevilla
Spain

Manuel Félix Ángel
Chemical Engineering
Universidad de Sevilla
Sevilla
Spain

Editorial Office
MDPI AG
Grosspeteranlage 5
4052 Basel, Switzerland

This is a reprint of the Topic, published open access by the journals *Foods* (ISSN 2304-8158), *Materials* (ISSN 1996-1944), *Polymers* (ISSN 2073-4360), *Sensors* (ISSN 1424-8220), and *Applied Sciences* (ISSN 2076-3417), freely accessible at: https://www.mdpi.com/topics/ZF9OF53756.

For citation purposes, cite each article independently as indicated on the article page online and as indicated below:

Lastname, A.A.; Lastname, B.B. Article Title. *Journal Name* **Year**, *Volume Number*, Page Range.

ISBN 978-3-7258-3839-4 (Hbk)
ISBN 978-3-7258-3840-0 (PDF)
https://doi.org/10.3390/books978-3-7258-3840-0

Cover image courtesy of Sergio Muñoz

Contents

applied sciences

MDPI

Editorial

Scientific Advances in STEM: Synergies to Achieve Success (Volume 3)

Yadir Torres [1,*], Ana M. Beltrán [1] and Manuel Felix [2]

1 Departamento de Ingeniería y Ciencia de los Materiales y del Transporte, Escuela Politécnica Superior, Universidad de Sevilla (US), 41011 Seville, Spain; abeltran3@us.es
2 Departamento de Ingeniería Química, Escuela Politécnica Superior, Universidad de Sevilla (US), 41011 Seville, Spain; mfelix@us.es
* Correspondence: ytorres@us.es

1. Introduction

"Scientific Advances in STEM: Synergies to Achieve Success, Volume 3" emphasizes the importance of interdisciplinary collaboration in Science, Technology, Engineering, and Mathematics (STEM) for driving innovation and societal benefits. It explores advanced materials, technologies, and their transformative impact on industries, medicine, agriculture, and sustainability. This volume addresses challenges like climate change and resource management with innovative solutions, emphasizing the integration of diverse knowledge to tackle global issues. By addressing collaborative efforts in computing, artificial intelligence, and environmental science, it highlights the interconnections of STEM disciplines, inspiring readers to value research for novel discoveries and practical applications.

2. Overview of the Published Articles

This collection demonstrates critical advancements in various STEM fields, including a machine learning framework to monitor machining processes, enhancing efficiency and reducing industrial costs [1]. In relation to research about heat transfer improvements through nanofluids, subsequently showcasing significant advancements in cooling systems and energy technologies [2]. Innovations in material science are also highlighted, analyzing the mechanical properties of advanced structural materials for aerospace and civil engineering applications [3]. Biocompatibility and biomedical innovations are also addressed [4], emphasizing 3D-printed nanocomposites for medical implants, tissue engineering and formability in medical-grade polymers [5], and polymer curing in dental applications [6]. Environmental sustainability is explored [7], focusing on biodegradable polymers derived from invasive seaweed species; the evaluation of food preservation techniques using cassava-based edible films is also analyzed [8]. Advances in structural engineering are addressed [9], analyzing the mechanical performance of injection-molded carbon fiber composites.

Industrial applications are also covered [10,11], where chloride permeability in alkali-activated slag concretes under high temperatures and the recovery of antimony from industrial waste are investigated. Hardware and sensor implementation are also addressed in this collection [12,13]. Thus, hyperspectral imaging for agricultural monitoring, Field Programable Gate Array (FPGA)-based cryptohardware protection, advancing cybersecurity, and advances in wearable sensor technologies for health monitoring are reported.

These works illustrate the integration of research across disciplines to address global challenges and drive innovation.

Received: 13 February 2025
Accepted: 16 February 2025
Published: 11 March 2025

Citation: Torres, Y.; Beltrán, A.M.; Felix, M. Scientific Advances in STEM: Synergies to Achieve Success (Volume 3). *Appl. Sci.* **2025**, *15*, 3030. https://doi.org/10.3390/app15063030

Author Contributions: The authors equally contributed to the paper. All authors have read and agreed to the published version of the manuscript.

Conflicts of Interest: The authors declare no conflicts of interest.

References

1. Gavrilović, T.; Despotović, V.; Zot, M.-I.; Trumić, M.S. Prediction of Flotation Deinking Performance: A Comparative Analysis of Machine Learning Techniques. *Appl. Sci.* **2024**, *14*, 8990. [CrossRef]
2. Zou, F.; Yao, S. Multi-Objective Optimization of Crashworthiness of Shrink Tube Energy Absorption Structure. *Appl. Sci.* **2024**, *14*, 7347. [CrossRef]
3. Yu, H.; Luan, X.; Zheng, G.; Hao, G.; Liu, Y.; Xing, H.; Liu, Y.; Fu, X.; Liu, Z. Analysis and Experiment of Thermal Field Distribution and Thermal Deformation of Nut Rotary Ball Screw Transmission Mechanism. *Appl. Sci.* **2024**, *14*, 5790. [CrossRef]
4. Hortigon, B.; Rodriguez-Mayorga, E.; Santiago-Espinal, J.A.; Ancio, F.; Gallardo, J.M. Influence of Materials of Moulds and Geometry of Specimens on Mechanical Properties of Grouts Based on Ultrafine Hydraulic Binder. *Materials* **2024**, *17*, 1645. [CrossRef] [PubMed]
5. Rosa-Sainz, A.; Silva, M.B.; Beltrán, A.M.; Centeno, G.; Vallellano, C. Assessing Formability and Failure of UHMWPE Sheets through SPIF: A Case Study in Medical Applications. *Polymers* **2023**, *15*, 3560. [CrossRef] [PubMed]
6. Díaz-Rodríguez, S.D.; Tarano-Artigas, O.; Herrera-Kao, W.; Cauich-Rodríguez, J.V.; Cervantes-Uc, J.M.; Rosa-Sainz, A.; La Serna, A.A.; Veranes-Pantoja, Y. Effect of the Silanization of Aerosil OX50 in the Properties of Light-Cured Dental Composites. *Appl. Sci.* **2024**, *14*, 2453. [CrossRef]
7. Santana, I.; Felix, M.; Bengoechea, C. Feasibility of Invasive Brown Seaweed Rugulopteryx okamurae as Source of Alginate: Characterization of Products and Evaluation of Derived Gels. *Polymers* **2024**, *16*, 702. [CrossRef] [PubMed]
8. Otálora González, C.M.; Felix, M.; Bengoechea, C.; Flores, S.; Gerschenson, L.N. Development and Characterization of Edible Films Based on Cassava Starch Modified by Corona Treatment. *Foods* **2024**, *13*, 468. [CrossRef] [PubMed]
9. Wang, X.; Bao, Z.; Gao, D.; Huang, S.; Huang, L.; Chen, Q.; Zhao, H.; Han, W.; Xu, Y. Mechanical Properties of Weld Lines in Injection-Molded Carbon Fiber-Reinforced Nylon (PA-CF) Composites. *Polymers* **2023**, *15*, 2476. [CrossRef] [PubMed]
10. Zhou, B.; Ma, Q.; Guo, R.; Li, P. Chloride Permeability of Alkali-Activated Slag Concretes after Exposure to High Temperatures. *Materials* **2024**, *17*, 1028. [CrossRef] [PubMed]
11. Díaz Gutiérrez, E.; Maldonado Calvo, J.A.; Gallardo Fuentes, J.M.; Paúl Escolano, A. Effect of pH Hydrolysis on the Recovery of Antimony from Spent Electrolytes from Copper Production. *Materials* **2023**, *16*, 3918. [CrossRef] [PubMed]
12. Domínguez-Cid, S.; Larios, D.F.; Barbancho, J.; Molina, F.J.; Guerra, J.A.; León, C. Identification of Olives Using In-Field Hyperspectral Imaging with Lightweight Models. *Sensors* **2024**, *24*, 1370. [CrossRef] [PubMed]
13. Potestad-Ordóñez, F.E.; Casado-Galán, A.; Tena-Sánchez, E. Protecting FPGA-Based Cryptohardware Implementations from Fault Attacks Using ADCs. *Sensors* **2024**, *24*, 1598. [CrossRef] [PubMed]

applied sciences

MDPI

Article

Prediction of Flotation Deinking Performance: A Comparative Analysis of Machine Learning Techniques

Tamara Gavrilović [1], Vladimir Despotović [2], Madalina-Ileana Zot [3] and Maja S. Trumić [1,*]

1 Technical Faculty in Bor, University of Belgrade, 19210 Bor, Serbia
2 Bioinformatics and AI Unit, Department of Medical Informatics, Luxembourg Institute of Health, 1445 Strassen, Luxembourg; vladimir.despotovic@lih.lu
3 Faculty of Mechanical Engineering, Politehnica University Timisoara, 300222 Timisoara, Romania; madalina.zot@upt.ro
* Correspondence: majatrumic@tfbor.bg.ac.rs

Abstract: Flotation deinking is one of the most widely used techniques for the separation of ink particles from cellulose fibers during the process of paper recycling. It is a complex process influenced by a variety of factors, and is difficult to represent and usually results in models that are inconvenient to implement and/or interpret. In this paper, a comprehensive study of several machine learning methods for the prediction of flotation deinking performance is carried out, including support vector regression, regression tree ensembles (random forests and boosting) and Gaussian process regression. The prediction relies on the development of a limited dataset that assumes representative data samples obtained under a variety of laboratory conditions, including different reagents, pH values and flotation residence times. The results obtained in this paper confirm that the machine learning methods enable the accurate prediction of flotation deinking performance even when the dataset used for training the model is limited, thus enabling the determination of optimal conditions for the paper recycling process, with only minimal costs and effort. Considering the low complexity of the Gaussian process regression compared to the aforementioned ensemble models, it should be emphasized that the Gaussian process regression gave the best performance in estimating fiber recovery (R^2 = 97.77%) and a reasonable performance in estimating the toner recovery (R^2 = 86.31%).

Keywords: deinking; flotation; paper recycling; machine learning; support vector regression

Citation: Gavrilović, T.; Despotović, V.; Zot, M.-I.; Trumić, M.S. Prediction of Flotation Deinking Performance: A Comparative Analysis of Machine Learning Techniques. *Appl. Sci.* **2024**, *14*, 8990. https://doi.org/10.3390/app14198990

Academic Editors: Yadir Torres Hernández, Ana María Beltrán Custodio and Manuel Félix Ángel

Received: 2 August 2024
Revised: 15 September 2024
Accepted: 25 September 2024
Published: 6 October 2024

1. Introduction

The flotation process has been used in mineral processing plants to separate valuable minerals from ore. In this process, three phases are combined in the flotation pulp: solid (mineral particles), liquid (water) and gaseous (air). Mineral particles are separated from the pulp based on the difference in their surface hydrophobicity. The ones that are easily wetted by water are called hydrophilic, while particles with a limited affinity for wetting are called hydrophobic [1,2]. Flotation is a key process in many paper recycling plants as well and was introduced successfully to the paper recycling industry in the 1980s. Generally, the deinking process is based on the separation between hydrophobic inks and hydrophilic paper fibers. In modern paper recycling plants, the process of removing unwanted particles from the pulp can have as many as three times more steps than in the mineral processing industry due to the fact that a higher quality product is needed to compete with virgin paper [3]. Given that the heterogeneity of the feed would affect the different steps of the recycling process, and those steps can change the quality of the paper, it is necessary to control the operating parameters in all steps.

A variable that can affect the final quality of the product but which will not affect the reduction of plant capacity or require significant operating costs is called a practical control variable. Variables that have a significant impact on the process but at the same time cause disturbances in the process and must also be optimized are considered non-practical control variables [4]. There are a number of possible control variables for deinking plants and many researchers have investigated their influence on the deinking process [3,5–10]. In addition to chemical additives, such as sodium hydroxide, sodium silicate, hydrogen peroxide and surfactant, the pulp temperature in the phase of disintegration, i.e., pulp formation and in the flotation phase, pulp formation time and flotation time, flotation consistency and pH value in the flotation are considered as process variables. Process variables that are often not adjustable in practice such as pulp consistency, flotation air flow rate and foam height are not considered as good modeling variables. The deinking process in industrial practice is continuous, with a constant flow of input, acceptance and rejection, and the quality of the recycled paper is the main control variable in the deinking process [3].

Predictive modeling is based on analyzing relationships between input variables to make predictions about continuous output variables. In supervised machine learning, these relationships are learned from the data, during a training process. The trained model can then be applied to previously unseen input data not used during the training process, thereby allowing the inference of implicit properties about the modeled process from the data. Recent research in the literature has focused on machine learning applications that have been developed to estimate and adjust parameters in flotation processes. Only a limited number of studies report the application of machine learning to ink removal processes. Artificial Neural Networks (ANNs) are used to model and predict flotation behavior in industrial paper recycling process [4], while Labidi et al., 2007 [11] propose a model to predict the deinking efficiency based on an ANN. Verikas et al., 2024, developed a method for monitoring ink removal based on neural network color image analysis [12]. ANNs were used by many authors to model and simulate the quality characteristics of pulp and paper [13] and the macroscopic mechanical properties of minerals [14]. Multivariate Nonlinear Regression (MNLR), Radial Basis Function Neural Networks (RBFNNs) and Recurrent Neural Networks (RNNs) were used to predict flotation performance [15]. Chehreh Chelgania et al., 2018 [16] used SVR to model coal flotation.

Szmigiel et al., 2024 [17], reviewed the research over the last ten years. They presented the work of many authors who approached different challenges in this mechanism using different models that have been developed and adapted for a specific flotation problem. They identified different categories of models, such as "Predictive Models for Evaluation and Recovery", "Models Developed to Evaluate the Importance of Flotation Parameters" and "Analysis of Flotation Foam Images with Machine Learning". The last category was divided into "Image Extraction of Foam", "Size Bubbles and Distribution Analysis", "Flotation Performance Predictions and Feature Importance Analysis Based on Ash Images" and "Ash Image Analysis and Predictions for Ash Content in the Coal Flotation Process". Generally, researchers have presented potential solutions for the mineral beneficiation process using machine learning and artificial intelligence techniques, but limited efforts were made for the prediction of grade and recovery in the flotation of other materials.

In this study, an attempt has been made to estimate Fe grade and cellulose recovery in the froth flotation products in laboratory conditions. This study performs a comparative analysis of different machine learning methods for the prediction and modeling of deinking flotation performance. To the best of our knowledge, there are no previously reported studies that use Gaussian Process Regression (GPR) and Regression Tree Ensembles for this purpose.

The remainder of this paper is organized as follows. In Section 2, a brief overview of machine learning techniques used in this paper is given. The dataset developed for model training and testing is also presented in Section 2, while experimental results are discussed in Section 3. Concluding remarks are given in Section 4.

2. Materials and Methods

2.1. Machine Learning Techniques for Prediction of Flotation Deinking Performance

2.1.1. Support Vector Regression

Support Vector Machines (SVMs) are a supervised machine learning method originally developed for solving binary classification problems. While the output variable is discrete in classification, it is continuous in regression (real number). Therefore, it is not possible to give an exact prediction as in classification, and an error (deviation) ε is introduced [18]. SVM for regression problems is usually denoted as Support Vector Regression (SVR).

Suppose a training dataset $D = \{(x_i, y_i), i = 1, \ldots, N\}$ which consists of N training pairs, where $x_i \in R^n$ is the n-dimensional vector denoting the model's inputs and $y_i \in R$ are the observed responses to these inputs (model's outputs). The goal of SVR is to determine a function $f(x)$ that deviates from y_i by a value not greater than ε for each training data point x_i.

$$f(x) = \langle \omega, x \rangle + b; \omega \in R^n, b \in R \tag{1}$$

where ω is the weight, b is the bias and $\langle \omega, x \rangle$ denotes the dot product. Values ω and b are determined from the training data by maximizing the margin $2/\|\omega\|$, or equivalently by minimizing $\frac{1}{2}\|\omega\|^2$, where the factor $\frac{1}{2}$ is used for mathematical convenience only [19]:

$$argmin \frac{1}{2}\|\omega\|^2 + C\sum_{i=1}^{N}(\xi_i + \xi_i^*); \text{ subjectto} \begin{cases} y_i - f(x_i) \le \varepsilon + \xi_i \\ f(x_i) - y_i \le \varepsilon + \xi_i^* \\ \xi_i, \xi_i^* \ge 0 \end{cases} \tag{2}$$

where constant $C > 0$ defines the amount of error larger than ε that is tolerated and ξ_i, ξ_i^*, are the error tolerances. The solution of (2) is found using Lagrange multipliers with the dual set of variables. To obtain the dual formula, a Lagrange function is constructed from the primal function by introducing non-negative Lagrange multipliers $\alpha_i, \alpha_i^*, \eta_i, \eta_i^*$ for each training data point x_i.

$$\omega = \sum_{i=1}^{N}(\alpha_i - \alpha_i^*) \cdot x_i$$

$$b = y_i - \langle \omega, x_i \rangle - \varepsilon, 0 < \alpha_i < C \tag{3}$$

$$b = y_i - \langle \omega, x_i \rangle + \varepsilon, 0 < \alpha_i^* < C$$

The function used to predict new values then becomes:

$$f(x) = \sum_{i=1}^{N}(\alpha_i - \alpha_i^*) \cdot \langle x_i, x \rangle + b \tag{4}$$

where the bias b is defined in (5).

When the linear model is not adequate, the Lagrange dual formulation can be extended to nonlinear functions by replacing the dot product $\langle x_i, x \rangle$ with a nonlinear kernel function $\langle x_i, x \rangle = \varphi(x_i) \cdot \varphi(x)$, where each data point x_i is mapped to a higher-dimensional space using the transform $\Phi : x_i \to \varphi(x_i)$. The solution to the optimization problem for the nonlinear case becomes:

$$\omega = \sum_{i=1}^{N}(\alpha_i - \alpha_i^*) \cdot \varphi(x_i)$$

$$f(x) = \sum_{i=1}^{N}(\alpha_i - \alpha_i^*) \cdot k(x_i, x) + b \tag{5}$$

Common kernel functions are the linear, polynomial, sigmoid and radial basis function (RBF). The RBF kernel is used in this paper [20]:

$$k\langle x_i, x \rangle = exp\left(-\gamma\|x_i - x\|^2\right) \tag{6}$$

where $\gamma > 0$ is the regularization parameter which determines the trade-off between the fitting error minimization and the smoothness of the estimated function.

2.1.2. Regression Tree Ensembles

Linear regression represents a global model, where a single formula describes the relations between the inputs and the outputs of the model over the entire data space. When there are many features interacting in nonlinear ways, it is very hard to design a single global model. An alternative approach is to divide the data space into smaller partitions, where the modelling of these interactions is easier to achieve. These partitions can be further divided into even smaller regions, until finally one obtains the data space cells where simple models can be applied. This is called recursive partitioning [21].

Regression trees use the tree to represent the recursive partition. It splits the input data space in partitions and assigns a prediction value to each partition. The terminal nodes of the tree, denoted as leaves, represent these partition cells. In order to determine to which leaf the input data belong to and to assign it the prediction value, the algorithm starts from the root node and asks successive binary questions. Depending on the outcome of the question, the sub-branch of the tree is chosen. Eventually, the algorithm arrives at the leaf node, where the prediction is made. This prediction is found as an average of all training data instances which reach that leaf node [21].

Suppose a training dataset $D = \{(x_i, y_i), i = 1, \ldots, N\}$ which consists of N training pairs, where $x_i \in R^n$ is the n-dimensional vector denoting model's inputs and $y_i \in R$ are the observed responses to these inputs (model's outputs). Suppose further a division of the input data space into M partitions R_i; $i = 1, 2, \ldots, M$, where the response is modelled as a constant c_i in each partition:

$$f(x) = \sum_{i=1}^{M} c_i I\{x \in R_i\} \tag{7}$$

where $I\{x \in R_i\}$ is a binary function that takes the value 0 or 1 depending on the outcome of the question at the tree's split point [21,22]. The constant c_i can be determined as the average of the responses y_i in the region R_i.

The greedy algorithm is used in order to determine the split point [21,23], which is very efficient, but might lead to poor decisions, especially in the lower tree branches, due to unreliable estimates based on the small number of samples. To overcome this issue, more regression trees can be combined in an ensemble, which represents a predictive model composed of a weighted combination of multiple regression trees. Different algorithms exist for ensemble learning, such as bagging, random forests and boosting [21].

Boosting, which is an ensemble technique where the predictors are created sequentially, is used in this paper. The rationale behind this is that each subsequent predictor learns from the mistakes committed by the previous predictors [21,22]. When gradient boosting is applied to regression tree ensembles, the first regression tree is the one that maximally reduces the loss function for the selected tree structure and the given training dataset. The residual (prediction error) is then calculated, which represents the mistake committed by the predictor model (the first regression tree). At the next step, a new tree is fitted to the residuals of the first tree. At each step, a new tree is added to the model, which is fitted to the residuals of the previous one. The residual values are usually multiplied by the learning rate (value less than 1) to avoid overfitting. The final model obtained by boosting is simply a linear combination of all trees (usually hundreds or thousands of trees).

The main idea of boosting is that, instead of using a complex single regression tree, which is easily over fitted, a much better fit is produced if many simple regression trees are trained iteratively, each of them improving the prediction performance of the previous ones [22]. Boosting algorithms play a crucial role in dealing with bias variance trade-off. Unlike bagging algorithms, which only control for high variance in a model, boosting controls both the aspects (bias and variance) and is considered to be more effective.

2.1.3. Gaussian Process Regression

Suppose a training dataset $D = \{(x_i, y_i), i = 1, \ldots, N\}$ which consists of N training pairs, where $x_i \in R^n$ is the n-dimensional vector denoting model's inputs and $y_i \in R$ are the observed responses to these inputs (model's outputs). Aggregating the column vectors of inputs in matrix X and the responses in the vector y, a training dataset becomes $D = (X, y)$. The goal of GPR is to predict the value of response, given the new (unseen) input vector and the training data, i.e., to determine the conditional distribution of the responses given the inputs [24].

Consider a standard linear regression model with Gaussian noise [24]:

$$y = f(X) + \varepsilon = X^T\omega + \varepsilon \tag{8}$$

where X is the input matrix, ω is the vector of weights, y is the vector of observed responses and $f(X) = X^T\omega$ is the function value which differs from the observed response y by error ε, that follows an independent, identically distributed Gaussian distribution with zero mean and variance σ^2, i.e., $\varepsilon \sim N(0,\sigma^2)$. The weights ω and the error variance σ^2 are estimated from the data [24].

Applying the Bayes' rule, the posterior distribution over the weights can be determined as [24,25]:

$$p(\omega|X,y) = \frac{p(y|X,\omega)p(\omega)}{p(y|X)} \tag{9}$$

where $p(y|X)$ is the normalizing term which is independent of weights ω and can be neglected. Assuming a zero mean Gaussian prior of the weights $p(\omega)$ with the covariance matrix Σ, i.e., $\omega \sim N(0,\Sigma)$, one obtains the posterior distribution $p(\omega|X,y)$ as a Gaussian with mean $\overline{\omega}$ and covariance matrix A^{-1} [24]:

$$p(\omega|X,y) \sim N\left(\overline{\omega} = \frac{1}{\sigma^2}A^{-1}Xy, A^{-1}\right) \tag{10}$$

where $A = \sigma^{-1}XX^T + \Sigma^{-1}$. To make a prediction for a new, unseen, test input x^*, one can average over all possible parameter values, weighted by their posterior probability, i.e., the distribution $f(x^*)$ at x^* is again Gaussian, with a mean given by the posterior mean of the weights multiplied by the test input [24,25]:

$$p(f(x^*)|x^*, X, y) \sim N\left(\frac{1}{\sigma^2}x^{*T}A^{-1}Xy, x^{*T}A^{-1}x^*\right) \tag{11}$$

When the linear regression model is not adequate, the input data points x_i can be mapped to a higher-dimensional space using the transform $\Phi : x_i \rightarrow \varphi(x_i)$. The model is further derived, same as in the linear case, substituting x everywhere with $\varphi(x)$. Equation (10) then becomes [24]:

$$p(f(x^*)|x^*, X, y) \sim N\left(\frac{1}{\sigma^2}\varphi(x^*)^T A^{-1}\varphi(X)y, \varphi(x^*)^T A^{-1}\varphi(x^*)^T\right) \tag{12}$$

This model represents a GPR model. Hence, a Gaussian process is completely defined by its mean function and covariance function. The choice of an adequate covariance function for a given dataset is very important. In our experiments, we use different covariance functions, such as exponential, squared exponential, Matérn and rational quadratic. Each of these covariance functions depend on the hyperparameters whose values also need to be tuned. For some of the covariance functions, hyperparameters are easy to interpret and can be used to also combine learning with automatic feature selection, i.e., to determine which inputs (features) are relevant and to exclude all the irrelevant ones from the learning process. For example, consider the covariance function [24]:

$$K(x_i, x_j) = \beta exp\left[-\frac{1}{2}\sum_{n=1}^{N}\left(\frac{x_i^n - x_j^n}{r_n}\right)^2\right] \tag{13}$$

where r_n denotes the length-scale of the covariance function along the input dimension n. It is obvious that if r_n is very large, the covariance function becomes independent of the n-th input; therefore, it can be considered irrelevant and can be removed from the inference. Such a covariance function implements automatic relevance determination (ARD). Exponential, squared exponential, Matérn and rational quadratic covariance functions with ARD are also considered in this paper.

It is important to emphasize that, when working with limited datasets, as in our case, selecting the right machine learning algorithms is crucial to maximize the model performance and generalization. The use of deep learning approaches was therefore avoided as the available data were not sufficient to train reliable deep learning models. Using regression tree ensemble approaches, such as random forests or boosting, aggregates decisions from multiple regression trees, helping to reduce the variance that might arise from small data samples and reducing overfitting. On the other hand, GPR relies on strong prior assumptions about the function being learned (encoded through the covariance function). These priors are especially valuable with limited data as they guide the model in the absence of sufficient empirical data. Finally, SVR handles nonlinear relationships in the data by using the kernel trick, which is particularly useful in cases of limited data, where simple linear models may not capture the underlying complexity, but adding too many parameters (like in neural networks) could lead to overfitting. SVR has a regularization parameter C that helps to balance the model complexity and the margin of error. When dealing with limited data, this regularization prevents the model from overfitting to noise or small fluctuations in the data, which is crucial when the data are scarce.

2.2. Dataset

Experimental specimens were obtained using IQ ECONOMY+ A4, 80 g/m^2 white paper and the HP LaserJet Q2610A toner. The paper was mechanically cut in a paper shredder, soaked in distilled water, and mixed to obtain cellulose fiber specimens. The toner was printed on precoated Q CONNECT A4 universal laser transparency film with polyvinyl alcohol [26] and disintegrated in a mechanical stirrer to obtain plate-shaped particles for toner specimens. The specimens of cellulose fibers and toner particles were further mixed to form a pulp, transferred to the Denver 1,6-L flotation cell and floated at the conditions specified below in Tables 1 and 2. The parameters which may have a significant effect on the deinking process, but are not used as the practical control variables, are summarized in Table 2.

Table 1. Ranges of deinking parameters used as the input model variables.

Process Control Variables	Range of Process Control Variables
Flotation pH	3–12
Surfactant in flotation cell: Oleic acid Oleic acid + $CaCl_2$	 0.1–7 kg/t 0.125–1.5 kg/t + 30 kg/t
Flotation time	1–20 min

The concentration of oleic acid, with or without $CaCl_2$, as a surfactant in flotation, pH value and retention time in flotation were used as parameters of the input model. The pH value is an important control parameter because it affects the function of surfactants in deinking flotation, particularly fatty acid soaps. The calcium concentration has been shown to affect the performance of deinking systems. The recovery of cellulose fibers and optical properties represents the trade-off between economy and quality that must be reconciled in any deinking operation, and flotation retention time is a consistent determinant of deinking performance [4,6]. The recovery of toner and the recovery of cellulose fibers were used as parameters of the output model [6]. Samples were extracted from the foam at 1, 2, 4, 10 and 20 min, and in order to calculate toner recovery in the rejected stream (E_t) and cellulose fiber recovery in the acceptable stream (E_m), float and sink products were filtered through the Buchner funnel, dried at room temperature and weighed, while the dried froth filter pads were then heated at 550 °C in a muffle furnace to determine the ash content by x-ray fluorescence (XRF), for E_t calculation. The printability of the prepared sink filter pads and hand sheet after deinking was checked via printing in a controlled environment using the monochrome laser-jet printer, HP 1018.

Table 2. Optimization variables for flotation deinking.

Optimization Variables	Range of Optimization Variables	Adopted Value
Pulping pH	7–10 [5,27–32]	8
Pulping time	4–60 min [6,31,33–37]	35 min
Pulping temperature	35–60 °C [4,9,27,33,35,38]	45 °C
Pulping consistency	8–18 wt % [9,33–37,39]	10 wt %
Flotation temperature	20–45 °C [4,27,35,40–43]	22 °C
Flotation consistency	0.5–1.5% [6,11,28,29,31,35–37,41,42]	1 wt %
Agitation speed	1000–1400 rpm [11,27,28,31,41,44]	1000 rpm
Airflow rate	225–775 L/h [9,11,35,43]	260 L/h

For each experiment, 100 experiments were performed, i.e., 100 pairs of input/output model parameters were created. Since a limited data set is used, 90% of all data were randomly selected for training the model and the remaining 10% were used for testing the prediction ability of the created model [44]. This ensures that the model has enough data to learn from, and using a cross-validation approach will allow a small amount of data to be used to train/test the model [10]. The data that were used for testing were not included in the training dataset [44].

2.3. Performance Measure

As a measure of performance, the Mean Squared Error (MSE) was used, which defines the mean squared deviation between the observed and predicted values of the output parameter [44]:

$$MSE(y, f(x)) = \frac{1}{N}\sum_{i=1}^{N}(y_i - f(x_i))^2 \quad (14)$$

where y_i represents the observed value of the output parameter and $f(x_i)$ is the predicted value obtained using the trained model. MSE is always non-negative, with values closer to zero defining better models [44].

Besides the MSE, the coefficient of determination R^2 was also used as a measure of performance, defined as [44]:

$$R^2(y, f(x)) = 1 - \frac{\sum_{i=1}^{N} (y_i - f(x_i))^2}{\sum_{i=1}^{N} (y_i - \overline{y})^2} \quad (15)$$

where $\overline{y}$ denotes the mean of y. Values of R^2 closer to one define better models. While MSE is an absolute measure of fit, R^2 represents a relative measure of fit [44].

The problem with the evaluation of any machine learning-based model is that it may result in adequate prediction on the data used for training the model but might not generalize well and fail to predict future unseen data. Cross-validation might be used to overcome this problem by dividing the data into two subsets: one for training a model and the other for model validation. The machine learning methods used in this paper were evaluated using 10-fold cross-validation, where the dataset is randomly partitioned into 10 subsets, 9 of them being used for training the model and the remaining one for model validation (testing). The cross-validation procedure is repeated 10 times, with each of the subsets used exactly once for validation, and the 10 obtained results are then averaged to produce a single estimation [44]. The optimal hyperparameters of the models were determined using the grid search, which exhaustively tries every combination of the provided hyperparameter values to select the best model [45].

3. Results

SVR was implemented using LIBSVM library [46] with the linear and RBF kernel functions. The optimal hyperparameters of the SVR model (C in case of the linear kernel; γ and C in case of the RBF kernel) are determined using the grid search. An example of hyperparameter tuning for the RBF kernel for estimation of cellulose fiber recovery in the sink product (E_m) is shown in Figure 1a, in case the surfactant is oleic acid, and in Figure 1b, in case the surfactant is oleic acid + $CaCl_2$. Hyperparameter C determines the tradeoff between the model complexity and the amount of error that can be tolerated. In general, a lower C tolerates a larger error at the cost of model accuracy, whereas a larger C increases the model complexity but enables better prediction. Note that in both subfigures, the model performance is extremely sensitive to the value of hyperparameter γ, which must be chosen very carefully. When γ is too large, the area of influence of the support vectors is too narrow, so the overfitting appears. On the other hand, when γ is too small, the hyperplane becomes too flat, again leading to poor performance on the test dataset.

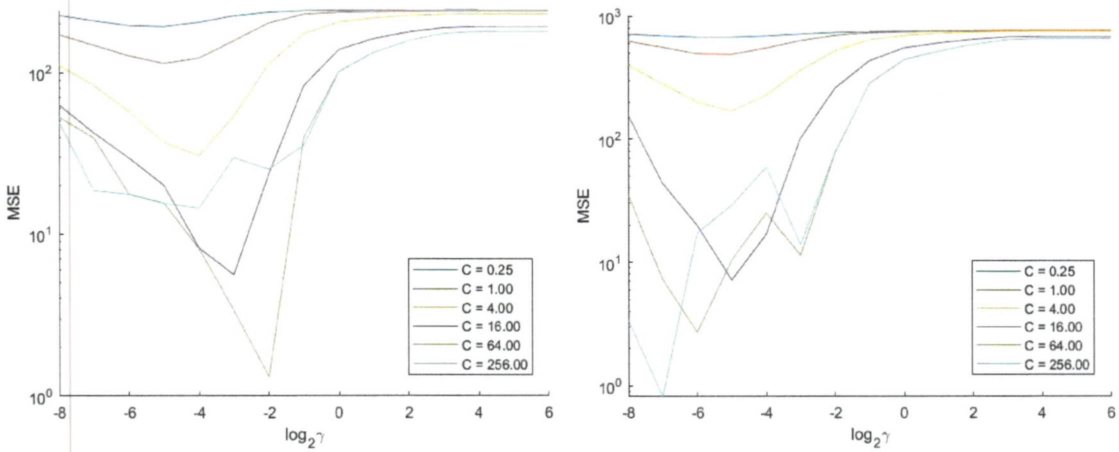

(a) (b)

Figure 1. Optimization of the hyperparameters γ and C in SVR with the RBF kernel for estimating cellulose fiber recovery in the sink product (E_m) when the surfactant is (**a**) oleic acid and (**b**) oleic acid + $CaCl_2$.

Ensembles of regression trees were realized using random forests and boosting. In case of random forests, the hyperparameters to be optimized were the number of trees and the minimum leaf size. A grid search was used for the optimization. An example of hyperparameter tuning for the estimation of cellulose fiber recovery in the sink product (E_m) is shown in Figure 2a in case the surfactant is oleic acid, and in Figure 2b, in case the surfactant is oleic acid + $CaCl_2$. In general, more trees usually lead to better estimates; however, note that after 100 trees in Figure 2a and 50 trees in Figure 2b, the MSE mostly stabilizes and there is no point to further increase the number of trees, since this would increase the model complexity. The minimum leaf size determines the smallest number of observations a node is allowed to have. If a child node should be created by splitting with fewer observations than the minimum leaf size, the node is not split. Note that the most accurate models are obtained for the smallest minimum leaf sizes (equal to two in our experiments). However, this leads to deeper trees; so, better performance comes at the cost of increased complexity.

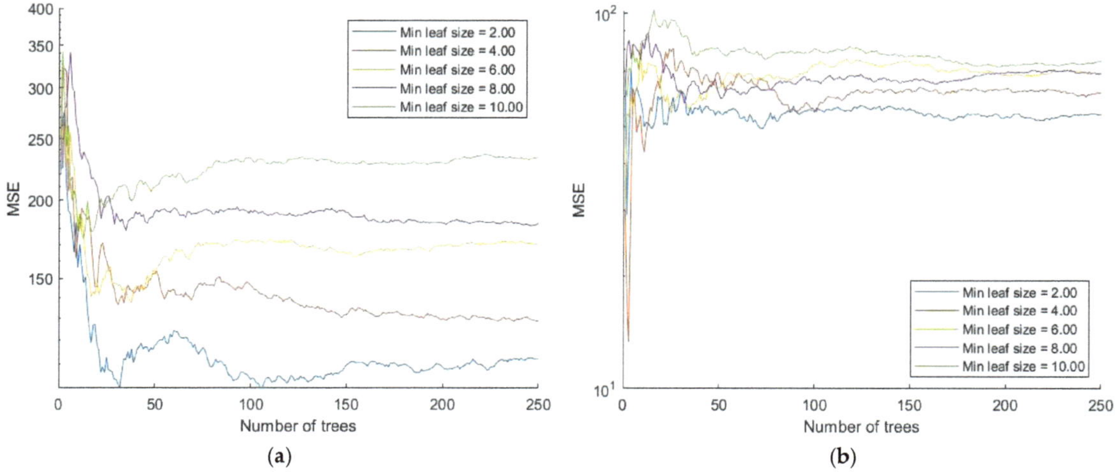

Figure 2. Optimization of the hyperparameters minimum leaf size and number of trees in random forests for estimating cellulose fiber recovery in the sink product (E_m) when the surfactant is (**a**) oleic acid and (**b**) oleic acid + $CaCl_2$.

In the case of boosting, the number of trees and the learning rate are optimized using a grid search. An example of hyperparameter tuning for the estimation of cellulose fiber recovery in the sink product (E_m) is shown in Figure 3a, in case the surfactant is oleic acid, and in Figure 3b, in case the surfactant is oleic acid + $CaCl_2$. As in the case of random forests, the MSE mostly stabilizes for more than 50 trees and there is no point to further increase the number of trees, since this would increase the model complexity. The learning rate is a number between 0 and 1, which multiplies the step magnitude in each gradient step and defines how quickly the prediction error is corrected in the subsequent tree of the model. In other words, shrinkage appears in each gradient step. A learning rate equal to 1 means there is no shrinkage. Small learning rates cause sample predictions to slowly converge towards the observed values and can improve the model's generalization ability. However, smaller learning rates require larger trees and might become computationally expensive. The learning rate equal to 0.25 is the most optimal in our experiments.

GPR is realized with exponential, squared exponential, Matérn 3/2, Matérn 5/2 and rational quadratic covariance functions. It is further combined with ARD to determine the feature importance, as shown in Figure 4. The length scale is the parameter that estimates the relevance of the input features to predict the model's response. A small length scale indicates a highly relevant feature and vice versa. The surfactant concentration for both surfactants is the most important feature, whereas the flotation time has the smallest impact on the overall model performance.

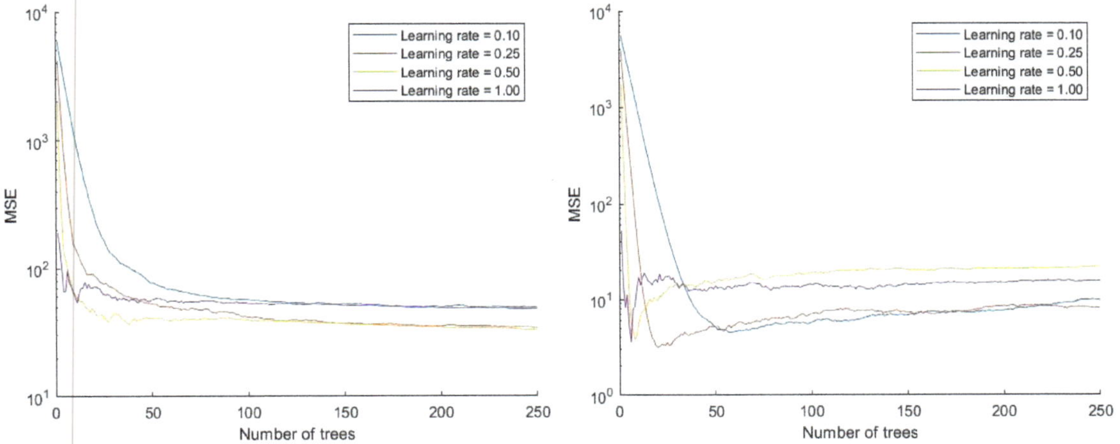

(a) (b)

Figure 3. Optimization of the hyperparameters learning rate and number of trees in boosting for estimating cellulose fiber recovery in the sink product (E_m) when the surfactant is (**a**) oleic acid and (**b**) oleic acid + $CaCl_2$.

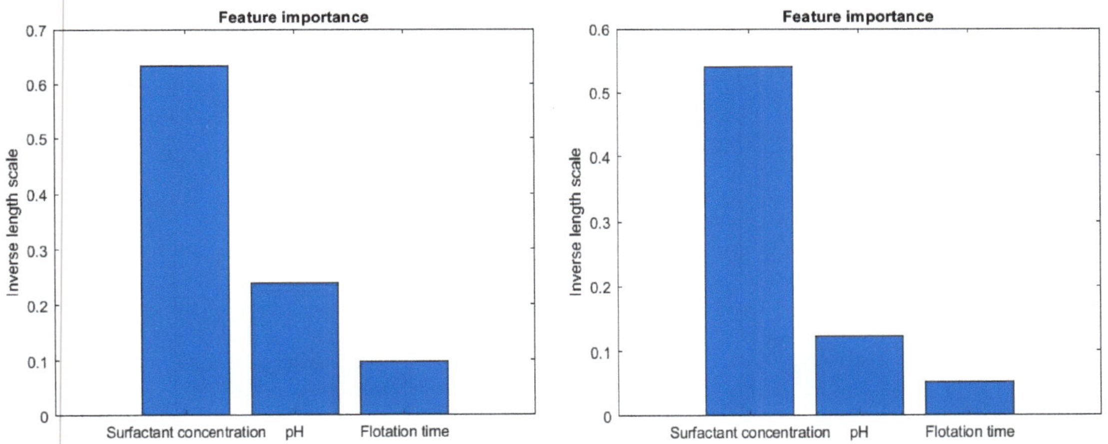

(a) (b)

Figure 4. Feature selection using ARD in GPR for estimating cellulose fiber recovery in the sink product (Em) when the surfactant is (**a**) oleic acid and (**b**) oleic acid + $CaCl_2$.

Tables 3 and 4 present the prediction results of the flotation deinking performance using various machine learning techniques. The fiber recovery in sink product (E_m) and the toner recovery in foam product (E_t) were used to estimate the performance of flotation deinking in Tables 3 and 4, respectively. Oleic acid and oleic acid with addition of $CaCl_2$ were used as the surfactants in both cases.

Table 3. Estimation of fiber recovery in the sink product (E_m); oleic acid and oleic acid + $CaCl_2$ were used as surfactants.

Models		Oleic Acid		Oleic Acid + $CaCl_2$	
		MSE	**R^2 [%]**	**MSE**	**R^2 [%]**
SVR	Linear	101.33	63.72	104.24	71.02
	RBF	20.31	93.56	30.97	93.37
Regression trees	Random forests	51.31	88.19	44.47	92.06
	Boosting	21.16	94.05	24.27	93.87
GPR	Exponential	24.06	94.87	27.67	93.34
	Squared exponential	11.85	97.32	19.72	95.43
	Matérn 3/2	14.03	97.66	19.73	95.95
	Matérn 5/2	12.64	97.77	20.21	95.73
	Rational quadratic	12.48	97.66	21.44	95.24

Table 4. Estimation of toner recovery in the foam product (E_t); oleic acid and oleic acid + $CaCl_2$ were used as surfactants.

Models		Oleic Acid		Oleic Acid + $CaCl_2$	
		MSE	**R^2 [%]**	**MSE**	**R^2 [%]**
SVR	Linear	82.20	49.22	56.24	43.91
	RBF	12.52	90.95	19.71	73.96
Regression trees	Random forests	31.80	84.01	29.12	63.96
	Boosting	7.33	93.90	7.95	88.33
GPR	Exponential	24.37	84.50	32.13	64.83
	Squared exponential	45.43	69.07	40.45	55.48
	Matérn 3/2	20.27	86.31	30.87	65.80
	Matérn 5/2	21.62	84.51	38.05	55.83
	Rational quadratic	21.57	84.94	35.98	60.88

The prediction results for E_m in Table 3 show that GPR with all covariance functions outperforms all other techniques by a large margin using both MSE and R^2 as performance measures for prediction. The best overall result for R^2 is obtained using the Matérn 5/2 covariance function when oleic acid was used as the surfactant (R^2 = 97.77%), whereas the Matérn 3/2 covariance function yields the best performance when $CaCl_2$ was added to oleic acid (R^2 = 95.95%). On the other hand, the squared exponential covariance function gives the best estimation for both surfactants when MSE was used as the performance measure. Boosting and SVR with the RBF kernel have comparable performances, which is 2–4% lower than GPR (measured by R^2), while the performance of SVR with the linear kernel is significantly worse, due to highly nonlinear dependencies between input data. Based on the presented results, it is shown that GPR using the squared exponential covariance function gave the best performance in the assessment of fiber recovery in the sink product considering MSE as the performance measure. The prediction results for E_t presented in Table 4 are, in general, lower than for E_m. However, boosting outperforms all other machine learning techniques significantly for both performance measures, which is especially evident when oleic acid with the addition of $CaCl_2$ was used as the surfactant, where the second-best results were 15% lower, measured by R^2. The dataset is limited (only 100 observations) and it is obviously not enough to capture the complex dependencies between the input features. However, boosting proved to be especially robust to our small-sample problem. The reason might be that boosting, as an ensemble method, can decrease the variance of a single estimate by combining more estimates from different models. Moreover, boosting, unlike random forests, can also reduce the bias by focusing on a weak single model and trying to decrease the prediction error in the next iteration, resulting in a model with higher stability.

4. Conclusions

A comprehensive comparative analysis of a variety of machine learning algorithms for the prediction of flotation deinking performance in the process of paper recycling is given in this paper. A dataset is created for training the models that assumes data samples obtained under a variety of experimental conditions, including different reagents, pH values and flotation residence times. The developed dataset is limited and includes only 100 representative observations, as the aim was to prove that it was feasible to learn reasonable models from "small data" and avoid running expensive, laborious and time-consuming experiments. In this way, it is possible to determine the optimal experimental conditions for the separation of toner particles and cellulose fibers in printed paper recycling using flotation deinking, with only minimal costs and effort.

The obtained results indicate that boosting proved to be especially robust to the small-sample problem under all analyzed conditions. On the other hand, GPR gave the best performance in the estimation of fiber recovery in the sink product, with R^2 = 97.77%, and also a reasonable performance in the estimation of toner recovery in the foam product, with R^2 = 86.31%. Another major advantage of GPR is its low complexity in comparison to ensemble models, such as random forests and boosting, which allows an efficient model training and inference.

This study is limited to selected variables that have been reported to have a significant effect on flotation. The scope of the database was limited to the laboratory scale. Machine learning for the optimization of such variables in the real conditions of the flotation process, as proposed in the paper, is only a theoretical approach at this time. Indeed, some machine learning applications still largely remain a relatively new area of research in mineral processing [10], especially in paper flotation.

Author Contributions: Conceptualization, V.D. and M.S.T.; methodology, V.D. and M.S.T.; validation, T.G. and V.D.; formal analysis, T.G., V.D., M.-I.Z. and M.S.T.; investigation, T.G., V.D., M.-I.Z. and M.S.T.; resources, T.G., V.D., M.-I.Z. and M.S.T.; data curation, T.G., V.D., M.-I.Z. and M.S.T.; writing—original draft preparation, T.G. and V.D.; writing—review and editing, M.-I.Z. and M.S.T.; visualization, T.G., V.D., M.-I.Z. and M.S.T.; supervision, V.D. and M.S.T. All authors have read and agreed to the published version of the manuscript.

Funding: This research was partly funded by the Ministry of Education, Science and Technological Development of the Republic of Serbia, within the funding of the scientific research work at the University of Belgrade, Technical Faculty in Bor [grant numbers 451–03-65/2024–03/200131].

Institutional Review Board Statement: Not applicable.

Informed Consent Statement: Not applicable.

Data Availability Statement: The original contributions presented in the study are included in the article, further inquiries can be directed to the corresponding author.

Conflicts of Interest: The authors declare no conflicts of interest.

References

1. Jovanovic, I.; Miljanović, I.; Jovanović, T. Soft computing-based modeling of flotation processes—A review. *Miner. Eng.* **2015**, *84*, 34–63. [CrossRef]
2. Mohammed, A.S.; Qarani, A.S. Dissolved Air Flotation (DAF): Operational Parameters and Limitations for Wastewaters Treatment with Cost Study. *Recycl. Sustain. Dev.* **2023**, *16*, 91–97. [CrossRef]
3. Mehari, A. Deinking of Black Toner Ink from Laser Printed Paper by Using Anionic Surfactant. Master's Thesis, Addis Ababa University, Addis Ababa, Ethiopia, 2017.
4. Pauck, W.J.; Venditti, R.; Pocock, J.; Andrew, J. Neural network modelling and prediction of the flotation deinking behaviour of industrial paper recycling processes. *Recycl. Nord. Pulp Pap. Res. J.* **2014**, *29*, 521–532. [CrossRef]
5. Costa, C.A.; Rubio, J. Deinking flotation: Influence of calcium soap and surface-active substance. *Miner. Eng.* **2005**, *18*, 59–64. [CrossRef]
6. Pauck, W.J.; Venditti, R.; Pocock, J.; Andrew, J. Using statistical experimental design techniques to determine the most effective variables for the control of the flotation deinking of mixed recycled paper grades. *Tappsa J.* **2012**, *2*, 28–41.

7. Husovska, V. Investigation of Recycled Paper Deinking Mechanisms. Ph.D. Thesis, Western Michigan University, Kalamazoo, MI, USA, 2013.
8. Abraha, M.; Kifle, Z. Deinking of Black Toner Ink from Laser Printed Paper by Using Anionic Surfactant. *Chem. Biomol. Eng.* **2019**, *4*, 23–30. [CrossRef]
9. Kumar, A.; Dutt, D. A comparative study of conventional chemical deinking and environment-friendly bio-deinking of mixed office wastepaper. *Sci. Afr.* **2021**, *12*, e00793. [CrossRef]
10. Gomez-Flores, A.; Heyes, G.W.; Ilyas, S.; Kim, H. Prediction of grade and recovery in flotation from physicochemical and operational aspects using machine learning models. *Miner. Eng.* **2022**, *183*, 107627. [CrossRef]
11. Labidi, J.; Pelach, M.A.; Turon, X.; Mutje, P. Predicting flotation efficiency using neural networks. *Intensification* **2007**, *46*, 314–322. [CrossRef]
12. Verikas, A.; Malmqvist, K.; Bacauskiene, M. Monitoring the De-Inking Process through Neural Network-Based Colour Image Analysis. *Neural Comput. Appl.* **2000**, *9*, 142–151. [CrossRef]
13. Laperrière, L.; Wasik, L. Modeling and simulation of pulp and paper quality characteristics using neural networks. *Tappi J.* **2001**, *84*, 59.
14. Lü, Q.; Liu, S.; Mao, W.; Yu, Y.; Long, X. A numerical simulation-based ANN method to determine the shear strength parameters of rock minerals in nanoscale. *Comput. Geotech.* **2024**, *169*, 106175. [CrossRef]
15. Nakhaei, F.; Irannajad, M. Application and comparison of RNN, RBFNN and MNLR approaches on prediction of flotation column performance. *Int. J. Min. Sci. Technol.* **2015**, *25*, 983–990. [CrossRef]
16. Chehreh Chelgania, S.; Shahbazib, B.; Hadavandi, E. Support vector regression modeling of coal flotation based on variable importance measurements by mutual information method. *Measurement* **2018**, *114*, 102–108. [CrossRef]
17. Szmigiel, A.; Apel, D.B.; Skrzypkowski, K.; Wojtecki, L.; Pu, Y. Advancements in Machine Learning for Optimal Performance in Flotation Processes: A Review. *Minerals* **2024**, *14*, 331. [CrossRef]
18. Vapnik, V. *The Nature of Statistical Learning Theory*; Springer: New York, NY, USA, 1995. [CrossRef]
19. Vapnik, V.; Golowich, S.; Smola, A. Support vector method for function approximation, regression estimation, and signal processing. In *Advances in Neural Information Processing Systems*; Mozer, M.C., Jordan, M.I., Petsche, T., Eds.; MIT Press: Cambridge, MA, USA, 1997; Volume 9, pp. 281–287.
20. Smola, A.J.; Schölkopf, B. A tutorial on support vector regression. *Stat. Comput.* **2004**, *14*, 199–222. [CrossRef]
21. Kovačević, M.; Ivanišević, N.; Petronijević, P.; Despotović, V. Construction cost estimation of reinforced and prestressed concrete bridges using machine learning. *Građevinar* **2021**, *73*, 1–13. [CrossRef]
22. Hastie, T.; Tibshirani, R.; Friedman, J. Boosting and Additive Trees. In *The Elements of Statistical Learning*; Springer Series in Statistics; Springer: New York, NY, USA, 2009. [CrossRef]
23. Cormen, T.H.; Leiserson, C.E.; Rivest, R.L.; Stein, C. *Introduction to Algorithms*, 3rd ed.; The MIT Press: Cambridge, MA, USA, 2009.
24. Despotovic, V.; Skovranek, T.; Schommer, C. Speech Based Estimation of Parkinson's Disease Using Gaussian Processes and Automatic Relevance Determination. *Neurocomputing* **2020**, *401*, 173–181. [CrossRef]
25. Rasmussen, C.E.; Williams, C.K.I. *Gaussian Processes for Machine Learning*; The MIT Press: Cambridge, MA, USA, 2006.
26. Lin, D.; Kuang, Y.; Chen, G.; Kuang, Q.; Wang, C.; Zhu, P.; Peng, C.; Fang, Z. Enhancing moisture resistance of starch-coated paper by improving the film forming capability of starch film. *Ind. Crops Prod.* **2017**, *100*, 12–18. [CrossRef]
27. Dorris, G.M.; Sayegh, N.N. The role of print layer thickness and cohesiveness on deinking of toner printed papers. *Tappi J.* **1997**, *80*, 181–191.
28. Azevedo, M.A.D.; Drelich, J.; Miller, J.D. The Effect of pH On Pulping and Flotation of Mixed Office Wastepaper. *J. Pulp Pap. Sci.* **1999**, *25*, 317–320.
29. Dorris, G.; Ben, Y.; Richard, M. Overview of flotation Deinking, Progress in paper recycling. *Tappi J.* **2011**, *20*, 3–43.
30. Gong, R. New Approaches on Deinking Evaluations. Ph.D. Thesis, Western Michigan University, Kalamazoo, MI, USA, 2013.
31. Yilmaz, U.; Tutus, A.; Sönmez, S. Effects of using recycled paper in inkjet printing system on colour difference. *Pigment. Resin Technol.* **2022**, *51*, 336–343. [CrossRef]
32. Muangnamsuk, R.; Chuetor, S.; Kirdponpattara, S. Development and Optimization of Chemical Deinking of Laser-Printed Paper. *Mater. Sci. Forum* **2023**, *1098*, 151–155. [CrossRef]
33. Behin, J.; Vahed, S. Effect of alkyl chain in alcohol deinking of recycled fibers by flotation process. *Colloids Surf. A Physicochem. Eng. Asp.* **2007**, *297*, 131–141. [CrossRef]
34. Jiang, C.; Ma, J. Deinking of waste paper: Flotation. In *Encyclopedia of Separation Science*; Academic Press: London, UK, 2000; pp. 2537–2544.
35. Tsatsis, D.E.; Valta, K.A.; Vlyssides, A.G.; Economides, D.G. Assessment of the impact of toner composition, printing processes and pulping conditions on the deinking of office waste paper. *J. Environ. Chem. Eng.* **2019**, *7*, 103258. [CrossRef]
36. Ghanbarzadeh, B.; Ataeefard, M.; Etezad, S.M.; Mahdavi, S. Optical and Printing Properties of Deinked Office Waste Printed Paper. *Prog. Color Color. Coat.* **2024**, *17*, 75–84.
37. Yilmaz, U. Investigation of deinking efficiencies of trigromi laserjet printed papers depending on the number of recycling. *Pigment. Resin Technol.* **2024**, *53*, 475–483. [CrossRef]

38. Ali, T.; McLellan, F.; Adiwinata, J.; May, M.; Evants, T. Functional and perfomance characteristics of solube silicate in deinking. Part I: Alkaline deinking of newsprint/magazine. *J. Pulp Pap. Sci.* **1994**, *20*, J3–J8.
39. Liphard, M.; Schereck, B.; Hornfeck, K. Surface chemical aspects of filler flotation in waste paper recycling. *Pulp Pap. Can.* **1993**, *94*, 218–222.
40. Luo, Q.; Deng, Y.; Zhu, J.; Shin, W.T. Foam control using a foaming agent spray: A novel concept for flotation deinking of waste paper. *Ind. Eng. Chem. Res.* **2003**, *15*, 3578–3583. [CrossRef]
41. Pathak, P.; Bhardwaj, N.K.; Singh, A.K. Optimization of Chemical and Enzymatic Deinking of Photocopier Waste Paper. *BioResources* **2011**, *6*, 447–463. [CrossRef]
42. Chandranupap, P.; Chandranupap, P. Enzymatic Deinking of Xerographic Waste Paper with Non-ionic Surfactant. *Appl. Sci. Eng. Prog.* **2020**, *13*, 136–145. [CrossRef]
43. Yilmaz, U.; Tutuş, A.; Sönmez, S. Fiber Classification, Physical and Optical Properties of Recycled Paper. *Cellul. Chem. Technol.* **2021**, *55*, 689–696. [CrossRef]
44. Despotović, V.; Trumić, M.S.; Trumić, M.Ž. Modeling and prediction of flotation performance using support vector regression. *Recycl. Sustain. Dev.* **2017**, *10*, 31–36. [CrossRef]
45. Khadka, K.; Chandrasekaran, J.; Lei, Y.; Kacker, R.N.; Kuhn, D.R. A Combinatorial Approach to Hyperparameter Optimization. In Proceedings of the 2024 IEEE/ACM 3rd International Conference on AI Engineering—Software Engineering for AI (CAIN), Lisbon, Portugal, 14–15 April 2024; pp. 140–149.
46. Chang, C.C.; Lin, C.J. LIBSVM: A Library for Support Vector Machines. *ACM Trans. Intell. Syst. Technol.* **2011**, *2*, 1–27. [CrossRef]

applied sciences

MDPI

Article

Multi-Objective Optimization of Crashworthiness of Shrink Tube Energy Absorption Structure

Fan Zou [1,2,3] and Shuguang Yao [1,3,*]

1 School of Traffic & Transportation Engineering, Central South University, Changsha 410083, China; zoufan@csu.edu.cn
2 School of Intelligent Manufacturing and Mechanical Engineering, Hunan Institute of Technology, Hengyang 421002, China
3 Key Laboratory of Traffic Safety on Track, Ministry of Education, Central South University, Changsha 410083, China
* Correspondence: ysgxzx@csu.edu.cn

Abstract: By means of material testing, truck testing and numerical simulation, the structural parameters of the shrink tube anti-climb device for high-speed trains were determined. The effects of cone angle, tube thickness, friction coefficient and axial length of the friction cone on the crashworthiness of the shrink tube were studied, and the main causes were analyzed. Using cone angle and tube wall thickness as input variables, and peak crush force, mean crash force and specific energy absorption as crashworthiness indexes, a proxy model was constructed using a radial basis function. The global response surface methodology was adopted to optimize the design of the shrink tube's structural parameters. The results showed that the crashworthiness of the shrink tube was positively correlated with the cone angle, the thickness of the shrink tube and the friction coefficient, and the influence decreased successively, while the influence was negatively correlated with the axial length of the friction cone, which had the least influence. Through the optimized design, the peak force of the shrink tube increased by only 5.41%, while the specific energy absorption increased by 31.03%. Additionally, the mean force was closer to the technical requirements of 600 kN, and the crashworthiness was optimized.

Keywords: vehicle engineering; crashworthiness; response surface methodology; shrink tube; multi-objective optimization; trolley test

Citation: Zou, F.; Yao, S. Multi-Objective Optimization of Crashworthiness of Shrink Tube Energy Absorption Structure. *Appl. Sci.* **2024**, *14*, 7347. https://doi.org/10.3390/app14167347

Academic Editor: Homer Rahnejat

Received: 19 July 2024
Revised: 18 August 2024
Accepted: 19 August 2024
Published: 20 August 2024

1. Introduction

Improving crashworthiness can reduce injuries [1,2]; therefore crashworthiness design is a necessary means and method to improve crashworthiness. In order to reduce passenger injuries from collision impacts, energy-absorbing structures are often designed and installed in unoccupied areas at the front end of the train [3–6], where collision energy is absorbed through the deformation of circular tubes. Thin-walled circular tubes can be compressed by molds to produce a specific deformation pattern with stable reaction forces and high specific energy absorption [7–9], which makes them ideal energy-absorbing structures and has led to their wide use in automotive, railway transportation and aerospace industries [10–13]. Some deformation modes of circular tubes have attracted a large number of scholars, such as expansion [14–19], outward turning [20–25] and splitting [26–30]. Li [31] found that the energy absorption efficiency of shrink circular tubes is significantly greater than that of expanded circular tubes for different dimensional parameters by means of numerical simulation combined with experimental methods. Compared with the long rod structure of the expanding tube, the shrink tube has a more compact structure, a more stable energy absorption process, a higher mean force and a greater energy absorption efficiency, making it suitable to be used as the energy-absorption structure for the train anti-climbing device.

The deformation mechanism of the shrink circular tube is similar to that of the expanding circular tube, but there are fewer studies on shrink tubes. Yao et al. [32] proposed a straight taper shrinking (STS) tube and used quasi-static experiments and finite element simulations to study the crashworthiness of the STS tube. They found that the STS tube could form two deformation modes, shrinking (S) and buckling (B), and that matching the structural parameters, such as taper angle and tube wall thickness, could lead to different deformation modes. Additionally, the cone angle had the most significant effect on the crashworthiness of STS tubes. Liu et al. [33] proposed a theoretical model for the shrinkage of metal tubes by classifying the deformation into three modes based on the relationship between the actual mold radius, the critical mold radius and the conical vertebral angle. By comparing the results with numerical simulations over a wide range of geometrical parameters, it was found that the range of applicability of the theoretical model was conical angles less than or equal to 40° and diameter–thickness ratio greater than or equal to 10, respectively. Almeida et al. [34] investigated the shrinkage process of round tubes as a forming method to produce round tubes with specific diameters. Alves et al. [35] addressed the forming limit of aluminum tubes with expansion and contraction and observed experimentally and through simulation the occurrence of toughness fractures, wrinkles and local buckling as phenomena. They found that the geometric parameters and lubrication conditions were the key factors for the success of the shrinkage process. Guan et al. [36] proposed a novel energy absorber for high-speed trains using a combined shrinkage round tube and found that the rate of increase in the maximum crushing force was much greater than the rate of increase in specific energy absorption as the wall thickness increased. They chose to use multi-objective particle swarm optimization to obtain a Pareto bound for the dual optimization objective. Tanaskovic et al. [37] found that by combining the simultaneous action of both the shrinkage and splitting modes, the combined mode energy absorption efficiency was about 60% higher than that of a single contraction absorber.

From the above review, it can be seen that the shrink tube has high mean force and specific absorption energy, light mass and stable deformation. They have been applied to energy-absorbing structures, but the application of shrink tubes in train anti-climb devices is rare. In this paper, in view of the narrow installation space at the end of the train and the requirement for a compact but large energy-absorbing stroke in the anti-climbing device, a kind of shrink tube anti-climbing device is designed for crashworthiness optimization. The effects of the cone angle, the thickness of the shrink tube, the coefficient of friction and the axial length of the friction cone surface on the crashworthiness of the shrink tube are investigated through methods of material tests and trolley collision tests combined with numerical simulation. Then, the main causes are analyzed and the GRSM (global response surface methodology) is used to obtain the optimal configuration of its structural parameters.

2. Geometry and Finite Element Modeling

2.1. Shrink Tube Anti-Climb Structure

The anti-climb device is located at the front end of the train, as shown in Figure 1a. In a collision accident, it meshes with the anti-climb device of the opposite train to prevent the car from climbing. The structure of this shrink tube anti-climbing device is shown in Figure 1c, which consists of four parts: a shrink tube, a flange, an anti-climbing tooth plate and a fixed plate. The flange plate is fixedly installed on the front end of the train through four bolts and the fixed plate securely connects the shrink tube to the flange plate. When the anti-climbing tooth plate undergoes an impact and pushes the shrink tube to move axially, the inner conical surface of the flange squeezes the shrink tube, causing it to deform. This generates friction with the outer surface of the shrink tube, absorbing the kinetic energy of the collision and generating a stable resistance force. The shrink tube is machined from 45# steel, the flange is machined from 40 Cr steel and other parts are machined from Q235 steel. Figure 1d shows a half-section view, where L is the maximum energy absorption stroke of the shrink tube. The size design of the shrink deformation part is shown in Figure 1b, where T represents the thickness of the shrink tube and R_1 and R_2 are the diameters of the shrink tube before and after deformation, respectively. The difference between the large diameter and the small diameter represents the amount of shrinkage. α is the cone angle of the shrink tube and S is the axial length of the friction cone surface, where S is set to fine-tune the cone angle within a certain range and is defined as the projected length of the tapered surface in the axial direction of the shrink tube. In the initial plan, the cone angle, thickness, major diameter and minor diameter of the shrink tube are 15°, 9 mm, 65 mm and 60 mm, respectively.

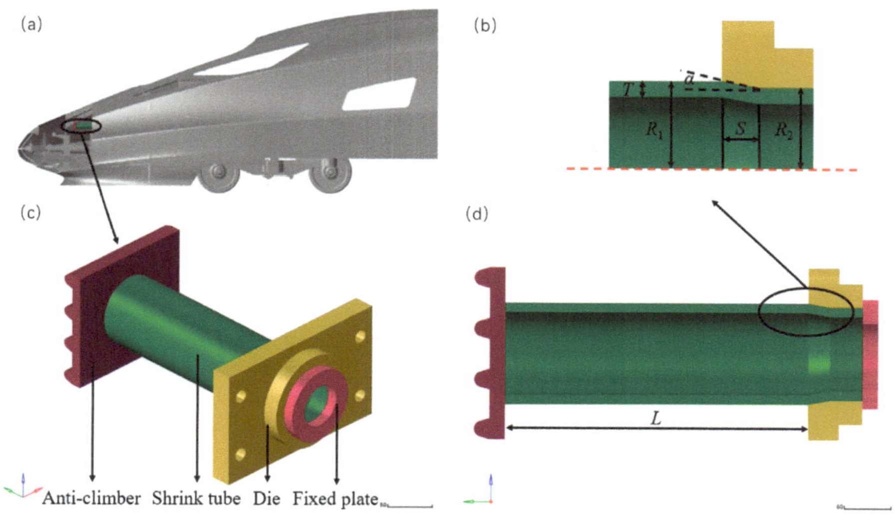

Figure 1. Structure description of the shrink tube anti-climb device: (**a**) installation diagram; (**b**) size detail drawing; (**c**) axonometric drawing; (**d**) semi-sectional view.

2.2. Finite Element Modeling

The LS-DYNA software (R12.1) was used to simulate the mechanical behavior of the shrink tube during the collision process. The shrink tube was a deformation component. The "Mat.24" material was used to simulate 45 steel, and the "Mat.20" material was used for other components. According to the material testing standard [38] (ISO 6892-1:2009, MOD.) [39], the MTS 647 hydraulic wedge clamping tensile testing machine was used to conduct a quasi-static tensile test to obtain the mechanical property curve of 45 steel. The test equipment and specimen clamping status are shown in Figure 2a. The test was conducted at room temperature and the tensile speed was 2 mm/min. The standard size of the material sample is shown in Figure 2b, and the fracture sample and the true stress-strain curve are shown in Figure 2c. Among them, the material parameters for 45 steel were as follows: a density of 7850 kg/m^3, a Poisson's ratio of 0.3, an elastic modulus of 206 GPa and a yield strength of 367 MPa.

The finite element model was consistent with the real test scenario. An anti-climb device was installed at the front end of the test trolley, which hit a fixed rigid wall at the required mass and speed. The finite element model was established, as shown in Figure 3. In order to take into account both calculation accuracy and calculation efficiency, a convergence analysis was performed, as shown in Table 1. The size of the shrink tube and the shrink mold flange was selected to be 3 mm for meshing, a size of 5 mm was used for other parts of the anti-climb device and a size of 30 mm was used for the trolley. The collision test finite element model after meshing had a total of 518,540 nodes and 454,685 elements. Automatic face-to-face contact was set between the contact surfaces of the components, with both the kinetic friction coefficient and static friction coefficient set to

0.1. The shrinkable circular tube was also set to have automatic single-surface contact, with both the kinetic friction coefficient and static friction coefficient set to 0.05 [40].

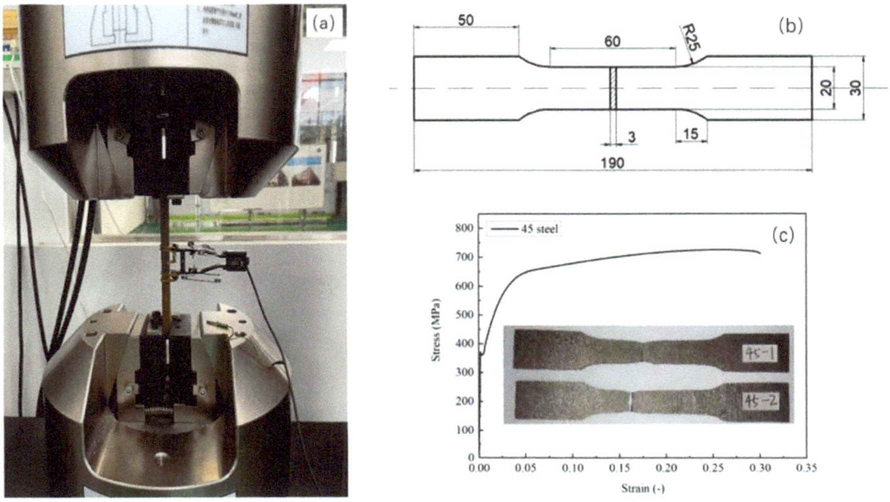

Figure 2. Quasi-static test: (**a**) test equipment; (**b**) standard size of material specimen; (**c**) test specimen and true stress-strain curve.

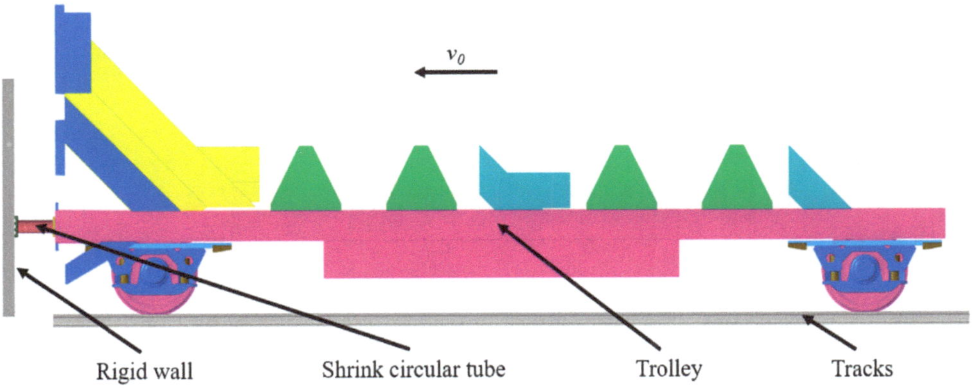

Figure 3. Trolley test collision model.

Table 1. Grid convergence analysis.

Mesh Size/mm	Mean Force/kN	Relative Error (%)
5 × 5 × 5	700	22.16
4 × 4 × 4	587	2.44
3 × 3 × 3	573	-

3. Test Verification and Crashworthiness Evaluation

3.1. Test Verification

A trolley test was conducted on the shrink tube anti-climb device at the High-speed Train Research Center of Central South University. The test scene is shown in Figure 4. There was a trigger speedometer under the front end of the test vehicle to record the collision speed. Three pressure sensors were arranged on the rigid wall to record changes in the impact force. A high-speed photography system was installed directly above and on the side of the impact position to record deformation images and displacement data during the impact. The mass of the test vehicle was 27 t and the impact speed was 3.64 m/s.

Figure 4. Test scenario.

The deformation sequence, as well as the force and displacement curves under test conditions and simulation conditions, are shown in Figure 5. It can be seen that the test curves and simulation curves were in good agreement, and the deformation process of the shrink tube was smooth and orderly. The final displacements of the test and simulation after the collision were 303.64 mm and 307.86 mm, respectively, with a relative error of 1.39%. The energy absorption values of the test and simulation were 175.55 kJ and 178.87 kJ, respectively, with a relative error of 1.89%. The average forces of the test and simulation were 578.15 kN and 581.01 kN, respectively, with a relative error of 0.49%. The results of the experimental mean force and the simulated mean force in this model were almost consistent. This indicates that the finite element model is reliable and effective and can be used for subsequent research.

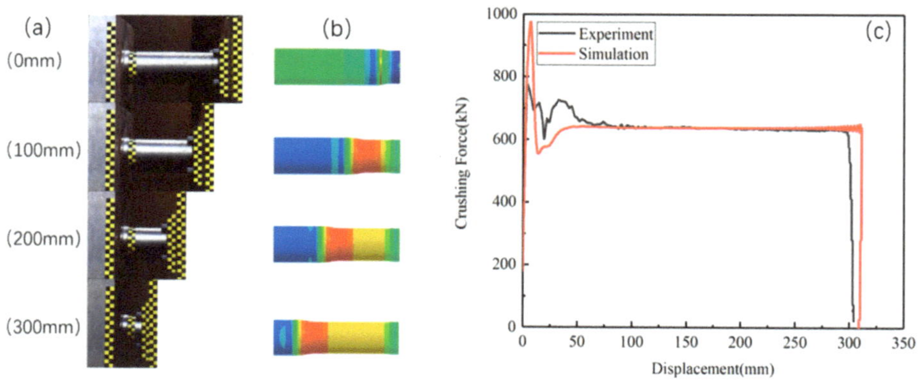

Figure 5. Shrink tube deformation plots and force versus displacement curves: (**a**) test deformation diagram; (**b**) simulation deformation diagram; (**c**) comparison of simulation and test force and displacement curves.

3.2. Evaluation Index

The energy-absorbing structure should have maximum energy absorption and minimum peak force based on a controllable deformation mode. This article uses the following five index parameters to evaluate the crashworthiness [41–43]: energy absorption (E_A), specific energy absorption (E_{SA}), mean force (F_{MC}), peak force (F_{PC}) and impact force efficiency (E_{IF}).

E_A represents the total energy absorbed by the structure during the deformation process, which can be expressed by the following formula:

$$E_A = \int_0^{S_{max}} F(s)ds \quad (1)$$

where $F(s)$ is the function of impact force and displacement, and s is the effective impact displacement.

E_{SA} is an important indicator for crashworthiness evaluation. It represents the amount of energy absorbed per unit mass and directly reflects the quality of the energy-absorbing structure design. Its formula is expressed as follows:

$$E_{SA} = \frac{E_A}{W_m} \quad (2)$$

where W_m represents the mass of the energy-absorbing circular tube.

F_{PC} refers to the maximum impact force during the energy absorption process, which generally occurs in the initial stage. The magnitude of the peak force has a great impact on the safety of the train structure and passenger comfort, and it is required to be as close to the mean force as possible.

F_{MC} refers to the average impact force under a given shrinkage displacement, and its formula is expressed as follows:

$$F_{MC} = \frac{E_A}{s} \quad (3)$$

E_{IF} refers to the uniformity of impact force and displacement response during the energy absorption process, which reflects the efficiency of the mean force. The calculation formula is as follows:

$$E_{IF} = \frac{F_{MC}}{F_{PC}} \times 100\% \quad (4)$$

4. Main Causes and Response Surface Analysis

4.1. Design of Experiments

DOE (design of experiments) is a methodology that is widely used in optimization design. The research goal of DOE is to define a series of tests, analyze the impact of the influencing factors and construct an approximation model that can be used as an alternative model to the computationally intensive real model. Analyzing the principle of energy absorption, the shrink tube anti-climbing device mainly relies on the shrinkage deformation of the shrink tube and the friction of the cone surface to absorb energy. The main factors influencing the energy absorption effect include the cone angle of the shrink tube α, the thickness of the shrink tube T, the coefficient of friction u and the axial length of the friction cone surface S. The specific range of values for these parameters is shown in Table 2. In this paper, we adopt the DOE method to screen the influencing factors on the input variables of the energy-absorbing structure of the shrink tube anti-climbing device, in order to improve the efficiency of the subsequent optimization design. According to the design parameters in Table 2, we generated 300 groups of sample points for simulation calculations using the full factorial method to determine the effects of the design parameters on peak force, mean force and specific energy absorption.

Table 2. Variable values.

Variable Names	Cone Angle $\alpha/^{\circ}$	Thicknesses T/mm	Friction Coefficient u	Axial Length S/mm
Variable values	5, 15, 25	3, 6, 9, 12	0.02, 0.04, 0.06, 0.08, 0.1	21, 23, 25, 27, 29

4.2. Analysis of the Main Causes

Analysis of the main causes was used to compare the extent to which each variable affects the outcome, generally ignoring the effect of other variables when examining the effect of one of them. Linear effects were calculated using a linear regression model, where the range of design variables was set to [0, 1] in proportion to the change and expressed as a linear effect. The slope of the linear regression model represents the degree of influence of the parameter on the results; the larger the slope, the greater the influence of the parameter on the results. In this paper, we analyzed the degree of influence of variables such as the shrink tube cone angle α, shrink tube thickness T, friction coefficient u, and axial length of the friction cone surface S on the three responses, peak shrink tube force F_{PC}, mean shrink tube force F_{MC} and the specific energy absorption E_{SA}. The results of the main cause analysis are shown in Figure 6. The peak force F_{PC} was positively correlated with the cone angle α, the shrinkage tube thickness T and the friction coefficient u. The degree of influence, in descending order, was the cone angle α, the shrinkage tube thickness T and the friction coefficient u, and it was negatively correlated with the axial length of the friction cone surface S. The influence was the least for the axial length of the friction cone surface S. The effects of the four variables on F_{MC} and F_{PC} were basically the same, with only the amplitude being slightly reduced. The cone angle α had the greatest effect on E_{SA}, and the gap with the other influencing factors was large. The friction coefficient u had a greater effect on E_{SA} than that of the thickness T, and the axial length of the friction cone surface S showed the same negative correlation with E_{SA}. Through the main cause analysis, it could be seen that the main factors affecting the energy absorption index were the cone angle α and the thickness of the shrinkage tube T. The friction coefficient was related to the lubrication method used in the actual working conditions, and the axial length of the friction cone had a small influence on the collision resistance index. In this paper, the outer surface of the shrink tube was evenly coated with lithium grease for lubrication. In order to improve efficiency, the cone angle α and the thickness of the shrinkage tube T were used as the input variables in the subsequent optimal design.

4.3. Response Surface Modelling and Agent Modelling

In order to obtain the three-dimensional cloud map of each output response, the two variables with less influence were set to the midpoint of their variation ranges. The other two design variables were both used as independent variables to construct each response surface, as shown in Figure 7, in accordance with the calculation results of the sample points in the experimental design. Figure 7 shows the response surface model of the peak force F_{PC}, mean force F_{MC} and specific energy absorption E_{SA} of the energy-absorbing structure, from which we can see the impact of each variable on the response index. Both the peak force and the mean force of the shrink tube increased with the increase in the cone angle α and thickness T. The degree of influence of the two factors was relatively close, with the influence of the two factors having a greater effect on the peak force than on the average force. The specific energy absorption increased positively with the influence of the two factors. It grew faster with the increase in the cone angle α and relatively slower with the increase in the thickness T. Within the variable design interval, the mean force increased by 476.42% with thickness and by 515.72% with the angle, and the specific energy absorption increased by 58.59% with thickness and by 531.19% with the angle.

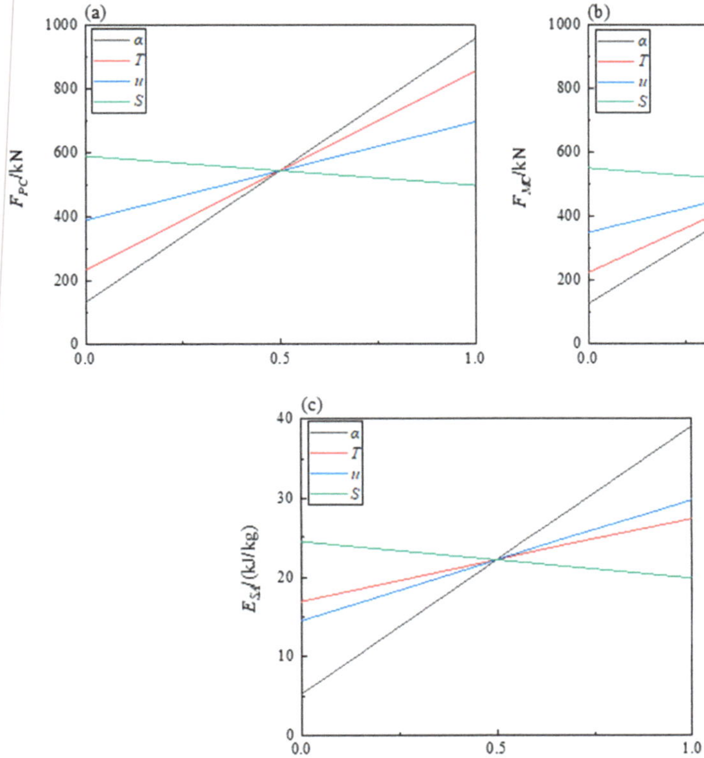

Figure 6. Main cause analysis: (**a**) effect on peak force; (**b**) effect on mean force; (**c**) effect on specific energy absorption.

Practical engineering problems are often complex, usually do not have an explicit functional relationship between design variables and target performance, and manifest themselves as multi-parameter, high-dimensional, strongly nonlinear problems, which are difficult to compute with a simplified physical model and often need to be fitted to discrete data through a proxy model. In order to obtain optimization results efficiently and accurately, proxy methods are used to fit discrete data and build efficient models to replace the actual model without reducing the accuracy of the model. These models are also referred to as proxy models. Among them, the radial basis function is an efficient surrogate method, especially effective in predicting E_{SA} with good accuracy. The mechanical behavior of the shrink tube anti-climb device during the impact energy absorption process is highly nonlinear; therefore, we used the radial basis function (RBF) to construct a proxy model for response surface analysis.

In order to evaluate the accuracy of the response model, corresponding error analysis and evaluation of the output indicators need to be performed using the coefficient of determination R^2, the mean absolute error E_{raa} and the root mean square error E_{rms} as evaluation parameters to verify the effectiveness of the proxy model [44]. These are calculated as follows:

$$R^2 = 1 - \frac{\sum (\hat{y}i - yi)^2}{\sum (\hat{y}i - \overline{y}i)^2} \tag{5}$$

$$E_{raa} = \frac{\frac{1}{N}\sum_{i=1}^{N} |yi - \hat{y}i|}{\sqrt{\frac{1}{N}\sum_{i=1}^{N} |yi - \overline{y}i|^2}} \tag{6}$$

$$E_{\mathrm{rms}} = \sqrt{\frac{\sum(\hat{y}i - yi)^2}{N}} \quad (7)$$

where $\hat{y}i$ is the predicted value of the proxy model at the point i; yi is the actual finite element calculation value of the point; $\overline{y}i$ is the average of the actual response values of all sample points; and N is the number of sample points. When R^2 is closer to 1 or E_{raa} and E_{rms} are smaller, it indicates that the accuracy of the proxy model is higher. The error analysis results are shown in Table 3. It can be seen that the R^2 of the agent model is above 0.99, indicating a high fitting accuracy and small error. The use of surrogate models can effectively reduce calculation time, and the high-precision fitting of F_{PC}, F_{MC} and E_{SA} ensures the validity of subsequent optimization results.

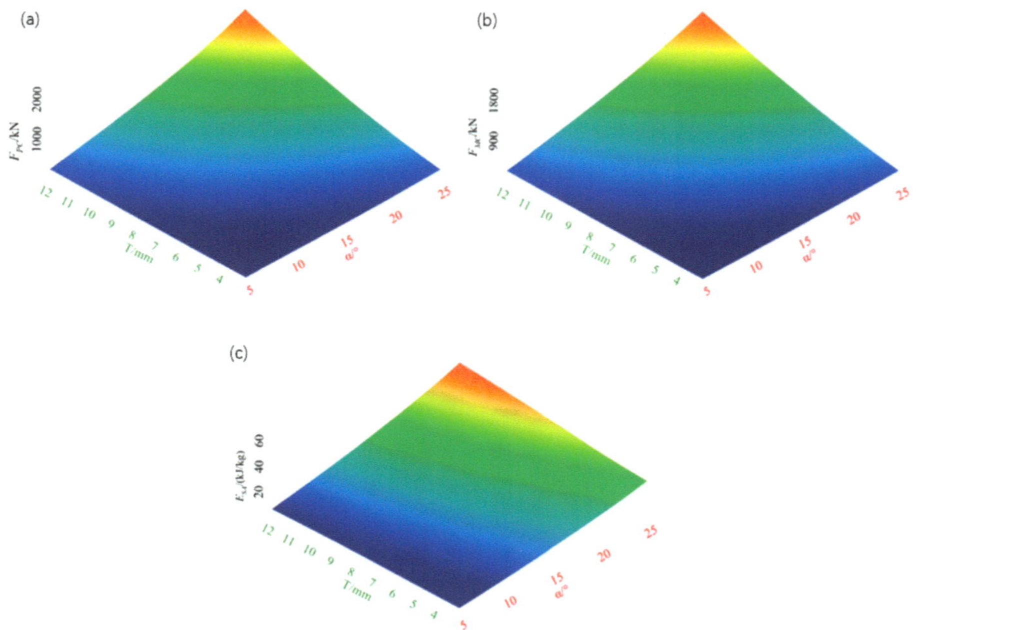

Figure 7. Response surface model diagram: (**a**) effect on peak force; (**b**) effect on mean force; (**c**) effect of specific energy absorption.

Table 3. Error analysis results.

Variant	R^2	E_{raa}	E_{rms}
F_{PC}	0.999	0.007	7.789
F_{MC}	0.991	0.052	18.986
E_{SA}	0.999	0.008	0.133

5. Multi-Objective Optimization

5.1. Optimal Algorithm

Commonly used optimization algorithms include ARSM (adaptive response surface method), GRSM (global response surface method), GA (genetic algorithm) and MOGA (multi-objective genetic algorithm). Among them, GRSM can generate a response surface based on very few data points for computational optimization and will only stop optimization when the solution reaches the optimal value, unlike other algorithms that stop calculation after reaching the convergence criterion. After each iteration, GRSM generates new sampling points in the unsampled intervals of the global design domain based on the distribution of existing sampling points and constructs a new DOE. In this paper, the specific energy absorption E_{SA}, peak force F_{PC} and mean force F_{MC} are taken as the optimization objectives, and the variables are designed for the cone angle α and the thickness of the shrinkage tube T. A series of sample points and their corresponding objective values of the sample points are generated through the design of experiments. The sample point data are then fitted using the Hyper-Study software (2019) to construct the agent model. Finally, the global response surface method (GRSM) is used to obtain the optimal solution for the structural parameters of the energy absorption device based on the agent model. The optimal solution of the structural parameter configuration of the device is obtained based on the proxy model using the GRSM. Due to the superior global search capability of the GRSM, computational optimization using response surfaces is well-suited for the optimal design of crashworthiness indexes for shrink tubes. The response surface fitted in this article has high accuracy, with errors generally less than 1%. This fitted response surface can be used for optimization analysis. Therefore, this article used the GRSM for multi-objective optimization, and the optimization design process is shown in Figure 8.

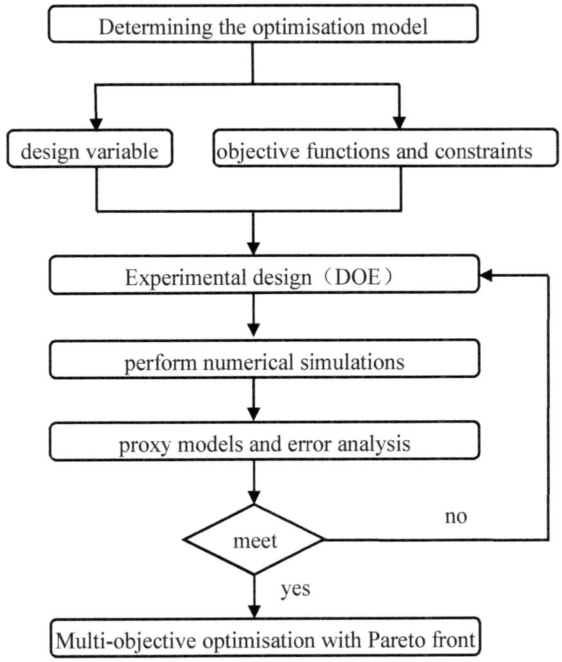

Figure 8. Multi-objective optimization process.

5.2. Optimization Objectives and Boundary Conditions

For this shrink tube anti-climb device, the technical requirements were as follows: an energy absorption of 228 kJ, an energy absorption stroke of 395mm, an average force range of 600 kN $\pm$ 7.5%, a small gap between the peak force and the average force and an original average force efficiency of more than 93%. The space for optimization of mean force efficiency was limited. In the crashworthiness design, the energy-absorbing structure is required to have the highest possible specific energy absorption E_{SA}. Therefore, the first objective of the optimization was to maximize the specific energy absorption E_{SA}, and at the same time, the peak force F_{PC} needed to be reduced as much as possible in order to avoid excessive deceleration during the collision, which could lead to heavy casualties among occupants. Hence, the second objective of the optimization was to minimize the peak force

F_{PC}. In addition, for this shrink tube design, the mean force is required to be as close as possible to 600 kN while meeting the first two targets. Therefore, the third target of the optimization was to take the absolute value of the difference between the mean force F_{MC} and 600 kN as the target of the mean force and to minimize it. Referring to the main cause analysis results, only two input variables, the cone angle α and the shrink tube thickness T, were considered in the optimal design. The friction coefficient u is related to the lubrication method used in the actual working conditions. This paper selected the original value of the friction coefficient u as 0.1 for the case of no lubrication, and the axial length of the friction cone S had a relatively small influence on the three indexes, and an intermediate value of 25 mm was used. In summary, this paper carried out the mathematical description of the multi-objective optimization problem and the boundary conditions, as shown in the following equation:

$$\begin{cases} \text{Max } E_{SA}(\alpha, T) \\ \text{Min } F_{PC}(\alpha, T) \\ \text{Min } |F_{MC} - 600| \\ 5^\circ \leq \alpha \leq 25^\circ \\ 3 \text{ mm} \leq T \leq 12 \text{ mm} \end{cases} \tag{8}$$

5.3. Optimization Results

When there are multiple objectives, a solution that is best for one objective may be worse for others due to the presence of conflicting and incomparable objectives. These solutions, which necessarily weaken at least one other objective function while improving another, are called Pareto solutions. The set of optimal solutions for a set of objective functions is called a Pareto front. In a multi-objective optimization problem, there is more than one objective function to be maximized or minimized, so the ultimate goal is not to seek a single optimal solution but to seek a Pareto front. This Pareto front needs to be selected from the set of Pareto optimal solutions according to the specifics of the problem. In order to obtain an ideal shrink tube energy-absorbing structure, it is required that the peak force F_{PC} is the minimum and the specific energy absorption E_{SA} is the maximum. In this paper, multi-objective optimal design was carried out using the GRSM algorithm with cone angle α and pipe wall thickness T as design variables, and the obtained Pareto front solution set is shown in Figure 9. Figure 9 shows that the peak force F_{PC} of the shrinkage tube and the specific absorption energy E_{SA} have a positive correlation, and as the specific absorption energy E_{SA} increases, the peak force F_{PC} of the shrinkage tube also increases. Therefore, the two objectives of the minimum peak force F_{PC} and the maximum specific absorption energy E_{SA} have mutual exclusivity, which makes it difficult to select the optimal structural parameters.

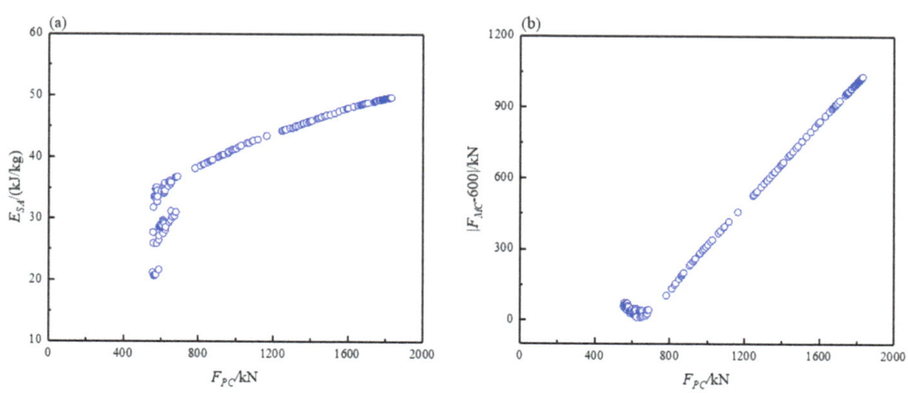

Figure 9. Pareto front: (**a**) Pareto front of specific energy absorption versus the peak force; (**b**) Pareto front of mean force target value and peak force.

Although the Pareto solution set has more advantages in solving optimal design solutions, the optimal solution must be selected to quantitatively analyze the optimization results. Combined with the technical requirements of the actual engineering problems on the mean force, a set of solutions with the mean force F_{MC} closest to 600 kN was selected as the optimal solution when the optimization objective three took the minimum value. The optimized parameters were a cone angle α value of 20.2°, a pipe wall thickness T value of 7.1 mm, a peak force F_{PC} of 658.16 kN, a specific energy absorption E_{SA} of 29.73 kJ/kg and a mean force F_{MC} of 613.23 kN. The error between the mean force and the design objective was 2.2%, which meets the requirement of a mean force error of less than or equal to 7.5%. In order to verify the reliability of the optimization prediction results, finite element simulation calculations were performed on the structural parameters obtained from the optimization results. The comparison between the optimization prediction results and the finite element simulation calculation results is shown in Table 4. It could be seen that the finite element calculation results were close to the optimization prediction results. The peak force error was 1.46%, the specific energy absorption error was 1.33%, and the mean force error was 1.31%. All errors were less than 1.5%, indicating that the optimization results have high accuracy and reliability.

Table 4. Comparison of simulation and optimization results.

Indicator	F_{PC}/kN	E_{SA}/(kJ/kg)	F_{MC}/kN
optimal value	658.16	29.73	613.23
simulation value	648.71	29.34	605.32
errors	1.46%	1.33%	1.31%

5.4. Comparison of Crashworthiness

The comparison between the optimized crashworthiness index and the original structure is shown in Table 5. In order to more intuitively show the improvement of crashworthiness, a radar chart of various crashworthiness indexes was drawn, as shown in Figure 10. The radar chart shows that, except for the average force efficiency, which remained almost unchanged, the other four indicators improved to varying degrees. The most important improvement was the specific energy absorption index, reaching 31.03%. The energy absorption, peak force and average force increased by 5.5%, 5.41% and 5.55%, respectively. The amplitudes were basically the same and the average force efficiency basically remained unchanged. This indicates that the crashworthiness of the optimized structure has been improved.

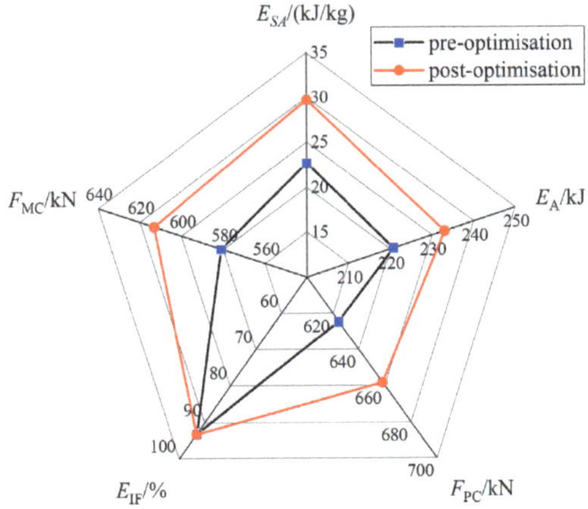

Figure 10. Radar map of crashworthiness indicators before and after optimization.

Table 5. Comparison of original design and optimization results.

Indicator	E_{SA}/(kJ/kg)	E_A/kJ	F_{PC}/kN	E_{IF}/%	F_{MC}/kN
original value	22.69	220.78	624.41	93.05	581.01
optimal value	29.73	233.03	658.16	93.17	613.23
elevation	31.03%	5.5%	5.41%	0.13%	5.55%

6. Conclusions

In this paper, a structural innovation for an anti-climbing energy absorption device with a shrink tube was presented, and its practical engineering application was demonstrated on the train. The effects of some structural parameters on the crashworthiness of the shrink tube were studied. A proxy model was constructed using a radial basis function, and the global response surface methodology was adopted to optimize the design of the shrink tube structure, so as to determine the optimal configuration scheme of its structural parameters. This article only selected some factors for optimizing the design variables. Further, different optimization methods can be used to conduct comparative research on all factors. The main conclusions of this study are summarized as follows:

(1) The effects of the cone angle, shrink tube thickness, friction coefficient and friction cone axial length on the crashworthiness of the shrink tube were studied, and the main causes were analyzed. It was found that the cone angle and thickness had the greatest impact on crashworthiness. Within the variable design interval, the mean force increased by 476.42% and 515.72% with the thickness and angle, respectively, and the specific energy absorption increased by 58.59% and 531.19% with the thickness and angle, respectively.

(2) The global response surface method was used to perform multi-objective optimization on the structural parameters of the shrink tube. Among the optimal structural parameters obtained, the cone angle was 20.2°, the tube wall thickness was 7.1 mm, the friction coefficient was 0.1 and the friction cone axial direction length was 25 mm. After optimization, the specific absorption energy was 29.73 kJ $(kg)^{-1}$, which was 31.03% higher than before optimization.

(3) The mean force obtained through surrogate model optimization was 613.23 kN. The mean force obtained by simulating the optimal structural parameters was 605.32 kN. The mean force error was 1.31%, which shows that the surrogate model has high accuracy.

Author Contributions: Conceptualization, S.Y.; Methodology, F.Z.; Software, F.Z.; Validation, F.Z.; Formal analysis, F.Z.; Investigation, F.Z.; Resources, S.Y.; Data curation, F.Z.; Writing—original draft preparation, F.Z.; Writing—review and editing, S.Y.; Visualization, F.Z.; Supervision, S.Y.; Project administration, S.Y.; Funding acquisition, S.Y. All authors have read and agreed to the published version of the manuscript.

Funding: This work was supported by the National Key Research and Development Program of China [2021YFB3703801].

Institutional Review Board Statement: Not applicable.

Informed Consent Statement: Not applicable.

Data Availability Statement: The original contributions presented in the study are included in the article, further inquiries can be directed to the corresponding author.

Conflicts of Interest: The authors declare no conflicts of interest.

References

1. Chen, Z.; Zhai, W.; Cai, C.; Sun, Y. Safety threshold of high-speed railway pier settlement based on train-track-bridge dynamic interaction. *Sci. China-Technol. Sci.* **2015**, *58*, 202–210. [CrossRef]
2. Wang, T.; Wang, L.; Wang, C.; Zou, X. Crashworthiness analysis and multi-objective optimization of a commercial vehicle frame: A mixed meta-modeling-based method. *Adv. Mech. Eng.* **2018**, *10*, 1687814018778480. [CrossRef]

3. Fang, J.; Gao, Y.; Sun, G.; Zhang, Y.; Li, Q. Crashworthiness design of foam-filled bitubal structures with uncertainty. *Int. J. Non-Linear Mech.* **2014**, *67*, 120–132. [CrossRef]
4. Xiao, Z.; Fang, J.; Sun, G.; Li, Q. Crashworthiness design for functionally graded foam-filled bumper beam. *Adv. Eng. Softw.* **2015**, *85*, 81–95. [CrossRef]
5. He, Q.; Ma, D.; Zhang, Z.; Yao, L. Mean compressive stress constitutive equation and crashworthiness optimization design of three novel honeycombs under axial compression. *Int. J. Mech. Sci.* **2015**, *99*, 274–287. [CrossRef]
6. Ko, Y.; Ahn, K.; Huh, H.; Choi, W.; Jung, H.; Kwon, T. Evaluation of Crash Energy Absorption Capacity of a Tearing Tube. *Exp. Appl. Mech.* **2011**, *6*, 647–654.
7. Liu, Y.; Qiu, X. A theoretical study of the expansion metal tubes. *Int. J. Mech. Sci.* **2016**, *114*, 157–165. [CrossRef]
8. Liu, Y.; Qiu, X.; Wang, W.; Yu, T.X. An improved two-arcs deformational theoretical model of the expansion tubes. *Int. J. Mech. Sci.* **2017**, *133*, 240–250. [CrossRef]
9. Yang, J.; Luo, M.; Hua, Y.; Lu, G. Energy absorption of expansion tubes using a conical–cylindrical die: Experiments and numerical simulation. *Int. J. Mech. Sci.* **2010**, *52*, 716–725. [CrossRef]
10. Yan, J.; Yao, S.; Xu, P.; Peng, Y.; Shao, H.; Zhao, S. Theoretical prediction and numerical studies of expanding circular tubes as energy absorbers. *Int. J. Mech. Sci.* **2016**, *105*, 206–214. [CrossRef]
11. Yao, S.; Yan, K.; Lu, S.; Xu, P. Prediction and application of energy absorption characteristics of thin-walled circular tubes based on dimensional analysis. *Thin-Walled Struct.* **2018**, *130*, 505–519. [CrossRef]
12. Hu, Y.; Liu, C.; Zhang, J.; Ding, G.; Wu, Q. Research on carbon fiber-reinforced plastic bumper beam subjected to low-velocity frontal impact. *Adv. Mech. Eng.* **2015**, *7*, 1687814015589458. [CrossRef]
13. Tang, Z.; Liu, F.J.; Guo, S.H.; Chang, J.; Zhang, J.J. Evaluation of coupled finite element/meshfree method for a robust full-scale crashworthiness simulation of railway vehicles. *Adv. Mech. Eng.* **2016**, *8*, 1687814016642954. [CrossRef]
14. Daxner, T.; Rammerstorfer, F.G.; Fischer, F.D. Instability phenomena during the conical expansion of circular cylindrical shells. *Comput. Methods Appl. Mech. Eng.* **2005**, *194*, 2591–2603. [CrossRef]
15. Fischer, F.D.; Rammerstorfer, F.G.; Daxner, T. Flaring—An analytical approach. *Int. J. Mech. Sci.* **2006**, *48*, 1246–1255. [CrossRef]
16. Shakeri, M.; Salehghaffari, S.; Mirzaeifar, R. Expansion of circular tubes by rigid tubes as impact energy absorbers: Experimental and theoretical investigation. *Int. J. Crashworthiness* **2007**, *12*, 493–501. [CrossRef]
17. Seibi, A.C.; Al-Hiddabi, S.; Pervez, T. Structural Behavior of a Solid Tubular Under Large Radial Plastic Expansion. *J. Energy Resour. Technol.* **2005**, *127*, 323–327. [CrossRef]
18. Seibi, A.C.; Barsoum, I.; Molki, A. Experimental and Numerical Study of Expanded Aluminum and Steel Tubes. *Procedia Eng.* **2011**, *10*, 3049–3055. [CrossRef]
19. Karrech, A.; Seibi, A. Analytical model for the expansion of tubes under tension. *J. Mater. Process. Technol.* **2010**, *210*, 356–362. [CrossRef]
20. Liu, Y.; Qiu, X.; Yu, T.X. A theoretical model of the inversion tube over a conical die. *Thin-Walled Struct.* **2018**, *127*, 31–39. [CrossRef]
21. Niknejad, A.; Moeinifard, M. Theoretical and experimental studies of the external inversion process in the circular metal tubes. *Mater. Des.* **2012**, *40*, 324–330. [CrossRef]
22. Rosa, P.; Baptista, R.; Rodrigues, J.; Martins, P. An investigation on the external inversion of thin-walled tubes using a die. *Int. J. Plast.* **2004**, *20*, 1931–1946. [CrossRef]
23. Rosa, P.A.R.; Rodrigues, J.M.C.; Martins, P.A.F. External inversion of thin-walled tubes using a die: Experimental and theoretical investigation. *Int. J. Mach. Tools Manuf.* **2003**, *43*, 787–796. [CrossRef]
24. Yu, X.; Qiu, X.; Yu, T.X. Analysis of the free external inversion of circular tubes based on deformation theory. *Int. J. Mech. Sci.* **2015**, *100*, 262–268. [CrossRef]
25. Yu, X.; Qiu, X.; Yu, T.X. Theoretical model of a metal tube under inversion over circular dies. *Int. J. Mech. Sci.* **2016**, *108–109*, 23–28. [CrossRef]
26. Huang, X.; Lu, G.; Yu, T.X. On the axial splitting and curling of circular metal tubes. *Int. J. Mech. Sci.* **2002**, *44*, 2369–2391. [CrossRef]
27. Moreno, C.; Beaumont, R.; Hughes, D.J.; Williams, T.; Dashwood, R. Quasi-static and dynamic testing of splitting, expansion and expansion-splitting hybrid tubes under oblique loading. *Int. J. Impact Eng.* **2017**, *100*, 117–130. [CrossRef]
28. Moreno, C.; Williams, T.; Beaumont, R.; Hughes, D.J.; Dashwood, R. Testing, simulation and evaluation of a novel hybrid energy absorber. *Int. J. Impact Eng.* **2016**, *93*, 11–27. [CrossRef]
29. Niknejad, A.; Rezaei, B.; Liaghat, G.H. Empty circular metal tubes in the splitting process—Theoretical and experimental studies. *Thin-Walled Struct.* **2013**, *72*, 48–60. [CrossRef]
30. Rouzegar, J.; Karimi, M. Numerical and experimental study of axial splitting of circular tubular structures. *Thin-Walled Struct.* **2016**, *105*, 57–70. [CrossRef]
31. Li, J.; Gao, G.; Guan, W.; Wang, S.; Yu, Y. Experimental and numerical investigations on the energy absorption of shrink circular tube under quasi-static loading. *Int. J. Mech. Sci.* **2018**, *137*, 284–294. [CrossRef]
32. Yao, S.; Li, Z.; Ma, W.; Xu, P. Crashworthiness analysis of a straight-tapered shrink tube. *Int. J. Mech. Sci.* **2019**, *157–158*, 512–527. [CrossRef]
33. Liu, Y.; Qiu, X. A theoretical model of the shrinking metal tubes. *Int. J. Mech. Sci.* **2018**, *144*, 564–575. [CrossRef]

34. Almeida, B.P.P.; Alves, M.L.; Rosa, P.A.R.; Brito, A.G.; Martins, P.A.F. Expansion and reduction of thin-walled tubes using a die: Experimental and theoretical investigation. *Int. J. Mach. Tools Manuf.* **2006**, *46*, 1643–1652. [CrossRef]
35. Alves, M.L.; Almeida, B.P.P.; Rosa, P.A.R.; Martins, P.A.F. End forming of thin-walled tubes. *J. Mater. Process. Technol.* **2006**, *177*, 183–187. [CrossRef]
36. Guan, W.; Gao, G.; Yu, Y. Crushing analysis and multiobjective crashworthiness optimization of combined shrinking circular tubes under impact loading. *Struct. Multidiscip. Optim.* **2021**, *64*, 1649. [CrossRef]
37. Tanaskovic, J.; Milkovic, D.; Lucanin, V.; Vasic Franklin, G. Experimental investigations of the shrinking–splitting tube collision energy absorber. *Thin-Walled Struct.* **2015**, *86*, 142–147. [CrossRef]
38. *ISO 6892-1:2009*; Metallic Materials—Tensile Testing. Part 1: Method of Test at Room Temperature. ISO: Geneva, Switzerland, 2009.
39. Fang, J.; Sun, G.; Qiu, N.; Kim, N.H.; Li, Q. On design optimization for structural crashworthiness and its state of the art. *Struct. Multidiscip. Optim.* **2017**, *55*, 1091–1119. [CrossRef]
40. Guan, W.; Gao, G. Crashworthiness analysis of shrink circular tube energy absorbers with anti-climbers under multiple loading cases. *Mech. Adv. Mater. Struct.* **2022**, *30*, 1453–1469. [CrossRef]
41. Gao, G.; Guan, W.; Li, J.; Dong, H.; Zou, X.; Chen, W. Experimental investigation of an active-passive integration energy absorber for railway vehicles. *Thin-Walled Struct.* **2017**, *117*, 89–97. [CrossRef]
42. Xie, S.; Zhou, H. Analysis and optimisation of parameters influencing the out-of-plane energy absorption of an aluminium honeycomb. *Thin-Walled Struct.* **2015**, *89*, 169–177. [CrossRef]
43. Aye, C.M.; Pholdee, N.; Yildiz, A.R.; Bureerat, S.; Sait, S.M. Multi-surrogate-assisted metaheuristics for crashworthiness optimisation. *Int. J. Veh. Des.* **2019**, *80*, 223–240. [CrossRef]
44. Peta, K.; Żurek, J. *Prediction of Air Leakage in Heat Exchangers for Automotive Applications Using Artificial Neural Networks*; Faculty of Mechanical Engineering and Management, Poznan University of Technology: Poznan, Poland, 2018.

applied sciences

MDPI

Article

Analysis and Experiment of Thermal Field Distribution and Thermal Deformation of Nut Rotary Ball Screw Transmission Mechanism

Hanwen Yu [1,*], Xuecheng Luan [1], Guiyuan Zheng [1], Guangchao Hao [1], Yan Liu [1], Hongyu Xing [1], Yandong Liu [1,*], Xiaokui Fu [2] and Zhi Liu [2]

[1] School of Mechanical and Electronic Engineering, Shandong Jianzhu University, Jinan 250101, China; 2023075139@stu.sdjzu.edu.cn (X.L.); zgy18764977064@163.com (G.Z.); haoguangchao23@sdjzu.edu.cn (G.H.); liuyan23@sdjzu.edu.cn (Y.L.); xinghongyu20@sdjzu.edu.cn (H.X.)

[2] Jinan Hope Wish Photoelectronic Technology Co., Ltd., Jinan 250101, China; fuxiaokui@chinahpws.com (X.F.); liuzhi@chinahpws.com (Z.L.)

* Correspondence: yuhanwen20@sdjzu.edu.cn (H.Y.); yandonliu@foxmail.com (Y.L.)

Abstract: This study designs a differential dual-drive micro-feed mechanism, superposing the two "macro feed motions" ("motor drive screw" and "motor drive nut") using the same transmission of "the nut rotary ball screw pair" structure. These two motions are almost equal in terms of speed and turning direction, thus the "micro feed" can be obtained. (1) Background: Thermal deformation is the primary factor that can restrict the high-precision micro-feed mechanism and the distribution of heat sources differs from that of the conventional screw single-drive system owing to the structure and motion features of the transmission components. (2) Discussion: This study explores the thermal field distribution and thermal deformation of the differentially driven micro-feed mechanism when two driving motors are combined at different speeds. (3) Methods: Based on the theory of heat transfer, the differential dual-drive system can be used as the research object. The thermal equilibrium equations of the micro-feed transmission system are established using the thermal resistance network method, and a thermal field distribution model is obtained. (4) Results: Combined with the mechanism of thermal deformation theory, the established thermal field model is used to predict the axial thermal deformation of the differential dual-drive ball screw. (5) Conclusions: Under the dual-drive condition, the steady-state thermal error of the nut-rotating ball screw transmission mechanism increases with the increase in nut speed and composite speed and is greater than the steady-state thermal error under the single screw drive condition. After reaching the thermal steady state, the measured thermal elongation at the end of the screw in the experiment is approximately 10.5 μm and the simulation result is 11.98 μm. The experimental measurement result demonstrates the accuracy of the theoretical analysis model for thermal error at the end of the screw.

Keywords: nut rotary ball screw pair; thermal field distribution; thermal deformation; differential dual-drive system; micro feed mechanism

Citation: Yu, H.; Luan, X.; Zheng, G.; Hao, G.; Liu, Y.; Xing, H.; Liu, Y.; Fu, X.; Liu, Z. Analysis and Experiment of Thermal Field Distribution and Thermal Deformation of Nut Rotary Ball Screw Transmission Mechanism. *Appl. Sci.* **2024**, *14*, 5790. https://doi.org/10.3390/app14135790

Academic Editors: Francesca Scargiali, Yadir Torres Hernández, Ana María Beltrán Custodio and Manuel Félix Ángel

Received: 24 May 2024
Revised: 24 June 2024
Accepted: 28 June 2024
Published: 2 July 2024

1. Introduction

Ultra-precision machining technology is an important research direction in the manufacturing equipment industry and is a fundamental technology for the development of cutting-edge technologies, the national defense industry, aviation, aerospace, microelectronics, optics, biology, medicine, and genetic engineering [1,2]. It is also a key technology that determines a country's comprehensive strength and international status. For most precision and ultra-precision optical instruments, scanning probe microscopes, semiconductor technology equipment, microelectromechanical system (MEMS) detection, and micro/nano machining machines, the thermal error is an important factor affecting the feed accuracy of the system, which can reduce the geometric and machining accuracy of the machine tool [3,4]. The thermal deformation error resulting from heat accounts for 40–70% of the total error of the machine tool [5]. The temperature rise and thermal deformation of the screw nut are caused by the relative motion friction and heat generation between the contact parts, and they are the primary heat sources of the system. Moreover, the thermal error of the screw, particularly the axial thermal error, makes a significant impact on the final machining accuracy. Therefore, investigating the temperature field distribution of the system and the variation law of axial thermal error of the ball screw becomes a vital direction for enhancing the transmission accuracy of the feed system.

The research results obtained using the thermal resistance network method include the following representative articles: Han L [6] established a temperature-distribution-related thermal resistance network model based on the thermal resistance network as well as the spherical wall heat conduction principle to improve the thermal design of the chip package structure. Li DT [7] proposed a new simulation approach according to the thermal resistance network approach and the chamber model, with the consideration of impacts of heat exchange, leakage, and component deformation and temperature. Sun ZH [8] presented a lumped-parameter thermal resistance network model for a permanent-magnet spherical motor to track its temperature transients. Meng QY [9] proposed thermal contact resistance (TCR) between contact surfaces and established the motorized spindle mathematical model by incorporating different fractal parameters. Zhu ZY [10] proposed an equivalent thermal resistance network approach for evaluating the temperature increase in an axial permanent-magnet magnetic bearingless flywheel machine, and they established an equivalent thermal network model based on the theory of baseline heat transfer. Bao YJ [11] developed a composite material thermal resistance network model by incorporating heat transfer direction and fiber ply angle based on a series–parallel resistance computing approach. In addition, they described their results, which were determined through thermal response tests for unidirectional carbon fiber laminates. Zhan ZQ [12] developed an implicit thermal network method for the accurate and efficient calculation of temperature, which mainly focused on converting the observed time-varying variables that served as the heat source and thermal resistance to latent variables to construct an implicit thermal equilibrium equation. To obtain real-time dynamic characteristics of ball bearings, Li TJ [13] presented a new model for calculating point-contact dynamic friction within blended lubrication, and they established a time-varying thermal contact resistance model to fit between the ball and the ring, between the housing and the ring, and between the shaft and the ring based on heat transfer and fractal theories. Yang YB [14] proposed a computationally volumetric heat source model according to a semi-analytical thermal modeling method, which was applied in modeling the thermal response during a selective laser melting process.

The following representative articles examine the feed system's temperature field distribution. Wu HY [15] derived a positive temperature field for ideal heat source heat transfer as well as a heat-transfer-induced negative temperature field by examining the heat transfer of a ball screw and the heat generation mechanism; they also established a dynamic temperature field mathematical model for the ball screw feed system with regard to position and time in line with the temperature field superposition principle. Liu JL [16] proposed an approach for optimizing thermal boundary conditions, such as thermal loads, thermal contact resistance, and the convective heat transfer coefficient, thus improving the accuracy of the conventional transient thermal characteristics analysis model in simulating the ball screw feed drive system. Sheng X [17] developed an analytical method for solving the temperature field distribution within a finite cylinder body resulting from periodic-motion heat sources and constant strength and proposed two heat source models, namely multiple and single. Taking a 250 kW permanent-magnet governor that was studied and independently fabricated as an example, Wang L [18] aimed to solve the existing issue of the inability to accurately calculate temperature distribution and transmission torque by a magnetic–thermal unilateralism coupling field. Liu WZ [19] proposed an optimization method for temperature field distribution to solve the issue of temperature influencing the stability and accuracy of a laser multi-degree-of-freedom measurement system. Sun

SB [20] conducted model testing, field measurement, and numerical analysis to investigate a tunnel's original temperature field distribution under higher rock temperatures.

Many scholars have achieved significant results in the study of thermal error in feed systems. Gao XS [21] proposed a thermal error compensation approach for ball screws according to the thermal expansion principle and extreme gradient boosting algorithm. Su DX [22] proposed a new thermal error modeling approach for the ball screw feed system based on FEM, where the grid distribution was changed at various time steps to load a moving heat source as the nut in this model. For investigating how thermal expansion affects the ball screw feed system in a precision machine tool, Yang JC [23] conducted theoretical modeling and experiments for thermally induced error. Li Y [24] established a thermal error difference equation model for describing transient change relations of thermal key point temperatures with ball-screw shaft elongation, and it was isolated from thermal characteristic experimental data in line with the linear superposition principle of thermal and geometric errors to constitute positioning error. Regarding the nut as a moving heat source, Liu HL [25] adopted a new approach for making the heat source moving process more closely resemble real nut movement, and they used a finite difference approach to simulate the thermal error and temperature field of the ball screw feed system in diverse working environments. For investigating how thermal expansion affects the ball screw feed drive system in a precision boring machine tool, Shi H [26] carried out theoretical modeling and experiments to examine heat generation features and thermally induced error. Rong R [27] proposed a screw thermal error iterative prediction model according to the finite difference equation. With regard to the ball screw, its thermal deformation belongs to the positioning error incurred by the quasi-static process. Cao L [28] constructed an adaptive reduced-order model (AROM) of real-time error prediction and compensation under time-varying and various operating conditions. Tanaka S [29] developed a wireless multi-point temperature sensor system based on a built-in temperature sensor array, installed it at the real machining center, and used it to measure the ball screw temperature.

In the above papers, there is almost no research on the thermal field distribution and thermal deformation of the nut-rotating ball screw transmission mechanism. However, in this study, a dual-drive micro-feed mechanism with the nut-rotating ball screw pair as the transmission component underwent changes in the distribution of heat sources owing to changes in the driving mode and transmission component structure. Under the action of multiple heat sources, a temperature field model of the nut-rotating ball screw pair can be established based on the thermal conduction equation of the screw. Subsequently, thermal field analysis is performed through the network topology structure, and a thermal network model of the dual-drive feed system is constructed. In addition, the thermal balance equation of the thermal node is established, while the steady-state temperature field distribution of the system is numerically solved. Based on the established temperature field model and the theory of mechanical thermal deformation, this study predicts the axial thermal error of the nut-rotating ball screw, explores the differences in thermal dynamic characteristics of the single/dual-drive system with the numerical analysis method, and performs thermal experiments on the dual-drive experimental platform to verify the analysis results.

The thermal error compensation method models the thermal error generated during the machining process of the machine tool and manually sets the thermal error compensation amount in the coordinate system [30]. This method can achieve high-precision positioning at low cost without changing the original structure of the machine. The core of the thermal error compensation method is to establish a nonlinear mapping between known physical quantities and thermal errors and to activate the tool center point set through the CNC system [31]. This mapping relationship needs to be constructed using the thermal field model and thermal error data established in this study.

2. Temperature Field and Thermal Error Modeling of Nut-Rotating Ball Screw Transmission Mechanism

Taking the nut-rotating ball screw transmission mechanism as the research object, a thermal resistance model for conduction and convection heat transfer of various components of the system was established based on heat transfer theory. Through thermal balance analysis of thermal nodes, the thermal balance equation system of the feed system was established using the thermal resistance network method. The Newton–Raphson method was used to solve the thermal balance equation system and the temperature field distribution model of the feed system was obtained. Based on the theory of mechanical thermal deformation, the axial thermal error of the nut-rotating ball screw can be predicted using the established temperature field model.

2.1. Nut Rotating Ball Screw Precision Transmission Mechanism

Figure 1 displays a sketch map for the micro-feed mechanism with a nut-rotating ball screw pair as a transmission component [32]. The meanings represented by the numbers are as follows: 1-Base, 2-Guide rail, 3-Nut servo motor, 4-Slide block, 5-Master synchronous belt wheel, 6-Nut motor mounting plate, 7-Screw servo motor, 8-Motor transmission seat, 9-Ball screw, 10-Slave synchronous belt wheel, 11-Rotating nut, 12-Table, and 13-Support bearing seat.

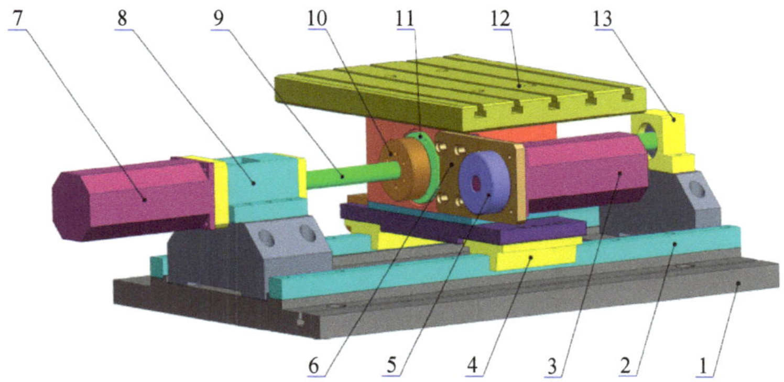

Figure 1. Nut rotating ball screw precision transmission mechanism.

Power and displacement are transmitted to the table through a nut-rotating ball screw pair. The screw servo motor triggers the rotation of the ball screw via coupling, the nut servo motor drives the nut to rotate through a synchronous belt. The ball screw adopts a "fixed support" installation method, with a fixed end using a diagonal contact ball bearing and a supporting end using a radial ball bearing, and the table moves back and forth on the rolling guide rail.

The CNC motion controller allocates motion instructions for the screw servo motor and nut servo motor according to a certain algorithm based on the given motion requirements of the table. The table's linear motion speed along the axial direction driven solely by the screw servo motor is indicated as V1, while that along the axial direction driven solely by the nut servo motor is represented by V2. Under the dual servo motor drive, the differential synthesis speed of the table approaches zero, that is $\Delta = V1 - V2 \approx 0$ avoids the crawling phenomenon caused by low-speed table motion under driving by an individual servo motor, allowing the differential table to achieve high-precision micro feed motion that a traditional servo system cannot achieve.

After repeated calculation of the system parameters, the critical crawling velocity of the table with a single screw drive system is obtained as approximately 2.5 mm/s. Several simulations were performed under the same parameters and the critical crawling velocity of the differential dual-drive system reached approximately 1.5 mm/s. By analyzing the output speed of the table under the constant velocity condition and the variable velocity condition of two types of drives, the conclusions are that the differential dual-drive system has better micro-feed performance at low speed and quicker responsiveness than the single-drive system [33].

2.2. Heat Conduction Analysis of Nut-Rotating Ball Screw Pairs

Thermal conductivity analysis on the nut-rotating ball screw feed system ultimately focuses on the screw itself. Due to the fact that the machine tool's feed accuracy is mostly affected by thermal deformation in the screw axis direction, assuming the uniform temperature distribution on the screw cross-section, with the screw being regarded as a one-dimensional thermal conductor with only temperature gradients in the axis direction. Therefore, the thermal conductivity equation of the one-dimensional rod is as follows.

$$\frac{\partial^2 T(x,t)}{\partial x^2} = \frac{1}{\alpha_C}\frac{\partial T(x,t)}{\partial t} + \frac{4h}{\kappa d_0}\left[T(x,t) - T_f\right] \quad (1)$$

where $T(x,t)$ stands for temperature function on the screw, representing temperature change at time t at position x from the heat source; α_C is the thermal diffusivity, $\alpha = \kappa/\rho c$; κ indicates thermal conductivity; ρ represents screw density; c indicates specific heat capacity; h suggests a convective heat transfer coefficient between the external environment and the screw surface; d_0 stands for lead screw's nominal diameter; T_f indicates ambient temperature.

As shown in Figure 2, there are four major heat sources in the nut-rotating ball screw feed system: heat source Q_{B1} near the motor end screw bearing, heat source Q_{B2} at the nut component bearing, heat source Q_N for the screw nut pair, and heat source Q_{B3} far away from the motor end screw bearing. The screw motor is connected to the screw through a diaphragm coupling, and an insulation pad is installed inside the coupling. Therefore, the heat source at the screw motor has a negligible impact on the system temperature field. In addition, due to the fact that the nut motor transmits motion to the nut through a synchronous belt and has a large transmission distance, the heat source at the nut motor has no effect on the screw's thermal deformation. According to the superposition principle, the heat source of the nut bearing and that of the screw nut pair can be superimposed to form $Q_{NB2} = Q_N + Q_{B2}$. During operation, the nut component shows forward and backward movements along the screw axis direction within the effective travel range of the screw. Therefore, Q_{NB2} is regarded as a ring-shaped fixed heat source on the screw's cylindrical surface.

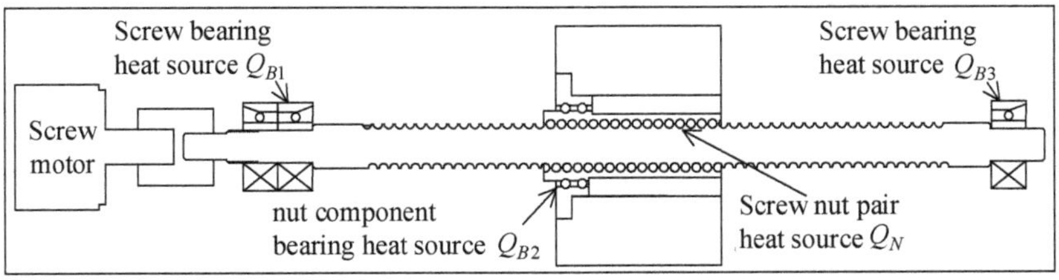

Figure 2. Schematic diagram of heat source distribution in the transmission system.

In the case of combined action of multiple heat sources, as Equation (1) belongs to the second-order linear partial differential equation, based on the linear superposition principle, the screw's temperature response at a given time and position can be determined by adding screw's temperature responses under each single heat source, which is displayed below:

$$T_{total}(x,t) = \Sigma T_i(x,t) \quad (2)$$

in which $T_1(x,t)$ stands for temperature response resulting from heat source Q_{B1}; $T_2(x,t)$ represents temperature response resulting from heat source Q_{NB2}; $T_3(x,t)$ represents temperature response resulting from heat source Q_{B3}; $T_{total}(x,t)$ represents the sum of the screw's temperature response under multiple heat sources.

2.3. Analysis of Temperature Field of the Nut-Rotating Ball Screw Pair under the Action of Heat Source

The screw temperature field at an equal distance from the heat source on both sides of the heat source exhibits symmetrical distribution, and the temperature response $T_2(x,t)$ caused by the heat source Q_{NB2} is equal within the range of the nut stroke, that is, $\partial T_2(x,t)/\partial x = 0$. The screw bearing can be detected at the screw terminal, therefore, the temperature response caused by the screw varies at different points. In this regard, it is of great necessity to investigate the temperature response on one side.

The screw temperature field is the solution to heat conduction Equation (1), but its analytical solution cannot be obtained. In the non-stationary stage of rapid heating or cooling of the screw, the screw's internal thermal conductivity greatly increases relative to its convective heat transfer efficiency in the air. At this time, $\kappa >> h$. Therefore, in the non-stationary stage of screw temperature change, the convective heat transfer term in the thermal conduction equation is negligible. The screw transient thermal conduction equation is obtained as follows:

$$\frac{\partial^2 T(x,t)}{\partial x^2} = \frac{1}{\alpha_C}\frac{\partial T(x,t)}{\partial t} \tag{3}$$

2.3.1. Constant Heat Source Temperature Response

Using one side of the screw as a constant power heat source and the boundary condition, the screw temperature field caused by the constant heat source is studied. Assuming that the initial temperature of the system is consistent with the environment, the initial condition is $T(x,0) = T_f$; If the temperature of a constant heat source is T_m, then the boundary condition is $T(0,t) = T_m$; The screw transient temperature field distribution is obtained by solving Equation (3):

$$T_t(x,t) = \Gamma_{\max}\left[1 - erf(\frac{x}{2\sqrt{\alpha_C t}})\right] + T_f \tag{4}$$

where $T_t(x,t)$ stands for the screw's transient temperature value; $\Gamma_{\max} = T_m - T_f$ represents the screw's temperature elevation after reaching thermal stability; $erf(x)$ is defined in mathematics as an error function:

$$erf(x) = \frac{2}{\sqrt{\pi}}\int_0^x e^{-\delta^2}d\delta \tag{5}$$

Equation (4) indicates that after a sufficiently long period of time, when the screw is insulated from the air, the steady-state temperature values at each point on the screw are equal to the heat source temperature T_m. However, this situation is not in line with actual working conditions, thus it is necessary to establish a steady-state temperature field model for the screw and to solve the screw's steady-state temperature values at each point while taking into account convective heat transfer. When the screw temperature field reaches the steady state, the temperature will not change with time at each point, that is, $\partial T(x,t)/\partial t = 0$. The screw steady-state heat conduction equation can be obtained from Equation (1):

$$\frac{\partial^2 T(x,t)}{\partial x^2} = \frac{4h}{\kappa d_0}\left[T(x,t) - T_f\right] \tag{6}$$

The screw steady-state temperature field distribution acquired through solving Equation (6) is:

$$T_s(x,t) = C_1 e^{-\sqrt{\frac{4h}{\kappa d_0}}x} + C_2 e^{\sqrt{\frac{4h}{\kappa d_0}}x} + T_f \tag{7}$$

where $T_s(x,t)$ stands for the screw's steady-state temperature value; C_1 and C_2 represent coefficients for the screw's temperature rise, in degrees, and temperature distribution, respectively.

When taking convective heat transfer into account, the screw's steady-state temperature rise value is $\Gamma_{\max} = T_s(x,t) - T_f$. The theoretical model of the screw temperature field under a constant heat source on one side can be obtained by combining Equations (4), (5) and (7):

$$T_c(x,t) = \left(C_1 e^{-\sqrt{\frac{4h}{\kappa d_0}}x} + C_2 e^{\sqrt{\frac{4h}{\kappa d_0}}x} \right) \left(1 - \frac{2}{\sqrt{\pi}} \int_0^{\frac{x}{2\sqrt{\alpha t}}} e^{-\delta^2} d\delta \right) + T_f \tag{8}$$

The first of the two terms in Equation (8) represents the screw's temperature rise distribution after steady state, while the second term represents the temperature changes at different points on the screw at different times.

2.3.2. Periodic Variation in Heat Source Temperature Response

When a periodic heat source is applied to the screw on one side, the screw's temperature at each point is always in a non-steady state process of rising or falling, thus convective heat transfer between the air and screw can be ignored. Boundary conditions are:

$$T(x,t)|_{x=0} = T_0 + T_1 \sin(\omega t - \varphi) \tag{9}$$

$T_0, T_1, \omega, \varphi$ are all constants.

According to the boundary conditions in Equation (9), we solve Equation (3) to acquire the screw temperature response as follows:

$$T_p(x,t) = T_0 + T_1 e^{-x\sqrt{\frac{\omega}{2\alpha_C}}} \sin(\omega t - x\sqrt{\frac{\omega}{2\alpha_C}} - \varphi) \tag{10}$$

From Equation (10), the screw temperature response at each point has the same frequency as the heat source, while the temperature amplitude at each point is:

$$A_m = T_0 + T_1 e^{-x\sqrt{\frac{\omega}{2\alpha_C}}} \tag{11}$$

From Equation (11), it can be seen that the temperature amplitude of the screw decreases exponentially with the increase in distance x, and the phase angles of each point are:

$$\varphi(x) = x\sqrt{\frac{\omega}{2\alpha_C}} + \varphi \tag{12}$$

From Equation (12), it can be seen that the phase angle at each point of the screw is different. As the distance x increases, the phase angle also increases, indicating that the temperature changes at each point of the screw relative to the heat source have a time lag.

2.3.3. Any Heat Source Temperature Response

The heat source of machine tools is variable under different working conditions, and there may be periodic or non-periodic changes in the heat source when processing multiple parts or processes. Due to the continuous variation in the heat source of the time domain, which satisfies Dirichlet boundary conditions, the heat source function can be expanded into a Fourier function with time t as the independent variable in the time domain. That is, the boundary conditions for any heat source are:

$$\begin{aligned} T(x,t)|_{x=0} &= \tfrac{a_0}{2} + \sum_{k=1}^{\infty} (a_k \cos kt + b_k \sin kt) \\ &= T_0 + T_1 \cos(\omega t - \varphi_1) + T_2 \cos(\omega t - \varphi_2) + \cdots + T_n \cos(\omega t - \varphi_n) + \cdots \end{aligned} \tag{13}$$

Under any changing heat source, temperature values at various positions change in real time. At this time, the system's internal heat conduction dominates, thus convective heat transfer between the air and the screw can be ignored. Boundary condition Equation (13) of any heat source is substituted into Equation (3) to obtain the screw temperature response function caused by any heat source:

$$T_a(x,t) = T_0 + \sum_{n=1}^{\infty} T_n e^{-x\sqrt{\frac{n\omega}{2\alpha_C}}} \cos(n\omega t - x\sqrt{\frac{n\omega}{2\alpha_C}} - \varphi_n) \tag{14}$$

2.4. Thermal Resistance Network Analysis of Nut-Rotating Ball Screw Transmission Mechanism

Using the thermal resistance network method to explore the steady-state temperature field distribution of the system exhibits the advantages of short simulation time, high computational efficiency, and strong real-time performance. The research object is subdivided using temperature nodes, and there is conduction or convective heat transfer resistance between each temperature node, forming a thermal resistance network. Based on the principle of thermoelectric analogy, Kirchhoff's law is followed to establish a thermal balance equation for each temperature node.

2.4.1. Heat Transfer Analysis

In this study, the transmission component used in the dual-drive mechanism is the DIR-type nut-rotating ball screw pair, which is a device integrating the single nut ball screw with the supporting bearing. In other words, the outer ring of the supporting bearing in the nut assembly can be integrated with the flange in the nut assembly. In addition, the inner ring of the supporting bearing is integrated with the outer ring of the rotating nut. The heat source distribution of this mechanism shows the difference with that of conventional mechanisms with the nut and the screw being driven, respectively. According to thermodynamic research and achievements on bearings and conventional screw drive mechanisms, both domestically and internationally, thermal analysis is performed on the dual-drive mechanism.

The dual-drive nut-rotating ball screw pair is an open heat transfer system. During the operation, there is a large amount of heat generated due to power loss and heat dissipation. According to the law of energy conservation, the entire dual-drive ball screw system ultimately reaches a thermal equilibrium state. In mechanical transmission systems, heat transfer mainly occurs in the form of thermal conduction and convection, and thermal radiation can be ignored. Additionally, part of the heat generated by power loss is transferred to other components in contact through thermal conduction, while another part is transferred to the internal air or external environment of the system through thermal convection.

Taking the dual-drive nut-rotating ball screw pair as the research object, the heat transfer of the ball screw pair system under the action of the screw nut heat source and the nut component bearing heat source is first analyzed, as displayed in Figure 3. Similarly, the effects of other heat sources are explored.

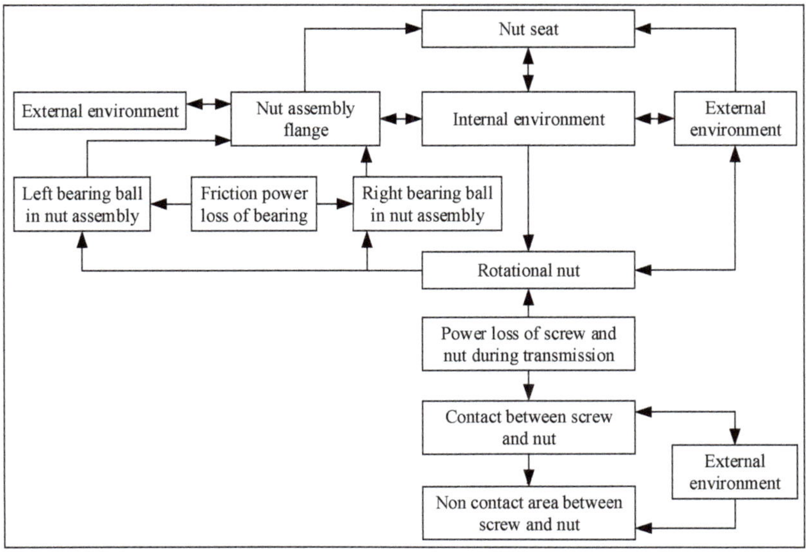

Figure 3. Heat transfer circuit diagram of the dual-drive nut-rotating ball screw pair.

2.4.2. Hot Node Layout

According to the thermal resistance network approach, we should arrange thermal nodes as densely as possible in important parts of the system and components that require specific temperature values. Through the heat transfer analysis of the dual-drive nut-rotating ball screw pair, it is found that the heat generated due to power loss significantly affects the entire ball screw pair's temperature field. Therefore, temperature nodes need to be arranged at the heat source point. In addition, temperature nodes should also be arranged at the parts in the ball screw pair that have thermal convection with the internal air or external environment.

From Figure 4, the temperature node arrangement of the dual-drive ball screw pair is shown under the screw nut heat source and the nut-component-bearing heat source. Figures 5 and 6, respectively, show the temperature node arrangement of the screw when the left-bearing heat source is applied and the temperature node arrangement of the screw when the right-bearing heat source is applied.

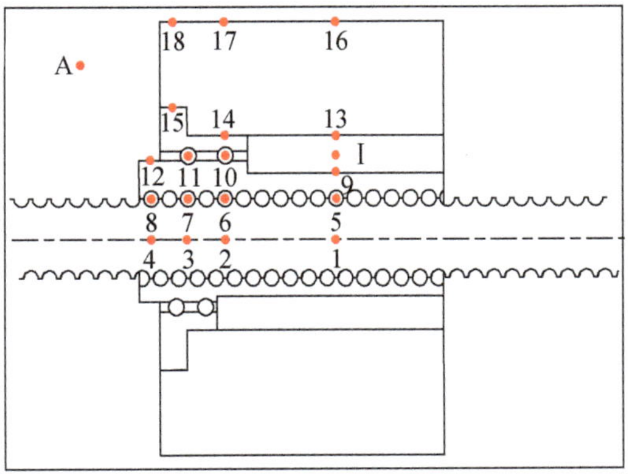

Figure 4. Temperature node layout diagram of the dual-drive nut-rotating ball screw pair.

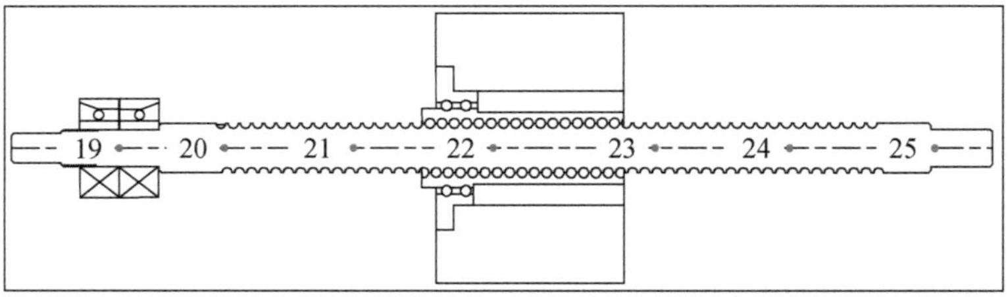

Figure 5. Layout diagram of the screw temperature nodes for the left-bearing heat source.

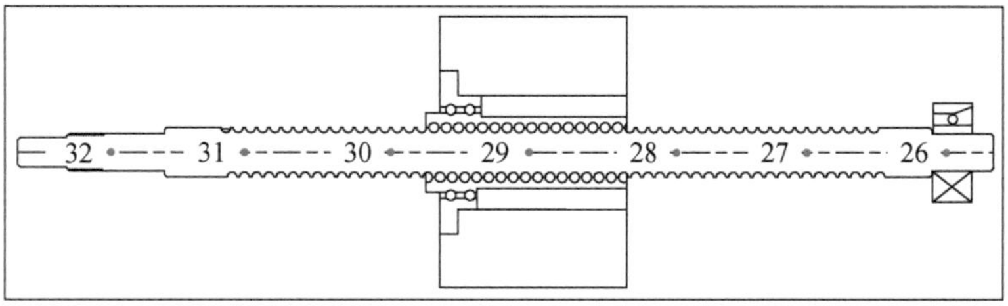

Figure 6. Layout diagram of the screw temperature nodes for the right-bearing heat source.

The red "•" in Figures 4–6 represent the temperature node positions arranged in the system. Table 1 lists the serial numbers and temperature symbols for each node.

Table 1. Meaning of temperature nodes for dual-drive nut-rotating ball screw pairs.

Node	Location of Each Temperature Node	Node Temperature
1	The temperature of the screw section at a distance of 25 mm from the right end of the rotating nut	T_1
2	The temperature of the screw section at the right support ball section in the nut assembly	T_2
3	The temperature of the screw section at the left support ball section in the nut assembly	T_3
4	The temperature of the screw section at a distance of 4 mm from the left end of the rotating nut	T_4
5	Surface temperature of the nut inner ring at a distance of 25 mm from the right end of the rotating nut	T_5
6	The surface temperature of the nut inner ring at the right support ball section in the nut assembly	T_6
7	The surface temperature of the nut inner ring at the left support ball section in the nut assembly	T_7
8	Surface temperature of the nut inner ring at a distance of 4 mm from the left end of the rotating nut	T_8
9	Surface temperature of the nut outer ring at a distance of 25 mm from the right end of the rotating nut	T_9
10	The average temperature of the right support ball in the nut assembly	T_{10}
11	The average temperature of the left support ball in the nut assembly	T_{11}
12	Surface temperature of the nut outer ring at a distance of 4 mm from the left end of the rotating nut	T_{12}
13	Surface temperature of the nut seat inner ring at a distance of 25 mm from the right end of the rotating nut	T_{13}
14	The surface temperature of the nut seat inner ring at the right support ball section in the nut assembly	T_{14}
15	Surface temperature of the nut seat inner ring at the flange section in the nut assembly	T_{15}

Table 1. *Cont.*

Node	Location of Each Temperature Node	Node Temperature
16	Surface temperature of the nut seat outer ring at a distance of 25 mm from the right end of the rotating nut	T_{16}
17	The surface temperature of the nut seat outer ring at the right support ball section in the nut assembly	T_{17}
18	Surface temperature of the nut seat outer ring at the flange section in the nut assembly	T_{18}
I	Air temperature between nut seat and rotating nut	T_I
19, 32	The temperature of the screw section at the left bearing section of the screw	T_{19}, T_{32}
20, 31	The temperature of the screw section at a distance of 40 mm from the bearing section on the left side of the screw	T_{20}, T_{31}
21, 30	The temperature of the screw section at a distance of 100 mm from the bearing section on the left side of the screw	T_{21}, T_{30}
22, 29	The temperature of the screw section at a distance of 160 mm from the bearing section on the left side of the screw	T_{22}, T_{29}
23, 28	The temperature of the screw section at a distance of 220 mm from the bearing section on the left side of the screw	T_{23}, T_{28}
24, 27	The temperature of the screw section at a distance of 280 mm from the bearing section on the left side of the screw	T_{24}, T_{27}
25, 26	The temperature of the screw section at the bearing section on the right side of the screw	T_{25}, T_{26}
A	External environmental temperature	T_A

2.4.3. Establishment of a Thermal Resistance Network

Temperature difference is the driving force for heat transfer, and thermal resistance is its resistance. The ratio of the two is the heat transfer amount (heat flux). According to heat transfer analysis of the dual-drive nut-rotating ball screw pair and arrangement of heat nodes, the thermal resistance network model is constructed for the dual-drive mechanism, as shown in Figures 7–9.

In Figures 7–9, the symbol "⊗" represents the system's heat source point. R represents thermal resistance between two temperature nodes, where the subscript "C" represents heat transfer between the two components through thermal conduction, and the subscript "V" represents heat convection between the components and the internal air or external environment of the system. The arrow represents the heat flow direction of two temperature nodes.

2.4.4. Heat Balance Equation

Analogous to Kirchhoff's law, the temperature difference between two temperature nodes is equivalent to the voltage in the circuit, and the thermal resistance between temperature nodes is equivalent to the resistance in the circuit. Therefore, the ratio of temperature difference to thermal resistance is equivalent to the current in the circuit. When the system reaches thermal equilibrium, the heat flow entering and exiting the temperature node is equal.

For the convenience of analyzing the system's temperature field, before listing the system thermal balance equation, the following assumptions and simplifications are made for the dual-drive nut-rotating ball screw system:

(1) The dual drive nut rotating ball screw pair has reached a thermal equilibrium state;
(2) Due to the temperature difference between various components in the mechanical transmission process being less than 200 °C, the influence of thermal radiation is ignored;
(3) The material of the components in the nut-rotating ball screw pair is isotropic, thus the heat flow direction does not affect the magnitude of thermal resistance;
(4) The thermal conductivity of each component in the system remains constant during temperature changes;

(5) Contact thermal resistance across various components in the nut-rotating ball screw pair is neglected;
(6) The external environment in which the system is located is at a constant temperature of 25 °C;

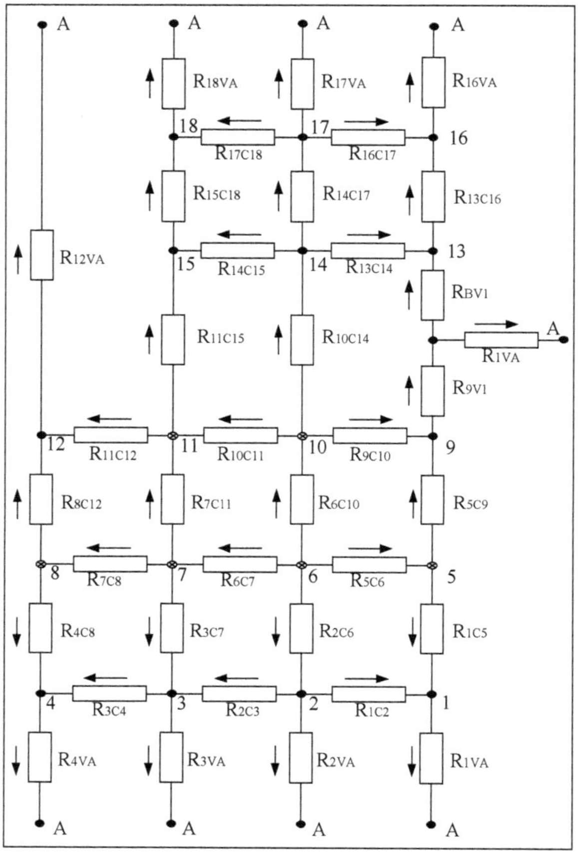

Figure 7. Thermal resistance network model of the dual-drive nut-rotating ball screw pair.

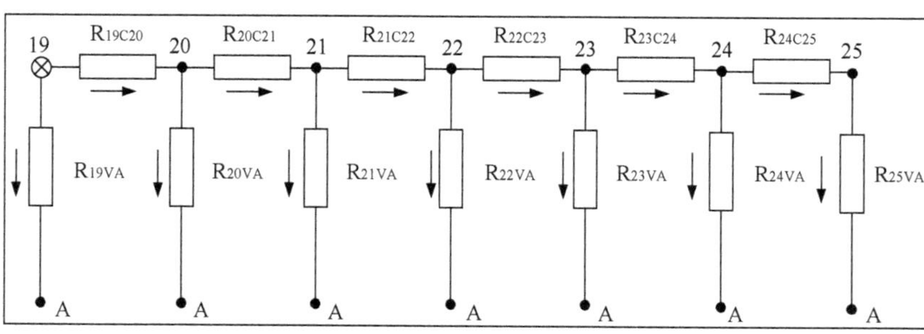

Figure 8. Thermal network model of the screw under the action of the screw's left-bearing heat source.

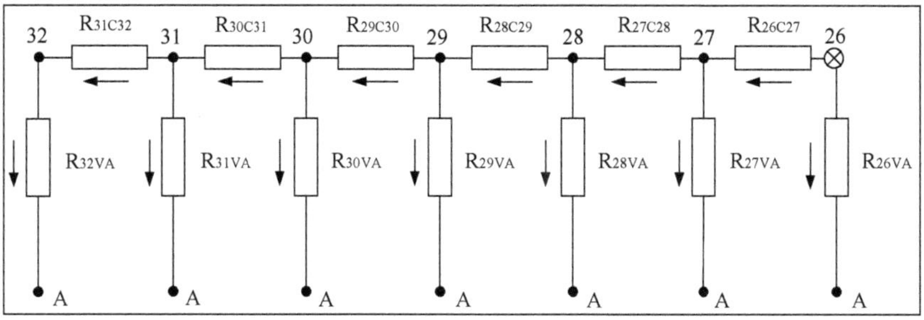

Figure 9. Thermal network model of the screw under the action of the screw's right-bearing heat source.

In line with Kirchhoff's law and the above assumptions, the thermal balance equations of every temperature node in the double drive nut rotating ball screw transmission system are listed as follows:

$$\left(\frac{1}{R_{1VA}}+\frac{1}{R_{1C2}}+\frac{1}{R_{1C5}}\right)T_1-\frac{1}{R_{1C2}}T_2-\frac{1}{R_{1C5}}T_5-\frac{1}{R_{1VA}}T_A=0 \quad (15)$$

$$\frac{1}{R_{1C2}}T_1-\left(\frac{1}{R_{1C2}}+\frac{1}{R_{2C3}}+\frac{1}{R_{2C6}}+\frac{1}{R_{2VA}}\right)T_2+\frac{1}{R_{2C3}}T_3+\frac{1}{R_{2C6}}T_6+\frac{1}{R_{2VA}}T_A=0 \quad (16)$$

$$\frac{1}{R_{2C3}}T_2-\left(\frac{1}{R_{2C3}}+\frac{1}{R_{3C7}}+\frac{1}{R_{3C4}}+\frac{1}{R_{3VA}}\right)T_3+\frac{1}{R_{3C4}}T_4+\frac{1}{R_{3C7}}T_7+\frac{1}{R_{3VA}}T_A=0 \quad (17)$$

$$\frac{1}{R_{3C4}}T_3-\left(\frac{1}{R_{3C4}}+\frac{1}{R_{4C8}}+\frac{1}{R_{4VA}}\right)T_4+\frac{1}{R_{4C8}}T_8+\frac{1}{R_{4VA}}T_A=0 \quad (18)$$

$$\frac{1}{R_{1C5}}T_1-\left(\frac{1}{R_{5C6}}+\frac{1}{R_{1C5}}+\frac{1}{R_{5C9}}\right)T_5+\frac{1}{R_{5C6}}T_6+\frac{1}{R_{5C9}}T_9=-Q_5 \quad (19)$$

$$\frac{1}{R_{2C6}}T_2+\frac{1}{R_{5C6}}T_5-\left(\frac{1}{R_{2C6}}+\frac{1}{R_{5C6}}+\frac{1}{R_{6C7}}+\frac{1}{R_{6C10}}\right)T_6+\frac{1}{R_{6C7}}T_7+\frac{1}{R_{6C10}}T_{10}=-Q_6 \quad (20)$$

$$\frac{1}{R_{3C7}}T_3+\frac{1}{R_{6C7}}T_6-\left(\frac{1}{R_{6C7}}+\frac{1}{R_{3C7}}+\frac{1}{R_{7C8}}+\frac{1}{R_{7C11}}\right)T_7+\frac{1}{R_{7C8}}T_8+\frac{1}{R_{7C11}}T_{11}=-Q_7 \quad (21)$$

$$\frac{1}{R_{4C8}}T_4+\frac{1}{R_{7C8}}T_7-\left(\frac{1}{R_{7C8}}+\frac{1}{R_{4C8}}+\frac{1}{R_{8C12}}\right)T_8+\frac{1}{R_{8C12}}T_{12}=-Q_8 \quad (22)$$

$$\frac{1}{R_{5C9}}T_5-\left(\frac{1}{R_{5C9}}+\frac{1}{R_{9C10}}+\frac{1}{R_{9VI}}\right)T_9+\frac{1}{R_{9C10}}T_{10}+\frac{1}{R_{9VI}}T_I=0 \quad (23)$$

$$\frac{1}{R_{6C10}}T_6+\frac{1}{R_{9C10}}T_9-\left(\frac{1}{R_{6C10}}+\frac{1}{R_{9C10}}+\frac{1}{R_{10C11}}+\frac{1}{R_{10C14}}\right)T_{10}+\frac{1}{R_{10C11}}T_{11}+\frac{1}{R_{10C14}}T_{14}=-Q_{10} \quad (24)$$

$$\frac{1}{R_{7C11}}T_7+\frac{1}{R_{10C11}}T_{10}-\left(\frac{1}{R_{7C11}}+\frac{1}{R_{10C11}}+\frac{1}{R_{11C12}}+\frac{1}{R_{11C15}}\right)T_{11}+\frac{1}{R_{11C12}}T_{12}+\frac{1}{R_{11C15}}T_{15}=-Q_{11} \quad (25)$$

$$\frac{1}{R_{8C12}}T_8+\frac{1}{R_{11C12}}T_{11}-\left(\frac{1}{R_{8C12}}+\frac{1}{R_{11C12}}+\frac{1}{R_{12VA}}\right)T_{12}+\frac{1}{R_{12VA}}T_A=0 \quad (26)$$

$$\left(\frac{1}{R_{13C16}}+\frac{1}{R_{13C14}}+\frac{1}{R_{13VI}}\right)T_{13}-\frac{1}{R_{13C14}}T_{14}-\frac{1}{R_{13C16}}T_{16}-\frac{1}{R_{13VI}}T_{I}=0 \quad (27)$$

$$\frac{1}{R_{10C14}}T_{10}+\frac{1}{R_{13C14}}T_{13}-\left(\frac{1}{R_{10C14}}+\frac{1}{R_{13C14}}+\frac{1}{R_{14C15}}+\frac{1}{R_{14C17}}\right)T_{14}+\frac{1}{R_{14C15}}T_{15}+\frac{1}{R_{14C17}}T_{17}=0 \quad (28)$$

$$\frac{1}{R_{11C15}}T_{11}+\frac{1}{R_{14C15}}T_{14}-\left(\frac{1}{R_{11C15}}+\frac{1}{R_{14C15}}+\frac{1}{R_{15C18}}\right)T_{15}+\frac{1}{R_{15C18}}T_{18}=0 \quad (29)$$

$$\frac{1}{R_{13C16}}T_{13}-\left(\frac{1}{R_{13C16}}+\frac{1}{R_{16C17}}+\frac{1}{R_{16VA}}\right)T_{16}+\frac{1}{R_{16C17}}T_{17}+\frac{1}{R_{16VA}}T_{A}=0 \quad (30)$$

$$\frac{1}{R_{14C17}}T_{14}+\frac{1}{R_{16C17}}T_{16}-\left(\frac{1}{R_{14C17}}+\frac{1}{R_{16C17}}+\frac{1}{R_{17C18}}+\frac{1}{R_{17VA}}\right)T_{17}+\frac{1}{R_{17C18}}T_{18}+\frac{1}{R_{17VA}}T_{A}=0 \quad (31)$$

$$\frac{1}{R_{15C18}}T_{15}+\frac{1}{R_{17C18}}T_{17}-\left(\frac{1}{R_{15C18}}+\frac{1}{R_{17C18}}+\frac{1}{R_{18VA}}\right)T_{18}+\frac{1}{R_{18VA}}T_{A}=0 \quad (32)$$

$$\frac{1}{R_{9VI}}T_{9}+\frac{1}{R_{13VI}}T_{13}-\left(\frac{1}{R_{9VI}}+\frac{1}{R_{13VI}}+\frac{1}{R_{IVA}}\right)T_{I}+\frac{1}{R_{IVA}}T_{A}=0 \quad (33)$$

$$\left(\frac{1}{R_{19C20}}+\frac{1}{R_{19VA}}\right)T_{19}-\frac{1}{R_{19C20}}T_{20}-\frac{1}{R_{19VA}}T_{A}=Q_{19} \quad (34)$$

$$\frac{1}{R_{19C20}}T_{19}-\left(\frac{1}{R_{19C20}}+\frac{1}{R_{20C21}}+\frac{1}{R_{20VA}}\right)T_{20}+\frac{1}{R_{20C21}}T_{21}+\frac{1}{R_{20VA}}T_{A}=0 \quad (35)$$

$$\frac{1}{R_{20C21}}T_{20}-\left(\frac{1}{R_{20C21}}+\frac{1}{R_{21C22}}+\frac{1}{R_{21VA}}\right)T_{21}+\frac{1}{R_{21C22}}T_{22}+\frac{1}{R_{21VA}}T_{A}=0 \quad (36)$$

$$\frac{1}{R_{21C22}}T_{21}-\left(\frac{1}{R_{21C22}}+\frac{1}{R_{22C23}}+\frac{1}{R_{22VA}}\right)T_{22}+\frac{1}{R_{22C23}}T_{23}+\frac{1}{R_{22VA}}T_{A}=0 \quad (37)$$

$$\frac{1}{R_{22C23}}T_{22}-\left(\frac{1}{R_{22C23}}+\frac{1}{R_{23C24}}+\frac{1}{R_{23VA}}\right)T_{23}+\frac{1}{R_{23C24}}T_{24}+\frac{1}{R_{23VA}}T_{A}=0 \quad (38)$$

$$\frac{1}{R_{23C24}}T_{23}-\left(\frac{1}{R_{23C24}}+\frac{1}{R_{24C25}}+\frac{1}{R_{24VA}}\right)T_{24}+\frac{1}{R_{24C25}}T_{25}+\frac{1}{R_{24VA}}T_{A}=0 \quad (39)$$

$$\frac{1}{R_{24C25}}T_{24}-\left(\frac{1}{R_{24C25}}+\frac{1}{R_{25VA}}\right)T_{25}+\frac{1}{R_{25VA}}T_{A}=0 \quad (40)$$

$$\left(\frac{1}{R_{26C27}}+\frac{1}{R_{26VA}}\right)T_{26}-\frac{1}{R_{26C27}}T_{27}-\frac{1}{R_{26VA}}T_{A}=Q_{26} \quad (41)$$

$$\frac{1}{R_{26C27}}T_{26}-\left(\frac{1}{R_{26C27}}+\frac{1}{R_{27C28}}+\frac{1}{R_{27VA}}\right)T_{27}+\frac{1}{R_{27C28}}T_{28}+\frac{1}{R_{27VA}}T_{A}=0 \quad (42)$$

$$\frac{1}{R_{27C28}}T_{27}-\left(\frac{1}{R_{27C28}}+\frac{1}{R_{28C29}}+\frac{1}{R_{28VA}}\right)T_{28}+\frac{1}{R_{28C29}}T_{29}+\frac{1}{R_{28VA}}T_{A}=0 \quad (43)$$

$$\frac{1}{R_{28C29}}T_{28}-\left(\frac{1}{R_{28C29}}+\frac{1}{R_{29C30}}+\frac{1}{R_{29VA}}\right)T_{29}+\frac{1}{R_{29C30}}T_{30}+\frac{1}{R_{29VA}}T_{A}=0 \quad (44)$$

$$\frac{1}{R_{29C30}}T_{29}-\left(\frac{1}{R_{29C30}}+\frac{1}{R_{30C31}}+\frac{1}{R_{30VA}}\right)T_{30}+\frac{1}{R_{30C31}}T_{31}+\frac{1}{R_{30VA}}T_{A}=0 \quad (45)$$

$$\frac{1}{R_{30C31}}T_{30} - \left(\frac{1}{R_{30C31}} + \frac{1}{R_{31C32}} + \frac{1}{R_{31VA}}\right)T_{31} + \frac{1}{R_{31C32}}T_{32} + \frac{1}{R_{31VA}}T_A = 0 \quad (46)$$

$$\frac{1}{R_{31C32}}T_{31} - \left(\frac{1}{R_{31C32}} + \frac{1}{R_{32VA}}\right)T_{32} + \frac{1}{R_{32VA}}T_A = 0 \quad (47)$$

$$T_A = 25\ ^\circ\mathrm{C} \quad (48)$$

The thermal balance equation of the nut-rotating ball screw transmission system in matrix form is written as follows:

$$A_{m\times n}T_{n\times 1} = B_{m\times 1} \quad (49)$$

where $A_{m\times n}$ is the temperature coefficient matrix composed of the reciprocal of thermal resistance; $T_{n\times 1}$ stands for the node temperature matrix; $B_{m\times 1}$ indicates a matrix composed of heat source points, with non-heat source point values considered as zero.

2.5. Thermal Error Analysis of Nut-Rotating Ball Screw

Thermal deformation error is an important component of motion error in micro feed systems, and for precision feed control systems, it is necessary to consider research. The nut-rotating ball screw in the micro feed mechanism in this article utilizes the fixed support in one terminal, whereas the free extension is used in the other terminal for installation. Thermal elongation at the ball screw terminal represents the sum of the ball screw thermal error. Ignoring the thread of the ball screw and treating it as a smooth cylinder, the ball screw's thermal error principle can be observed in Figure 10.

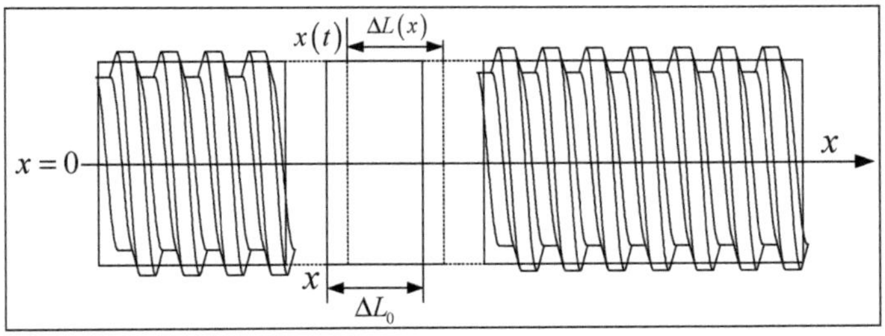

Figure 10. Schematic diagram of the thermal error of the screw.

Divide the screw into individual "micro units" ΔL_0. For the accurate prediction of thermal error, assume $\Delta L_0 \to 0$, and following a period of time when the heat source acts on it, the temperature changes to $T(x,t) - T_0$. Thermal error of the screw's "micro units" can be expressed as:

$$\Delta L(x) - \Delta L_0 = \alpha_T[T(x,t) - T_0]\Delta L_0 \quad (50)$$

where $\Delta L(x) - \Delta L_0$ stands for the "micro unit"'s thermal elongation and α_T indicates the screw's average linear expansion coefficient.

Thermal error at the screw terminal is acquired by adding all micro unit's thermal elongations, that is, the screw's total thermal error can be expressed as:

$$E = \int_0^L \alpha_T[T(x,t) - T_0]d\delta = \sum_{i=1}^{n} \alpha_T[T_i(x,t) - T_0]\Delta L_0 \quad (51)$$

3. Numerical Analysis and Simulation Calculation

Table 2 presents the thermodynamic parameters of the micro feed system for the dual-drive nut-rotating ball screw pair. The screw's fixed end adopts an angular contact ball bearing model of 7210AC, and the free support end adopts a radial ball bearing model of 6201. Based on the design parameters and empirical parameters of the nut-rotating ball screw precision transmission mechanism shown in Figure 1, an analysis program is written using Matlab R2016b numerical calculation software.

Table 2. Thermodynamic parameters of the nut-rotating ball screw precision transmission mechanism.

Parameters	Values	Unit
Ball screw diameter d_0	16	mm
Axial force F_a	500	N
Thermal conductivity κ	50	$W/(m \cdot K)$
Thermal diffusivity α_C	1.04×10^{-5}	m^2/s
Lubricant kinematic viscosity v_0	40	cst or mm^2/s
Average linear expansion coefficient of screw α_T	11.8×10^{-6}	mm/ °C

3.1. Thermodynamic Model Parameter Calculation of the Nut-Rotating Ball Screw Transmission Mechanism

As long as the thermal resistance value between nodes and heat generation at the heat source point are obtained, coefficients of matrices $A_{m\times n}$ and $B_{m\times 1}$ in Equation (49) can be determined. Then, numerical calculations can be carried out using Matlab to solve the specific temperature values of each node of the screw when the system is under single/dual driving conditions, respectively.

3.1.1. Heat Generation Calculation

According to a previous analysis, two heat source types are available for the nut-rotating ball screw transmission mechanism: (1) calculation of heat production from the screw support bearing and the bearing in the nut assembly; (2) Calculation of heat production at the screw–nut contact point.

1. Heat generation calculation for rolling bearings

Rolling bearings mainly generate heat through friction. The friction moment method is used to calculate heat produced through bearing under a given speed and friction moment. Its calculation formula is:

$$Q_B = 1.047 \times 10^{-4} n_B M_B \tag{52}$$

where Q_B indicates heat produced through bearing (W); n_B represents bearing speed (RPM); M_B stands for frictional torque acting on the bearing $(N \cdot mm)$.

Friction torque acting on rolling bearings is calculated based on the Palmgren empirical formula:

$$M_B = M_1 + M_2 \tag{53}$$

$$M_1 = f_1 P d_m \tag{54}$$

$$f_1 = \begin{cases} 0.0013 \times (P_0/C_0)^{0.33} & \text{Single-row installation} \\ 0.001 \times (P_0/C_0)^{0.33} & \text{Double-row installation} \end{cases} \tag{55}$$

$$P = \begin{cases} P_a + 0.1P_r & \text{Single-row installation} \\ 1.44P_a - 0.1P_r & \text{Double-row installation} \end{cases} \tag{56}$$

$$M_2 = \begin{cases} 10^{-7} f_0 (v_0 n_B)^{2/3} d_m^3 & v_0 n_B \geq 2000 \\ 160 \times 10^{-7} f_0 d_m^3 & v_0 n_B < 2000 \end{cases} \tag{57}$$

where M_1 is the friction torque during startup and low-speed operation $(N \cdot mm)$; M_2 represents friction torque associated with the bearing model, lubricating grease, and speed $(N \cdot mm)$; f_1 is a coefficient associated with the load and bearing type; P represents the external load of bearing (N); d_m accounts for the pitch diameter of bearing (mm); P_0 suggests equivalent static load of bearing (N); C_0 stands for the basic rated static load of bearing (N); P_a indicates axial load on bearing (N); P_r presents radial load on bearing (N); v_0 is lubricant viscosity (cst or mm^2/s); f_0 is an empirical coefficient associated with bearing model and lubricant, with $f_0 = 4$ for double-row installation and $f_0 = 2$ for single-row installation.

2. Calculation of heat generation at the contact point of the screw and the nut

The frictional heat produced through the nut-rotating ball screw pair can be determined by adding the heat generated by the rotating nut and that generated by the screw:

$$Q_{NS} = 1.047 \times 10^{-4}(n_N M_N + n_S M_S) \tag{58}$$

where Q_{NS} is heat produced at the nut screw pair contact point (W); n_N stands for the nut rotational speed (RPM); n_S indicates the screw rotational speed (RPM); M_N indicates the mut friction torques $(\mathrm{N} \cdot \mathrm{mm})$, which can be obtained from kinematic analysis; M_S stands for the screw friction torque $(\mathrm{N} \cdot \mathrm{mm})$, which can also be obtained from kinematic analysis.

3.1.2. Thermal Resistance Calculation

The size of thermal resistance represents the ability to hinder heat transfer. According to heat transfer theory, in a one-dimensional steady-state heat transfer process, thermal resistance is divided into conduction resistance and convective heat transfer resistance.

3. Calculation of conduction thermal resistance of thin-walled cylinders

$$R_C = \frac{\ln(d/D)}{2\pi l \kappa} \tag{59}$$

where d is the outer diameter of the thin cylinder (m); D is the inner diameter of a thin cylinder (m); κ is the thermal conductivity of the material $[\mathrm{W}/(\mathrm{m} \cdot \mathrm{K})]$; l is the axial length of a thin cylinder (m).

4. Calculation of Conduction Thermal Resistance of Solid Bodies

$$R'_C = \frac{l'}{A\kappa} \tag{60}$$

where l' stands for heat transfer length of solid body (m); A is thermal conductivity area vertical to the heat flow direction (m^2).

5. Calculation of convective heat transfer resistance

$$R_V = \frac{1}{\alpha S} \tag{61}$$

where α stands for convective heat transfer coefficient $[\mathrm{W}/(\mathrm{m}^2 \cdot \mathrm{K})]$ and S indicates the area of convective heat transfer (m^2).

3.1.3. Calculation of Convective Heat Transfer Coefficient

Based on the model for calculating thermal resistance, when the size of each component is determined, its thermal resistance value will be determined by the thermal conductivity and convective heat transfer coefficient. Within a certain temperature range, the thermal conductivity coefficient can be considered a constant value, which is only related to the properties of the material. Therefore, the calculation of the convective heat transfer coefficient will be a key factor for calculating thermal resistance value and temperature distribution.

6. Calculating the convective heat transfer coefficient of rotating components with the external environment

$$\mathrm{Re} = \frac{\omega \times d^2}{v_f} \tag{62}$$

$$Nu = 0.133\mathrm{Re}^{2/3}\mathrm{Pr}^{1/3}\left(\mathrm{Re} < 4.3 \times 10^5, 0.7 < \mathrm{Pr} < 670\right) \quad (63)$$

$$\alpha = \frac{Nu \cdot \lambda}{d} \quad (64)$$

where Re indicates the Reynolds number; ω represents the component's rotational angular velocity (rad/s); d is the diameter of the convective heat transfer part in the rotating component (m); v_f is the air kinematic viscosity at 25 °C; $v_f = 15.5 \times 10^{-6}\mathrm{m}^2/\mathrm{s}$; Nu is the Nusselt number; Pr stands for Prandtl number, while Pr of air is $\mathrm{Pr} = 0.72$; α suggests convective heat transfer coefficient of the rotating component with external air $[\mathrm{W}/(\mathrm{m}^2 \cdot \mathrm{K})]$; λ indicates thermal conductivity of the external air, at 25 °C, take $\lambda = 2.54 \times 10^{-2}\mathrm{W}/(\mathrm{m} \cdot \mathrm{K})$.

7. Calculating natural convection heat transfer coefficient

The convective heat transfer coefficient between components in a stationary state and the external environment is calculated as follows:

$$Gr = \frac{g\beta l^3 \Delta t}{v_f^2} \quad (65)$$

$$Nu = C(Gr \cdot \mathrm{Pr})^n \quad (66)$$

$$\alpha = \frac{Nu \cdot \lambda}{l} \quad (67)$$

where G_r is the Grayshev quasi number; g is the gravitational acceleration, taken as $g = 9.8\ \mathrm{m/s}^2$; β is the coefficient of expansion of the external air, taken as $\beta = 1/273$; l is the length of heat flux (m); Δt is the temperature difference between the stationary component and the external environment, taken as $\Delta t = 10$ °C; C, n are both constants, depending on the liquid state and heat source of the fluid.

The specific values are shown in Table 3. The convective heat transfer coefficient in actual calculation is generally taken as 3–10 times the theoretical calculation result, while in this article, the convective heat transfer coefficient is taken as 10 times the theoretical calculation value.

Table 3. The values of C and n.

C	n	Scope of Application	Location of Heat Exchange Surface	Flow Pattern
0.59	1/4	$10^4 \leq (Gr \cdot \mathrm{Pr}) \leq 10^9$	Vertical flat wall	laminar flow
0.12	1/3	$10^9 \leq (Gr \cdot \mathrm{Pr}) \leq 10^{12}$	Vertical flat wall	turbulence
0.54	1/4	$10^5 \leq (Gr \cdot \mathrm{Pr}) \leq 2 \times 10^7$	Horizontal wall facing downwards	laminar flow
0.14	1/3	$2 \times 10^7 \leq (Gr \cdot \mathrm{Pr}) \leq 3 \times 10^{10}$	Horizontal wall facing upwards	turbulence
0.27	1/4	$3 \times 10^5 \leq (Gr \cdot \mathrm{Pr}) \leq 3 \times 10^{10}$	Horizontal wall facing downwards	laminar flow

3.2. Node Temperature of Nut-Rotating Ball Screw under a Single Heat Source Condition for Bearing

Figures 11–16 present the temperature values of each node in the left bearing of the screw under the action of a single heat source. Obviously, with increasing distance from the heat source point, the steady-state temperature value of the node rapidly decays. At a specific point away from the heat source at this intensity, with the screw speed reaching a certain value, the temperature of that node will maintain a thermal equilibrium state. This indicates that when the screw speed reaches a certain value, the incoming heat and outgoing heat will always be consistent at that point. At nodes far from the heat source, with increasing screw speed, the temperature exhibits first a slight increase and then a decrease and tends toward room temperature because of an increase in convective heat transfer intensity between the screw speed and the air. As the temperature changes caused by heat sources are mainly concentrated near the heat source point, when considering temperature field changes, it becomes necessary to mainly investigate the temperature changes near the heat source point. The temperature field changes far away from the heat source point are relatively small. The variation pattern of the temperature field of the screw under the action of a single heat source on the right bearing of the screw is similar.

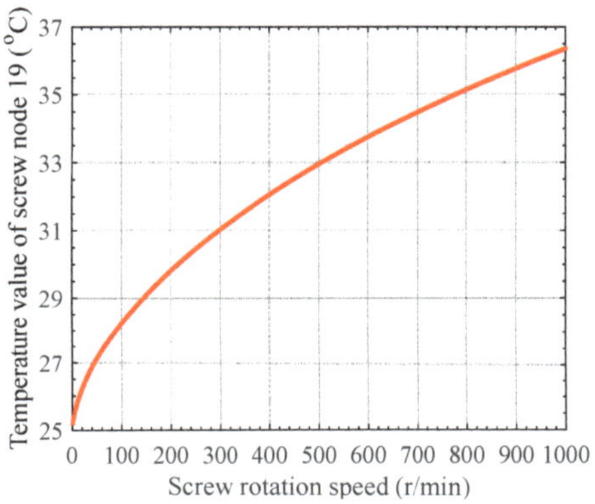

Figure 11. Temperature value of screw node 19 under single heat source.

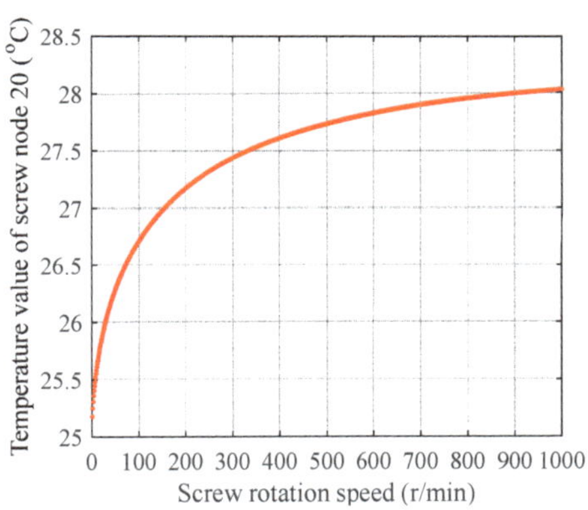

Figure 12. Temperature value of screw node 20 under single heat source.

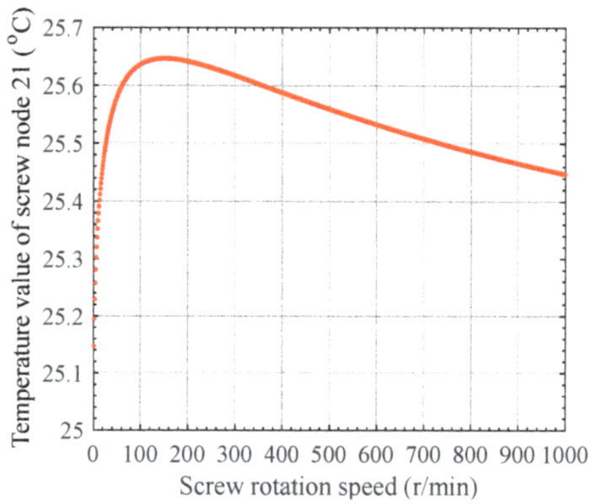

Figure 13. Temperature value of screw node 21 under single heat source.

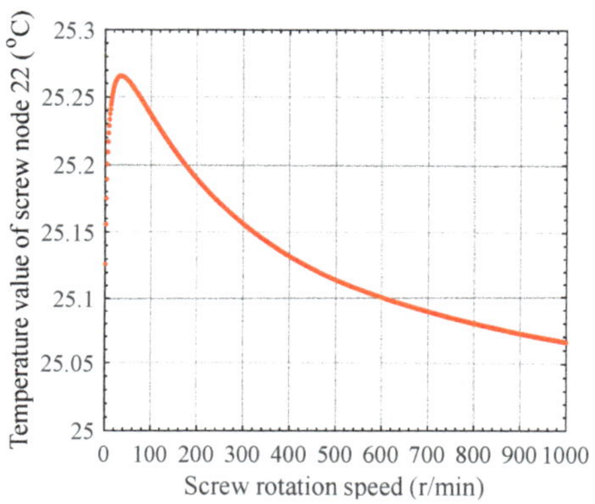

Figure 14. Temperature value of screw node 22 under single heat source.

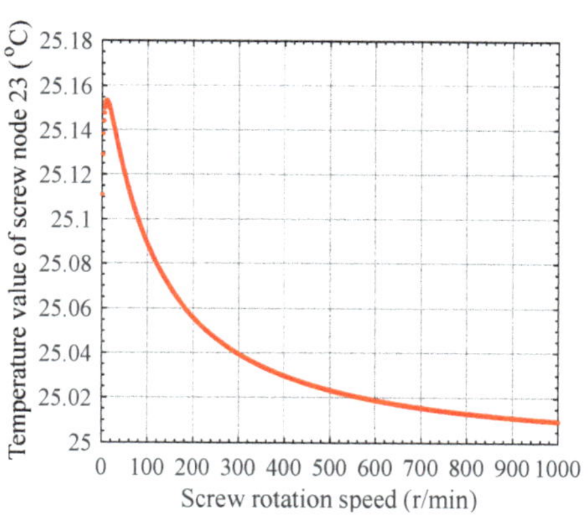

Figure 15. Temperature value of screw node 23 under single heat source.

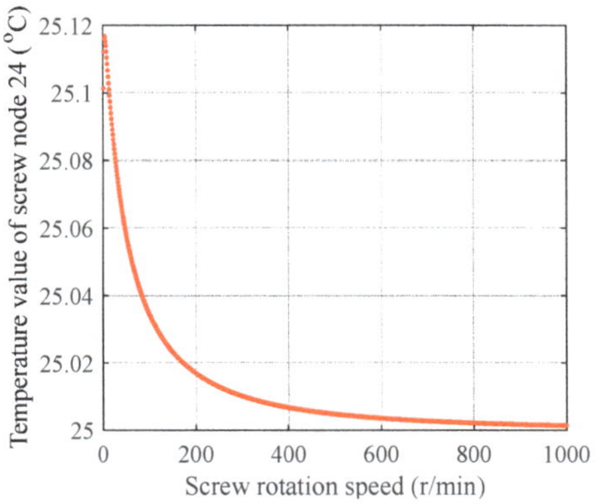

Figure 16. Temperature value of screw node 24 under single heat source.

3.3. Node Temperature of Screw under a Single Heat Source Condition for Nut–Screw Pair

Figures 17–20 show the temperature values of node 4, node 8, node 10, and node 11 of the screw under the single heat source condition for the nut–screw pair as a function of the nut rotation speed and composite rotation speed. When the single heat source is used on the screw–nut pair, the temperature of node 4 in nodes 1–4 remains relatively high. Owing to the linear motion of the nut component within the effective travel of the screw, the steady-state temperature value of the screw is the same as the temperature of node 4, which increases with increasing nut rotation speed and composite rotation speed. Node 8 is also the point with a higher temperature between nodes 5 and 8, and its variation pattern remains the same as that of node 4. Node 10 and Node 11 are two unique heat source points in the nut-rotating ball screw pair, and they are also the primary factors causing the heat source distribution of the nut-rotating ball–screw transmission mechanism to be different from that of the conventional ball screw pair. The temperature change with the system speed is the key to determining the temperature field distribution of the system.

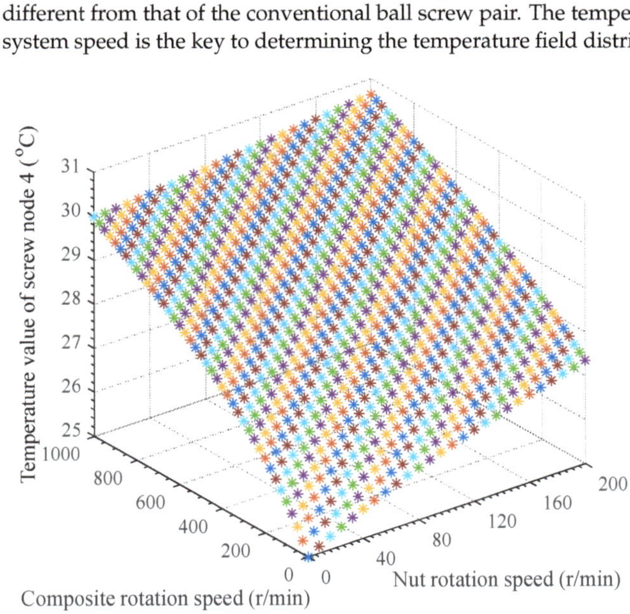

Figure 17. Temperature value of screw node 4 under single heat source.

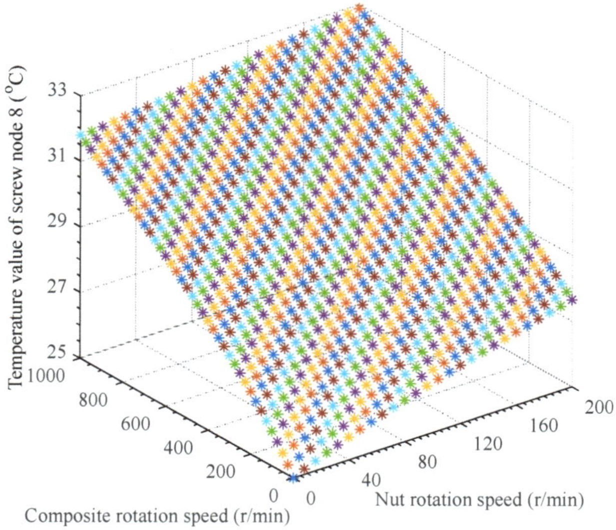

Figure 18. Temperature value of screw node 8 under single heat source.

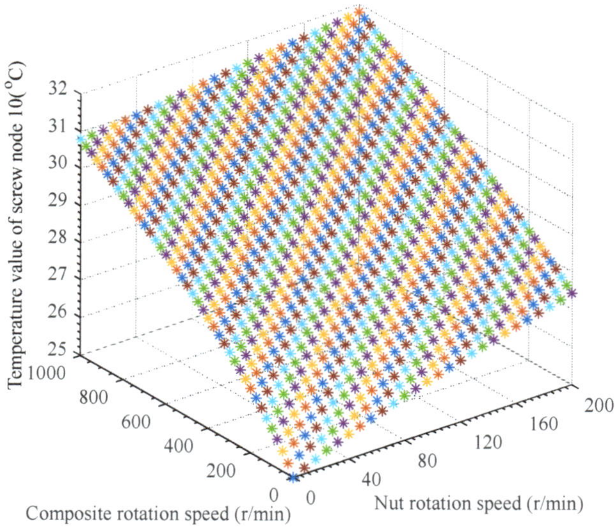

Figure 19. Temperature value of screw node 10 under single heat source.

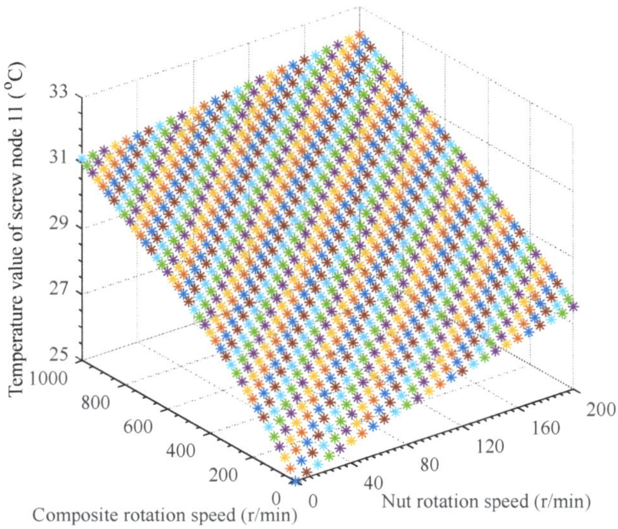

Figure 20. Temperature value of screw node 11 under single heat source.

3.4. Node Synthesis Temperature of Nut-Rotating Ball Screw under Multiple Heat Source Working Conditions

Figures 21–26 show the variation in node temperature values of the nut-rotating ball screw transmission mechanism under multiple heat source working conditions with variations in nut rotation speed and composite rotation speed. In line with the principle of linear superposition, the node temperature of the multiple heat sources is equal to the superposition of temperatures under the action of different single heat sources. Moreover, the composite temperature of the nodes is a prerequisite for addressing the thermal error of the screw.

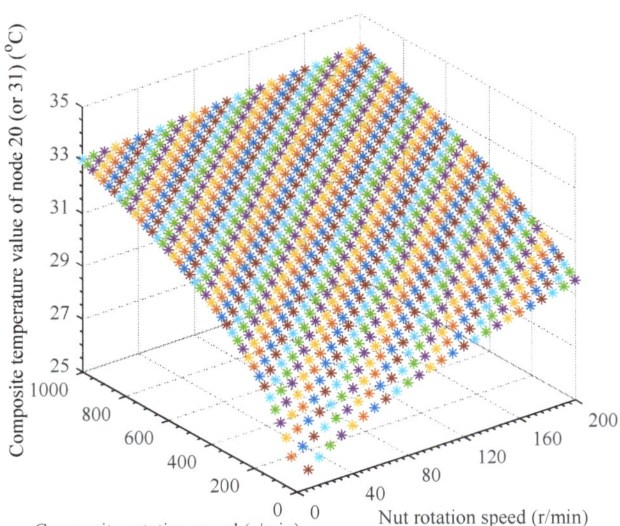

Figure 21. Synthetic temperature value of screw node 20 (or 31) under multiple heat sources.

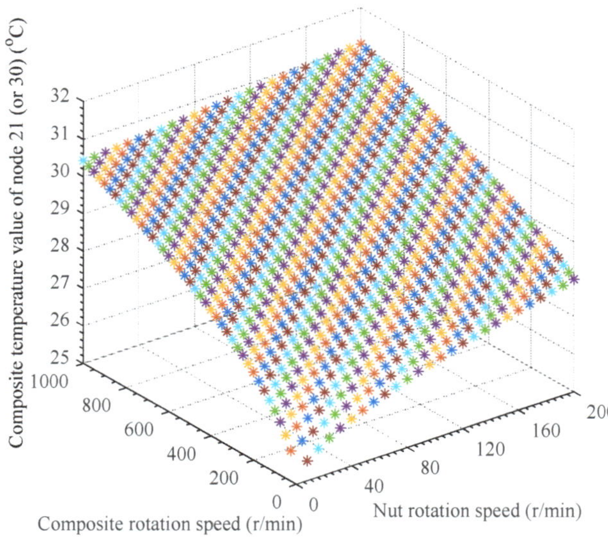

Figure 22. Synthetic temperature value of screw node 21 (or 30) under multiple heat sources.

When the nut rotation speed is $n_N = 0$RPM, the node temperature rises rapidly with the increasing screw speed. When the nut rotation speed and the composite rotation speed are $\Delta \neq 0$RPM, under the same feed rate, the node temperature value under the dual-drive condition is higher than that under the single-drive condition. When the composite rotation speed is $\Delta = 0$RPM, the composite temperature of the nodes is much higher with the increasing nut (or screw) speed compared with the node temperature of the screw single drive at the same speed. The distribution of its temperature field is much more complex than the screw's single-drive mechanism under the same parameters considering the unique transmission components and driving mode of the nut-rotating ball screw transmission mechanism.

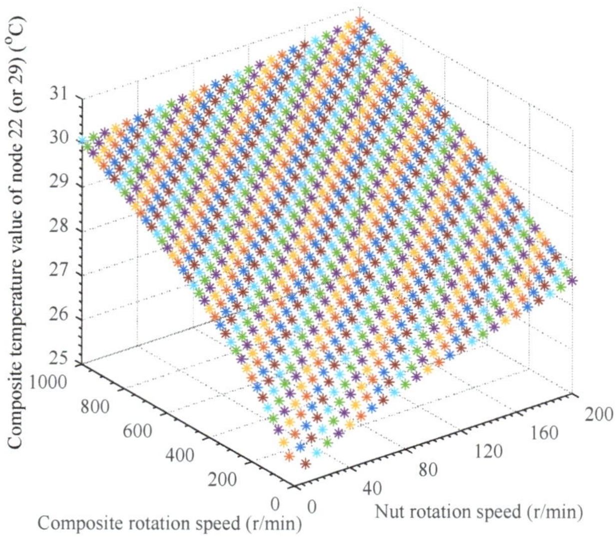

Figure 23. Synthetic temperature value of screw node 22 (or 29) under multiple heat sources.

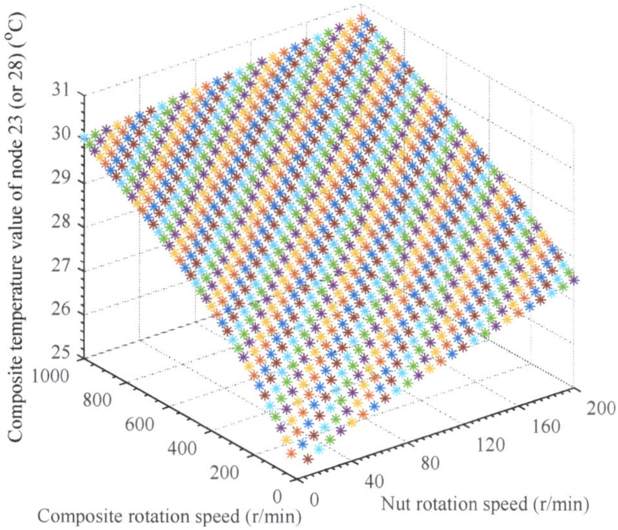

Figure 24. Synthetic temperature value of screw node 23 (or 28) under multiple heat sources.

3.5. Steady State Thermal Error of Nut-Rotating Ball Screw

Figures 27–32 display the steady-state thermal error of each node of the screw as a function of the nut rotation speed and the combined rotation speed. With the increasing distance from the fixed end of the screw, the steady-state thermal error of the screw also tends to increase. Because the steady-state thermal error of the screw can be obtained by integrating temperature with position, its value also increases with the increasing nut rotation speed and the composite rotation speed. The steady-state thermal error of each node under the dual-drive condition is greater than under the single-drive condition. Thermal deformation error is a vital component of motion error in micro-feed systems, and it is essential to consider research for precision feed control.

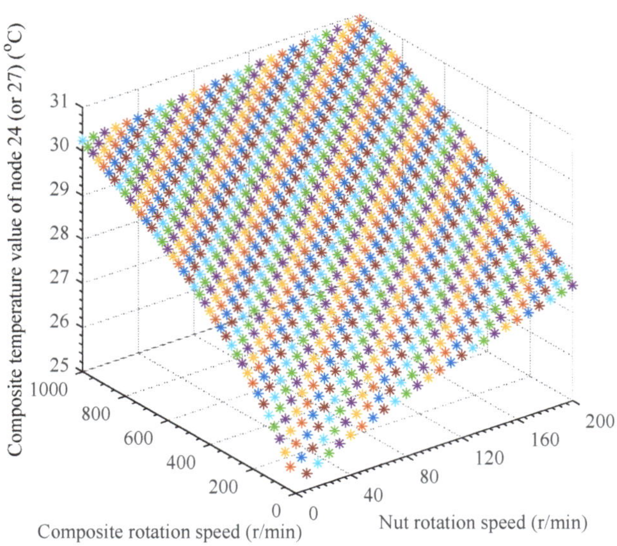

Figure 25. Synthetic temperature value of screw node 24 (or 27) under multiple heat sources.

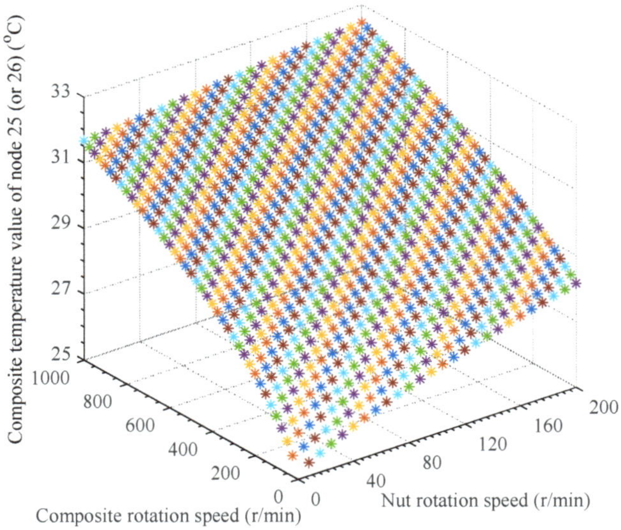

Figure 26. Synthetic temperature value of screw node 25 (or 26) under multiple heat sources.

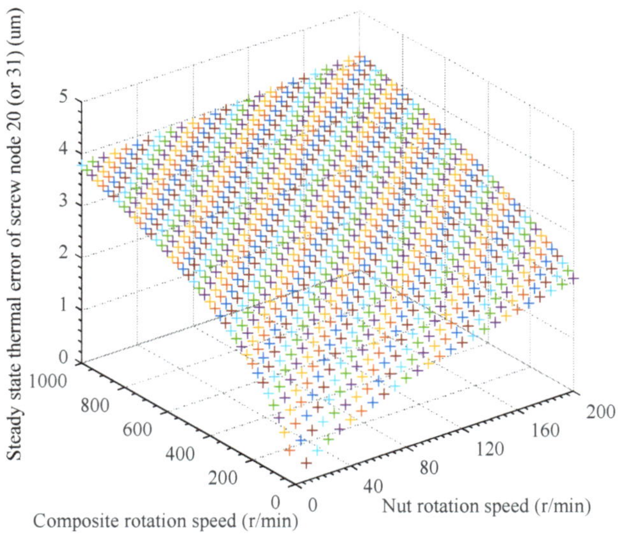

Figure 27. Steady-state thermal error of screw node 20 (or 31).

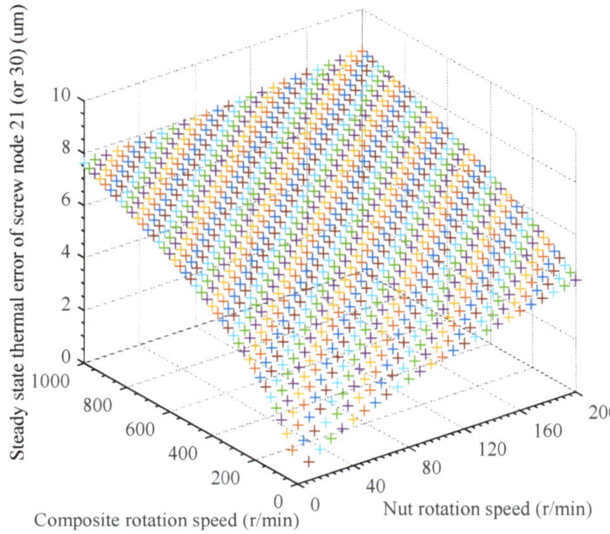

Figure 28. Steady-state thermal error of screw node 21 (or 30).

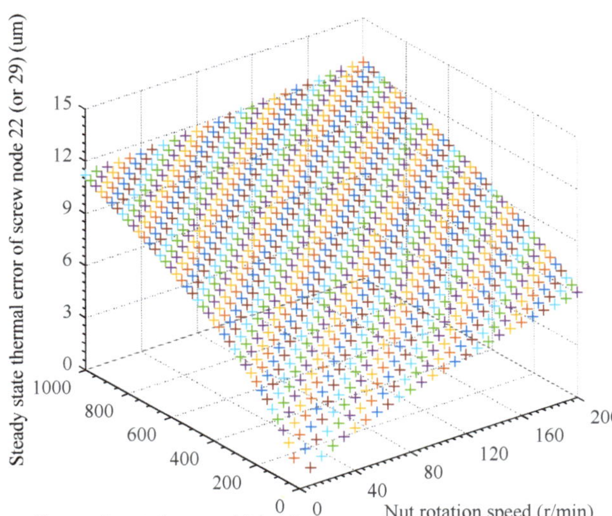

Figure 29. Steady-state thermal error of screw node 22 (or 29).

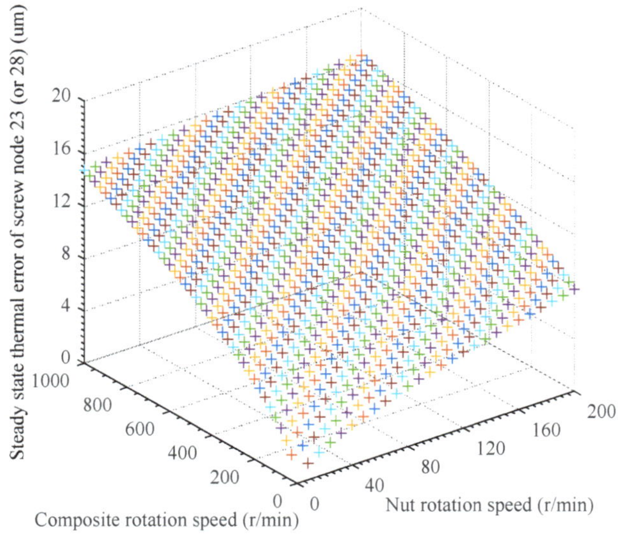

Figure 30. Steady-state thermal error of screw node 23 (or 28).

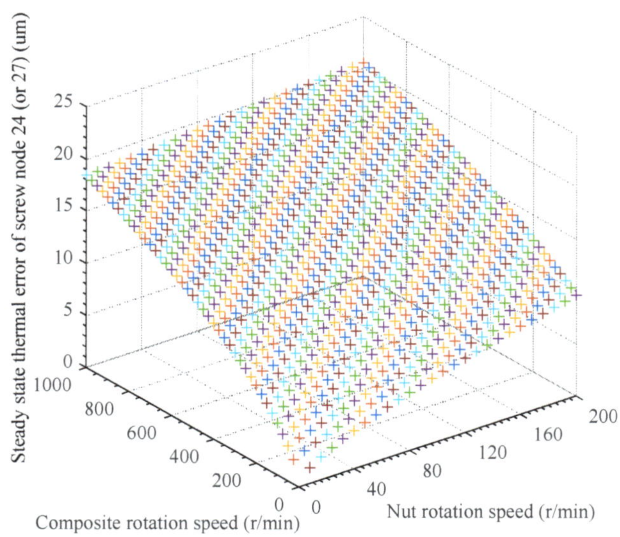

Figure 31. Steady-state thermal error of screw node 24 (or 27).

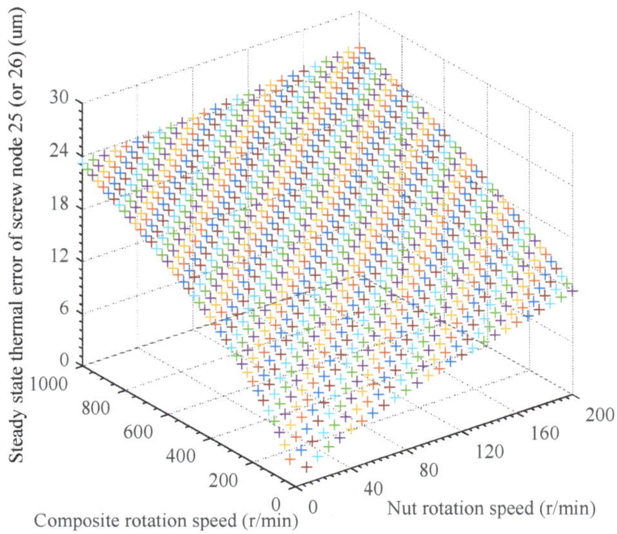

Figure 32. Steady-state thermal error of screw node 25 (or 26).

4. Temperature Rise and Thermal Elongation Testing Experiment for Nut-Rotating Ball Screw Transmission Mechanism

The temperature field distribution and thermal elongation exert a significant impact on the transmission accuracy of a feed system. Therefore, it is necessary to verify the established thermal resistance network model by experimentally measuring the temperature rise in the dual-drive feed system and the thermal elongation of the screw, providing a theoretical foundation for studying thermal error compensation.

4.1. Construction of Temperature Rise and Thermal Elongation Testing System

Figure 33 presents a schematic of the testing scenario. In most cases, the hardware of the detection system consists of temperature sensors (T1, T2, T3, and T4), transmitters, data acquisition cards, micro-displacement sensors (W1), etc. To detect the temperature of the proximal bearing, temperature sensor T1 is installed near the screw motor bearing, while temperature sensor T2 is installed far away from the screw motor bearing to detect the temperature of the remote bearing; in addition, temperature sensor T3 is installed at the flange of the nut assembly to detect the temperature of the nut assembly and temperature sensor T4 is adopted to detect the ambient temperature. Micro-displacement sensor W1 employs the Keens LK-031 laser micrometer to measure axial thermal elongation at the end of the screw, with a maximum measurement length of 1 mm, a transmission output of −5~5 V, and a resolution of 0.1 μm.

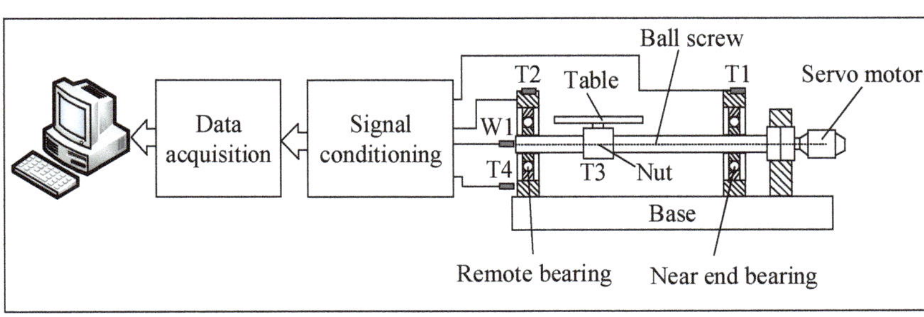

Figure 33. Schematic of the testing scenario.

Based on the temperature measurement requirements of the machine tool as well as the measurement range and accuracy requirements, the temperature sensor takes a Pt100 thermistor and a matching temperature transmitter, with a temperature measurement range of 0 °C–100 °C and a transmission output of 0–5 V. An advantech USB-4711A multi-channel data acquisition card with 16 analog input channels, 12-bit resolution, and a maximum sampling rate of 150 k/s is selected.

Figure 34 presents the experimental setup. According to a previous publication by the author of this paper [33], the minimum speed of both motors is no less than 60 RPM, and the speed difference between the two is no less than 36 RPM to ensure the stable output speed of the dual-drive system table. During the temperature rise and thermal deformation experiment, the nut motor speed was set to 180 RPM, and the screw motor speed was set to 216 RPM. At this speed, the synthetic feed rate of the table was 3 mm/s. The dual-drive system was operated for a sufficient amount of time. The temperature increases at the near-end bearing of the screw motor, the far-end bearing of the screw motor, the flange of the nut component, and the thermal deformation at the end of the screw were detected through temperature sensors, transmitters, data acquisition cards, and micro displacement sensors.

4.2. Experimental Results and Test Analysis

Figures 35–37 present the temperature changes at the near-end bearing of the screw motor, the far-end bearing of the screw motor, and the flange of the nut assembly under the dual-drive condition, indicated by a nut speed of 180 RPM and a screw speed of 216 RPM. As per Figure 35, after running for a sufficient period, the near-end bearing of the screw motor reaches a thermal steady state. The experimentally measured temperature rise value is approximately 7.5 °C, and the simulated steady-state temperature rise value of the near-end bearing of the screw motor is 6.21 °C. Figure 36 shows that the experimental temperature rise in the remote bearing of the screw motor after reaching thermal stability is approximately 4 °C, and the simulation shows that the steady-state temperature rise in the remote bearing of the screw motor is 3.2 °C. Figure 37 shows that the experimental temperature rise in the nut assembly flange after reaching thermal stability is about 3.8 °C, and the simulation demonstrates that the steady-state temperature increase in the nut assembly flange is 2.37 °C. The steady-state temperature rise experimental values obtained at the key points are consistent with the simulation results.

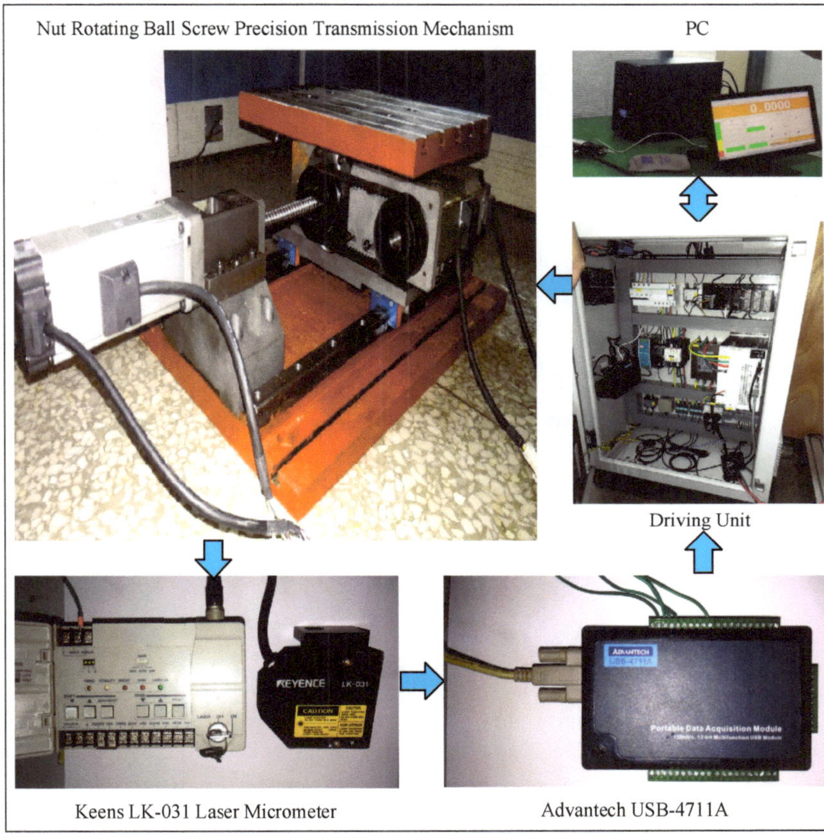

Figure 34. Experimental setup.

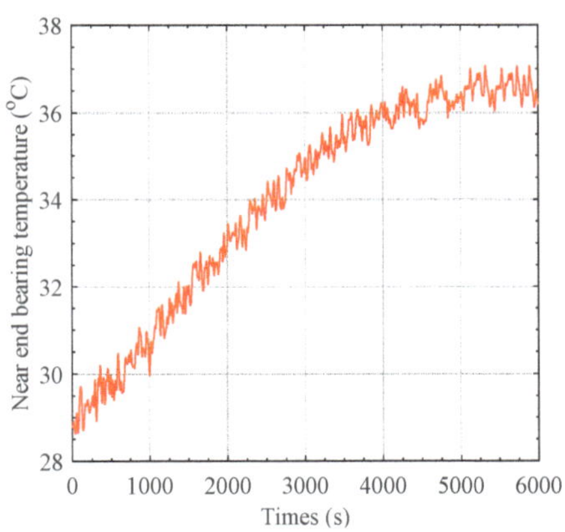

Figure 35. Temperature of the screw motor near end bearing under dual-drive condition.

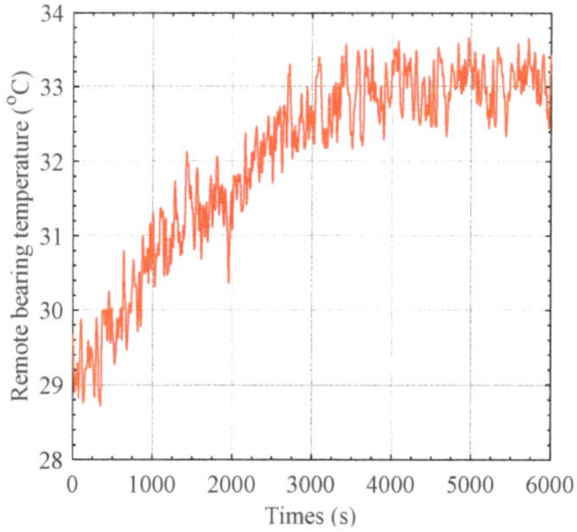

Figure 36. Temperature of the screw motor remote end bearing under dual-drive condition.

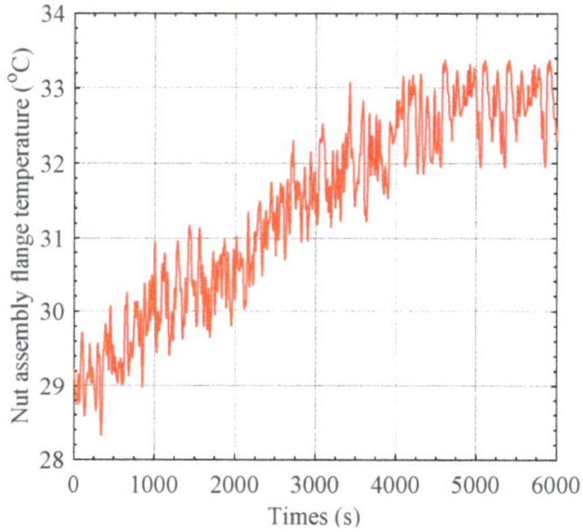

Figure 37. Temperature of the nut assembly flange under dual-drive condition.

Figure 38 shows that the thermal elongation at the end of the screw was experimentally measured under the dual-drive condition. After reaching the thermal steady state, the measured thermal elongation at the end of the screw in the experiment is approximately 10.5 μm, and the simulation result is 11.98 μm. The experimental measurement result demonstrates the accuracy of the theoretical analysis model for thermal error at the end of the screw.

Through system temperature rise and thermal deformation experiments, it can be seen that the temperature field distribution of the nut-rotating ball screw transmission mechanism is much more complex than the traditional screw single-drive system, and the temperature rise is also more obvious. This phenomenon is mainly caused by the complex structure and unique driving method of the nut-rotating ball screw pair. In the nut assembly, friction heat is generated between the nut and the ball, with friction heat generation being integrated into the rolling bearing inside the nut assembly. Installing the nut assembly in a relatively narrow space is not conducive to heat dissipation, which can easily increase the system temperature and thermal deformation. In this study, the thermal network method is used to establish a thermodynamic model of a nut-rotating ball screw transmission mechanism. The correctness of the theoretical model is demonstrated through the experiments, which is important for improving the transmission accuracy of the system.

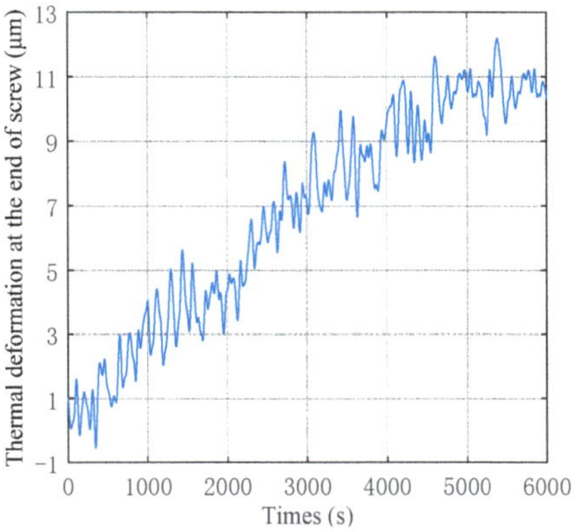

Figure 38. Thermal elongation at the end of the screw under dual-drive condition.

4.3. *Thermal Effects Optimization Measures*

The thermal deformation caused by temperature rise can affect the feed accuracy of the system, so the thermal effects of the system can be optimized and improved from the following three perspectives.

4.3.1. Reduce the Heat Generation of the System Heat Source

(1) Using fewer intermediate transmission links to reduce the number of heat sources;
(2) Using non-contact bearings, guide rails, and screws (such as hydraulic bearings, hydrostatic guide rails, and hydrostatic screws) or low-friction ceramic bearings;
(3) Reasonably adjusting the pre-tightening force of the bearing and screw and supplementing with appropriate lubricants;
(4) Improving the structure of the ball screw pair, such as by increasing the number of threaded heads, reducing the diameter of steel balls, using hollow balls, and improving the surface processing quality of the raceway.

4.3.2. Enhance the Cooling Capacity of the System

(1) Using hollow ball screw pairs as transmission components and performing forced circulation cooling on the screw pairs and bearings;
(2) Using oil–air micro-lubrication;
(3) Set a constant temperature working environment.

4.3.3. Compensating for Thermal Errors in the Transmission System

(1) Feedback interception method: By inserting a feedback loop into the servo system, thermal error compensation is achieved by adjusting the position of the tool holder;
(2) Zero drift method: Calculating the compensation amount based on the established thermal error model and then sending it to the CNC controller. Finally, the compensation amount is added at each position as a command signal to the servo loop for feedforward compensation.

5. Conclusions

(1) A thermal resistance network model of a nut-rotating ball screw transmission system was established based on heat transfer theory. By addressing the system's thermal balance equations, the distribution of its temperature field was analyzed. The results objectively reflected the temperature distribution in the system. The temperature distribution of the entire system basically conforms to the law of heat flow, indicating that the use of the thermal resistance network method for temperature field analysis of the multi-heat source system is effective. The prediction accuracy of the model can

be enhanced by arranging more temperature nodes and employing experimental data to identify and correct the model parameters.

(2) When a unilateral heat source functions, as the distance from the heat source point increases, the steady-state temperature value of the node rapidly decays. Under the condition that the screw speed reaches a certain value, there is a point where its temperature will maintain a thermal equilibrium state and not change with the increase in screw speed. At nodes far away from the heat source, the temperature is characterized by a slight increase followed by a decrease and tends towards the ambient temperature as the screw speed increases.

(3) The outer ring supporting the bearing in the nut assembly is integrated with the flange in the nut assembly, and the inner ring supporting the bearing can be integrated with the outer ring of the rotating nut. In addition, the nut and screw are driven separately. Due to the proposed structure and driving method, the dual-drive mechanism has varying heat source points from the conventional mechanism, which are also the main factors causing the different heat source distributions of the dual-drive feed mechanism.

(4) Under the action of multiple heat sources and the same feed rate, the node temperature value under the dual-drive condition is higher when compared with that under the single-drive condition, and the distribution of the temperature field is much more complex than that of the single-drive mechanism under the same parameters. Considering the fact that the steady-state thermal error of the screw can be acquired by integrating temperature with the position, the value of the steady-state thermal error also tends to increase with an increase in nut rotation speed and composite rotation speed. In addition, the steady-state thermal error of each node under the dual-drive condition is greater than that under the single-drive condition.

(5) The potential applications of the nut rotary ball screw transmission mechanism and the results of numerical analysis can be applied to advanced technology fields such as robotics, suspensions, powertrain, national defense, integrated electronics, optoelectronics, medicine, and genetic engineering so that the new system can have a lower stable speed limit and achieve precise micro-feed control.

Author Contributions: H.Y. was responsible for the conception and design, acquisition of data, analysis and interpretation of data, drafting the initial manuscript, and revising it critically for important intellectual content; X.L. and G.Z. collected the data and pictures, completed the follow-up information, and wrote the draft; G.H., Y.L. (Yan Liu) and H.X. were responsible for numerical analysis of the data and provided valuable suggestions for the manuscript; Y.L. (Yandong Liu) was responsible for the conception and design, interpretation of data, and reviewing all drafts of the manuscript; X.F. and Z.L. were responsible for the conception and design, and reviewing all drafts of the manuscript. All authors have read and agreed to the published version of the manuscript.

Funding: This research was funded by the Doctoral Research Fund Project of Shandong Jianzhu University (Grant No. X21030Z).

Institutional Review Board Statement: Not applicable.

Informed Consent Statement: Not applicable.

Data Availability Statement: The original contributions presented in the study are included in the article, and further inquiries can be directed to the corresponding author.

Conflicts of Interest: Authors Xiaokui Fu and Zhi Liu were employed by the company Jinan Hope Wish Photoelectronic Technology Co., Ltd. The remaining authors declare no conflicts of interest.

References

1. Delor, M.; Weaver, H.L.; Yu, Q.Q.; Ginsberg, N.S. Imaging material functionality through three-dimensional nanoscale tracking of energy flow. *Nat. Mater.* **2020**, *19*, 56–62. [CrossRef]
2. Li, Z.M.; Gao, X.Y.; Yang, J.K.; Xin, X.D.; Yi, X.Y.; Bian, L.; Dong, S.X. Designing Ordered Structure with Piezoceramic Actuation Units (OSPAU) for Generating Continual Nanostep Motion. *Adv. Sci.* **2020**, *7*, 2001155. [CrossRef]

3. Russo, M.; Zhang, D.; Liu, X.J.; Xie, Z.H. A review of parallel kinematic machine tools: Design, modeling, and applications. *Int. J. Mach. Tools Manuf.* **2024**, *196*, 104118. [CrossRef]
4. Weng, L.T.; Gao, W.G.; Zhang, D.W.; Huang, T.; Duan, G.L.; Liu, T.; Zheng, Y.J.; Shi, K. Analytical modelling of transient thermal characteristics of precision machine tools and real-time active thermal control method. *Int. J. Mach. Tools Manuf.* **2023**, *186*, 104003. [CrossRef]
5. Li, Y.; Yu, M.L.; Bai, Y.M.; Hou, Z.Y.; Wu, W.W. A review of thermal error modeling methods for machine tools. *Appl. Sci.* **2021**, *11*, 5216. [CrossRef]
6. Han, L.; Tong, Z.Y. A thermal resistance network model based on three-dimensional structure. *Measurement* **2019**, *133*, 439–443. [CrossRef]
7. Li, D.T.; He, Z.L.; Wang, C.; Sun, S.Z.; Ma, K.; Xing, Z.W. Simulation of dry screw vacuum pumps based on chamber model and thermal resistance network. *Appl. Therm. Eng.* **2022**, *211*, 118460. [CrossRef]
8. Sun, Z.H.; Xu, Y.F.; Wang, Q.J.; Xu, J.Z.; Li, G.L.; Wen, Y. Design and identification of lumped-parameter thermal network model for real-time temperature estimation of permanent-magnet spherical motors. *Therm. Sci. Eng. Prog.* **2023**, *44*, 102047. [CrossRef]
9. Meng, Q.Y.; Yan, X.X.; Sun, C.C.; Liu, Y. Research on thermal resistance network modeling of motorized spindle based on the influence of various fractal parameters. *Int. Commun. Heat Mass Transf.* **2020**, *117*, 104806. [CrossRef]
10. Zhu, Z.Y.; Zhang, W.; Li, Y.B.; Guo, J. Thermal analysis of axial permanent magnet flywheel machine based on equivalent thermal network method. *IEEE Access* **2021**, *9*, 33181–33188. [CrossRef]
11. Bao, Y.J.; Ma, Y.X.; Yang, Y.X.; Wang, J.L.; Cheng, D. Study on the effect of carbon fiber ply angle on thermal response of CFRP laminate based on the thermal resistance network model. *Polym. Compos.* **2023**, *44*, 8792–8804. [CrossRef]
12. Zhan, Z.Q.; Fang, B.; Wan, S.K.; Bai, Y.; Hong, J.; Li, X.H. An accurate and efficient implicit thermal network method for the steady-state temperature field. *Proc. Inst. Mech. Eng. Part C J. Mech. Eng. Sci.* **2024**, *238*, 1800–1810. [CrossRef]
13. Li, T.J.; Wang, M.Z.; Zhao, C.Y. Study on real-time thermal–mechanical–frictional coupling characteristics of ball bearings based on the inverse thermal network method. *Proc. Inst. Mech. Eng. Part J J. Eng. Tribol.* **2021**, *235*, 2335–2349. [CrossRef]
14. Yang, Y.B.; Zhou, X. A volumetric heat source model for thermal modeling of additive manufacturing of metals. *Metals* **2020**, *10*, 1406. [CrossRef]
15. Wu, H.Y.; Guan, Q.; Xi, C.F.; Zuo, D.W. Construction of dynamic temperature field model of ball screw based on superposition of positive and negative temperature fields. *Numer. Heat Transf. Part A Appl.* **2023**, *83*, 343–360. [CrossRef]
16. Liu, J.L.; Ma, C.; Wang, S.L.; Wang, S.B.; Yang, B.; Shi, H. Thermal boundary condition optimization of ball screw feed drive system based on response surface analysis. *Mech. Syst. Signal Process.* **2019**, *121*, 471–495. [CrossRef]
17. Sheng, X.; Lu, X.; Zhang, J.R.; Lu, Y.Q. An analytical solution to temperature field distribution in a thick rod subjected to periodic-motion heat sources and application in ball screws. *Eng. Optim.* **2021**, *53*, 2144–2163. [CrossRef]
18. Wang, L.; Jia, Z.Y.; Zhang, L. Investigation on the accurate calculation of the temperature field of permanent magnet governor and the optimization method of heat conduction. *Case Stud. Therm. Eng.* **2019**, *13*, 100360. [CrossRef]
19. Liu, W.Z.; Zhang, C.; Duan, F.J.; Fu, X.; Bao, R.J.; Yu, Z.X.; Gong, X. An optimization method of temperature field distribution to improve the accuracy of laser multi-degree-of-freedom measurement system. *Optik* **2022**, *269*, 169721. [CrossRef]
20. Sun, S.B.; Yan, S.H.; Cao, X.P.; Zhang, W. Distribution law of the initial temperature field in a railway tunnel with high rock temperature: A model test and numerical analysis. *Appl. Sci.* **2023**, *13*, 1638. [CrossRef]
21. Gao, X.S.; Zhang, K.; Zhang, Z.T.; Wang, M.; Zan, T.; Gao, P. XGBoost-based thermal error prediction and compensation of ball screws. *Proc. Inst. Mech. Eng. Part B J. Eng. Manuf.* **2024**, *238*, 151–163. [CrossRef]
22. Su, D.X.; Li, Y.; Zhao, W.H.; Zhang, H.J. Transient thermal error modeling of a ball screw feed system. *Int. J. Adv. Manuf. Technol.* **2023**, *124*, 2095–2107. [CrossRef]
23. Yang, J.C.; Li, C.Y.; Xu, M.T.; Zhang, Y.M. Analysis of thermal error model of ball screw feed system based on experimental data. *Int. J. Adv. Manuf. Technol.* **2022**, *119*, 7415–7427. [CrossRef]
24. Li, Y.; Fan, J.B.; Zheng, Y.N.; Gao, F.; Li, W.Q.; Hei, C.F. Thermal error compensation for a fluid-cooling ball-screw feed system. *Proc. Inst. Mech. Eng. Part B J. Eng. Manuf.* **2024**, 09544054231210963. [CrossRef]
25. Liu, H.L.; Rao, Z.F.; Pang, R.D.; Zhang, Y.M. Research on thermal characteristics of ball screw feed system considering nut movement. *Machines* **2021**, *9*, 249. [CrossRef]
26. Shi, H.; Ma, C.; Yang, J.; Zhao, L.; Mei, X.S.; Gong, G.F. Investigation into effect of thermal expansion on thermally induced error of ball screw feed drive system of precision machine tools. *Int. J. Mach. Tools Manuf.* **2015**, *97*, 60–71. [CrossRef]
27. Rong, R.; Zhou, H.C.; Huang, Y.B.; Yang, J.Z.; Xiang, H. Novel Real-Time Compensation Method for Machine Tool's Ball Screw Thermal Error. *Appl. Sci.* **2023**, *13*, 2833. [CrossRef]
28. Cao, L.; Park, C.H.; Chung, S.C. Real-time thermal error prediction and compensation of ball screw feed systems via model order reduction and hybrid boundary condition update. *Precis. Eng.-J. Int. Soc. Precis. Eng. Nanotechnol.* **2022**, *77*, 227–240. [CrossRef]
29. Tanaka, S.; Kizaki, T.; Tomita, K.; Tsujimura, S.; Kobayashi, H.; Sugita, N. Robust thermal error estimation for machine tools based on in-process multi-point temperature measurement of a single axis actuated by a ball screw feed drive system. *J. Manuf. Process.* **2023**, *85*, 262–271. [CrossRef]
30. Liu, H.Y.; Deng, H.G.; Feng, X.Y.; Liu, Y.D.; Li, Y.F.; Yao, M. Data-driven thermal error modeling based on a novel method of temperature measuring point selection. *Int. J. Adv. Manuf. Technol.* **2024**, *131*, 1823–1848. [CrossRef]

31. Peng, J.; Yin, M.; Cao, L.; Liao, Q.H.; Wang, L.; Yin, G.F. Study on the spindle axial thermal error of a five-axis machining center considering the thermal bending effect. *Precis. Eng.* **2022**, *75*, 210–226. [CrossRef]
32. Yu, H.W.; Feng, X.Y.; Sun, Q. Kinematic analysis and simulation of a new type of differential micro-feed mechanism with friction. *Sci. Prog.* **2020**, *103*, 0036850419875667. [CrossRef] [PubMed]
33. Yu, H.W.; Zhang, L.G.; Wang, C.; Feng, X.Y. Dynamic characteristics analysis and experimental of differential dual drive servo feed system. *Proc. Inst. Mech. Eng. Part C J. Mech. Eng. Sci.* **2021**, *235*, 6737–6751. [CrossRef]

materials MDPI

Article

Influence of Materials of Moulds and Geometry of Specimens on Mechanical Properties of Grouts Based on Ultrafine Hydraulic Binder

Beatriz Hortigon [1,*], Esperanza Rodriguez-Mayorga [2], Jose Antonio Santiago-Espinal [1], Fernando Ancio [1] and Jose Maria Gallardo [3]

1 Departamento de Mecánica de Medios Continuos y Teoría de Estructuras, Escuela Politécnica Superior, Universidad de Sevilla, 41011 Sevilla, Spain; jsespinal@us.es (J.A.S.-E.); ancio@us.es (F.A.)
2 Departamento de Estructuras de Edificación e Ingeniería del Terreno, Escuela Técnica Superior de Arquitectura, Universidad de Sevilla, 41012 Sevilla, Spain; espe@us.es
3 Departamento de Ingeniería y Ciencia de los Materiales y del Transporte, Escuela Técnica Superior de Ingeniería, Universidad de Sevilla, 41092 Sevilla, Spain; josemar@us.es
* Correspondence: bhortigon@us.es; Tel.: +34-954552828

Abstract: Ultrafine hydraulic binder grout injection is a technique utilised for repairing masonry, either to connect sections, seal joints, or fill voids due to its great capacity for penetration and higher mechanical strength than lime grout. In this research, the mechanical properties of ultrafine hydraulic cement grout are analysed considering the influence of the mould material for preparing the specimens and their geometry characteristics in the context of the specifications set out in several international standards. The test campaign to ascertain compressive and flexural strength in different circumstances is supplemented with a physical and chemical characterisation of both binder and fresh and hardened grout. Significant differences in mechanical properties between specimens prepared with absorbent or non-absorbent-water material are found due to the influence of drying shrinkage and decanting binder during the curing process. Furthermore, the slenderness of specimens is presented as an important factor in determining the compressive strength of mixtures.

Keywords: ultrafine hydraulic cement grout; injection of masonry; material of moulds; shape of specimens; compressive strength; flexural strength

Citation: Hortigon, B.; Rodriguez-Mayorga, E.; Santiago-Espinal, J.A.; Ancio, F.; Gallardo, J.M. Influence of Materials of Moulds and Geometry of Specimens on Mechanical Properties of Grouts Based on Ultrafine Hydraulic Binder. *Materials* **2024**, *17*, 1645. https://doi.org/10.3390/ma17071645

Academic Editor: John D. Clayton

Received: 5 March 2024
Revised: 29 March 2024
Accepted: 1 April 2024
Published: 3 April 2024

1. Introduction

Grout injections are one of the most common methods for the consolidation of soils. This method has been used for decades in geotechnical applications and always supplies quality results [1–3]. Grout injection can be executed in numerous ways, depending mainly on the tool employed to inject and the nature of the soil to be consolidated. A common method of grout injection uses the sleeve-port-pipe as the injection tool, also known by its French name tube-a-manchette (TAM). The procedure is suitable for the consolidation and stabilisation of soils since it enables the injection to be carried out in progressive stages or using reinjection [4]. In this way, in the first stages of the consolidation, binders with larger grain sizes than in the last stages must be used. TAM injections usually begin with cement-based grouts, in many cases coupled with bentonite and different proportions of water. The last stages of grout injected are based on binders with finer grain sizes, such as microfine and ultrafine cements and even chemicals, which provide their greater penetrability, strength, and stiffness.

Grout injection has also long been used as a consolidation technique for ancient masonries. The presence of voids is frequent in old masonries, either due to their own morphology or to the levels of degradation that this historical constructive system presents, thereby rendering it suitable for injection [5–7]. These types of masonry are normally injected with lime-based grouts since they are chemically compatible with the usual calcareous nature of ashlars and mortars that can be found in historical buildings. Despite this, several efforts have been focused on modifying the mechanical properties of lime grouts, inherently characterised by low values of strength and stiffness [8]. Admixtures with Portland cement or other additions have also been extensively researched and applied [9–11]. It is frequent that these trials use solely Portland cement as the binder. However, in order to attain injectable and masonry-compatible grout, additions are usually included, such as fly ash [12–14] and blast-furnace slag [13,15,16]. These additions not only improve the fluidity of the admixtures and reduce bleeding, but they also reduce the carbon footprint of the grouts and are more economical than exclusively cement-based grouts. As a disadvantage, compressive strength, flexural strength, and bond strength decrease, although other additives including pozzolans, such as metakaolin [17–19], perlite [20], and bentonite [12], do not affect the mechanical properties of the grouts. Recent interventions have demonstrated that certain types of ultrafine binders perform correctly when applied to historic masonries, such as the interventions in San Dionisio's Church [21], Santiago's Church [22,23], and Roman Theatre of Cadiz [24], all of them in Spain.

Full knowledge of the materials used in the repair is compulsory for the success of the intervention. Since ultrafine blast-furnace slag binder grouts are largely used for geotechnical purposes, this must be mechanically characterised when they are injected to repair masonry. This research has been conducted to test and document the values of compressive and flexural strength, along with other physical and chemical properties, of these kinds of grouts.

Significant differences can be found between several standards that refer to grouts regarding the procedure for the preparation of specimens to be tested and their geometry. The American Standard ASTM C1019 [25] deals only with grouts to inject masonries, while the European Standard EN 445 [26] addresses grouts for all purposes. Regarding the material of the moulds, the use of metallic moulds is established in EN 445 [26], and ASTM C1019 [25] offers two possible choices: (i) the use of stony or ceramic moulds in keeping with the masonry to be repaired, or otherwise another absorbent material such as wood, thereby reproducing the grout-masonry bond and (ii) when several specimens are simultaneously manufactured, the lateral faces of the specimens must be moulded with a material similar to masonry units, while the bases of the moulds and the internal partitions between specimens are of a non-absorbent material. The amount of water contained in the mixture is decisive in its final behaviour, and, given the diverse water absorption capacity of said materials, the mould material will be a key factor in the final mechanical properties of the hardened grout. In this context, shrinkage, in that it can induce the appearance of cracks and porosity, is one of the several factors that can affect mechanical properties. Plastic and drying shrinkage must be considered [27]. Plastic shrinkage depends on water loss due to evaporation and water suction from the subbase. Consequently, the use of absorbent moulds must be considered a potential cause of increased shrinkage. On the other hand, drying shrinkage is related to capillary forces as absorbed water is lost in 2.5 to 50 nm pores. Shrinkage increases as the water/binder ratio increases [28] and certain additives, such as blast furnace slag, are incorporated [29].

With regard to obtaining the mechanical properties of the grouts, the standard EN 445 [26] adheres to the mortar standard EN 196-1 [30]. Consequently, most research projects aim to ascertain these properties in accordance with these standards [20,31–33], and even standards for concrete are applied [12,14]. Nevertheless, significant differences can be observed in these standards concerning the shape of specimens when ascertaining compressive strength. While ASTM-C1019 [25], EN 196-1 [30], EN 1015-11 [34], and ASTM C942-15 [35] propose a parallelepiped shape, IS 4031-4 [36] establishes cuboid specimens. In this context, several research studies can be found regarding the inversely proportional relationship between the slenderness of concrete [37–39] or mortars based on Portland cement [40–42] specimens and their compressive strength.

Specifically, the compressive strength of mortars in joints can also be obtained using the Double Punch Test (hereinafter DPT). The standard DIN 18555-9 [43] proposes thin prism specimens for DPT (Table 1) hardened in moulds of masonry.

Table 1. Specimen sizes for the compressive test of cement-based mortars and grouts in accordance with different standards.

Standard	Mixture	Country	Specimen Size (mm)	Material of Mould
EN 445 [26] (Refers to EN 196-1 [30])	Grout	Europe	$40 \times 40 \times 80$ $40 \times 40 \times 160$	Steel
ASTM C942-15 [35]	Grout	United States	$40 \times 40 \times 160$	Steel
ASTM C1019 [25]	Grout	United States	Proportions of prism specimen H [1] = 2L [2]	Masonry units with non-absorbent mould bottom and optional spacers
EN 1015-11 [34]	Mortar	Europe	$40 \times 40 \times 80$ $40 \times 40 \times 160$	Metallic
IS 4031-4 [36]	Mortar	India	$70.6 \times 70.6 \times 70.6$	Cast iron/Mild Steel [44]
GB/T 17671 [45]	Mortar	China	$40 \times 40 \times 160$	Steel [46]
DIN 18555-9 [43] [3]	Mortar	Germany	$40 \times 40 \times 16$	Masonry units on all the mould faces

[1] Height of the prism; [2] Edge of the base, [3] Double-Punch Test.

Since sealing joints and cracks is one of the main applications of grouts, the results obtained from DPT are considered interesting for the determination of the compression strength of the mixture in this situation. With regard to obtaining flexural strength, the three-point flexural test is the proposed methodology, where researchers [12,31,47–49] and standards [30,34,45,50] employ similar shapes and sizes of specimens.

Consequently, this study investigates the effects of the material of the moulds and the shape of the specimens utilised in different standards regarding the testing of grouts made with 0.75 water/binder dosage, based on the flexural and compressive strength and other characteristics of the obtained samples. As an aid to discussing the obtained results, the physical characterisation of the specimens has also been carried out.

2. Materials and Methods

This section provides a detailed description of the experimental campaign (Table 2) accomplished to characterise grouts made of microfine ground granulated blast slag (hereinafter GGBS) furnace binders while paying special attention to the specific standard followed in the execution of the test. Hydraulic SPINOR® A12, from HOLCIM S.A:S., Paris, France, has been employed.

Table 2. Test for the evaluation of grouts.

Analysed Element	Property	Laboratory Test	Section
Binder	Chemical composition Particle size	SEM/EDX Laser diffraction	Section 2.1.1 Section 2.1.2
Cured blocks	Visual observations/Shrinkage Physico-chemical characterisation Physical characterisation	Observation of dimensional changes and cracks SEM/EDX CT	Section 2.3.1 Section 2.3.2 Section 2.3.3
Specimens	Density Flexural strength Compression strength	Measurement and weight Three-point flexural test Compression test/DPT	Section 2.4.2 Section 2.4.3 Section 2.4.3

2.1. Binder

2.1.1. Chemical Composition

According to the product specification, this binder is prepared by finely grinding a mixture of GGBS and clinker. The chemical characterisation provided by the manufacturer

(Table 3) gives a typical proportion of the components [51]. This group of binders is usually prepared from 30–70% GGBS and Portland cement mixtures, even though EN 197-1:2011 [52] permits broader ranges. The composition given earlier is an average of the typical compositions of each of the two substances as follows: CaO (64–67%), Al_2O_3 (4.4–5.2%), Fe_2O_3 (2.2–4.0%), SiO_2 (21.5–23.5%), MgO (0.7–0.86%) (Clinker), and CaO (30–50%), Al_2O_3 (8–24%), SiO_2 (28–38%), MgO (1–18%) (GGBS).

Table 3. Chemical composition (%) of SPINOR® A12.

CaO	Al_2O_3	Fe_2O_3	SiO_2	MgO
44	9.5	1.3	31	6.5

Scanning Electron Microscopy (SEM) has been carried out in both Zeiss EVO, from Carl Zeiss Microscopy, Jena, Germany, and FEI TENEO equipment, from FEI Company, Hillsboro, OR, USA, for the analysis of cured blocks and received binder, respectively. To this end, a small specimen of binder was adhered to a carbon ribbon and coated with a gold-sputtered film for observation. Morphological and chemical characterisation was carried out using up to 30 kV accelerating voltage and secondary electrons, back-scattered electrons, and OXFORD Energy Dispersive X-ray analysis (hereinafter EDX) detectors, from Oxford Instruments Pic, Abingdon, UK

2.1.2. Particle Size

The binder particle size was measured using a specific dry method for cements, with Malvern Mastersizer 2000 equipment, from Akribis Scientific Limited, Manchester, UK. The Mastersize 2000 measurement technique is based on laser diffraction to find a particular material size distribution. In particular, the equipment measures the intensity of light scattered when the laser beam goes through the sample dispersed in a fluid.

2.2. Fresh Grout

Grouts were prepared with a constant dosage water/binder ratio of 0.75, according to both the values recommended for cement-based grouts utilised in the repair of masonry and grouting [53]. A sole additive, superplasticizer Plast 355, from Sika AG, Baar, Swizerland,, was added at a percentage of 5%, in accordance with the manufacturer's guidelines, to prevent the flocculation of ultrafine particles and the formation of lumps. A Bunsen AGV-10-1321 high-turbulence mixer, from Bunsen S.A., Alcorcon, Spain, with a top speed of 2000 rpm, was employed to prepare the grouts. As a first step, water and the superplasticizer were mixed in a 3 L stainless steel container. Mixing was carried out for 1 min at a low rotary speed of 400 rpm. The binder was subsequently poured as the rotary speed was progressively increased up to 1800 rpm maintaining this value, once was reached, for 15 min. The whole process was carried out at a room temperature of 20 ± 2 °C and a relative humidity of over 50%. Fresh grouts were immediately poured inside the moulds.

In order to consider the influence of the moulds suggested in the standards, two different materials were used: methacrylate as a non-water-absorbent material, in accordance with EN 445 [26]; and medium-density fibreboard (hereinafter MDF) as an absorbent material in accordance with the first choice offered by ASTM C1019 [25]. These materials were combined in three types of moulds. Type I moulds were prepared entirely from methacrylate plates. Type II moulds were made from MDF 10 mm thick boards, and type III moulds combined both materials, also in accordance with [25] in its second choice. Type I and II moulds, measuring $40 \times 160 \times 110$ mm, were utilised to harden blocks individually. In type III moulds, methacrylate and MDF were combined in the lateral walls of the moulds, while internal methacrylate walls divided the internal cavity of the mould into three smaller cavities, two of which measured 40 mm × 40 mm × 110 mm and the third one $40 \times 80 \times 110$ mm (Figure 1a).

Figure 1. (a) type I moulds (methacrylate), type II moulds (MDF), and type III moulds (methacrylate and MDF); (b) Curing of fresh grouts inside the curing cabinet; (c) Curing cabinet.

The fresh grout, once poured into the moulds, up to a height of 105 mm (Figure 1b), was stored in a curing cabinet at 20 °C and 90% relative humidity for 48 h, at which point they were demoulded. The blocks were then returned to the curing cabinet for 26 additional days maintaining the temperature at 20 °C and the relative humidity at 100% (Figure 1c). In this way, type I, II, and III blocks were obtained for the analysis from the mould type I, II, and III moulds, respectively.

2.3. Cured Blocks

2.3.1. Visual Observations/Shrinkage Measurement

Differences between the diverse kinds of blocks after curing can be clearly observed. Density and visual aspects provide evidence of these differences. Visual observations reveal irregular surfaces where the block hardened in contact with MDF (type II and III moulds). More uniform blocks, hardened in contact with methacrylate (type I and III moulds), present certain changes in colour nuances that suggest the decanting of the binder (Table 4).

Table 4. Blocks obtained from moulds of type I, II and III.

Type Block	Mould Material	Visual Aspect	Dimensions (mm)
I	Methacrylate	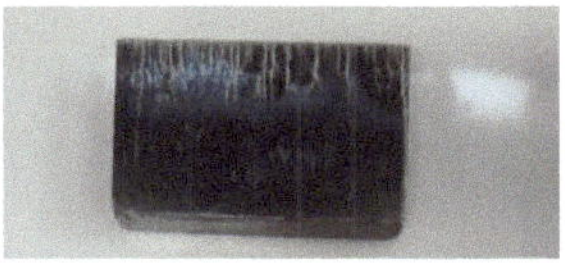	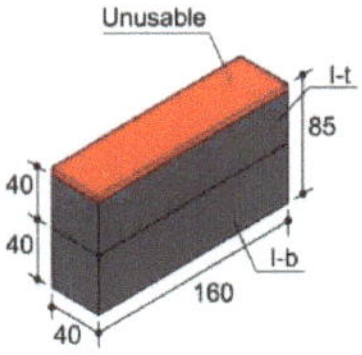
II	MDF	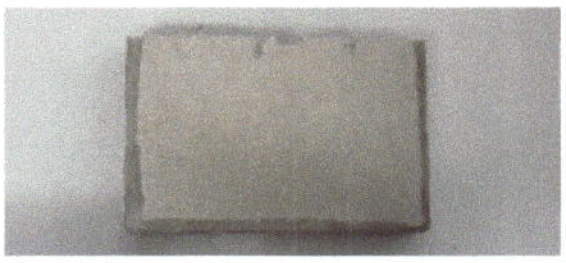	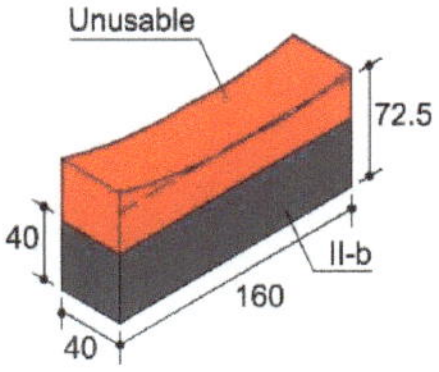
III	Methacrylate-MDF	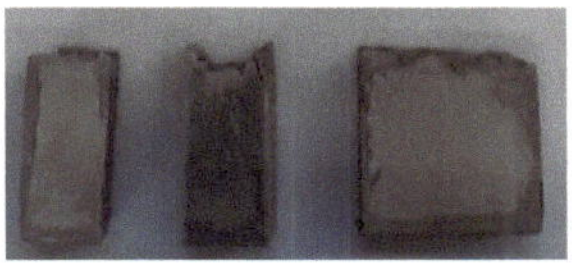	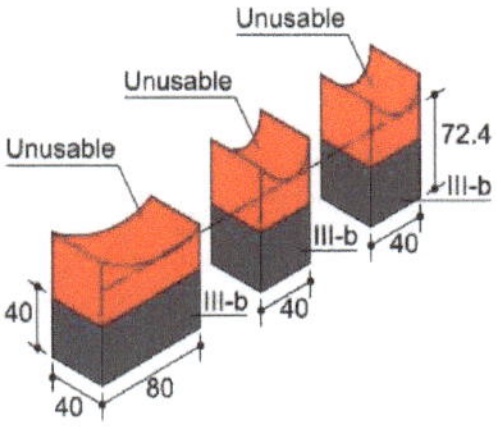

Another significant difference involves shrinkage, given a common fresh grout initial level of 105 mm. While the final height of type I blocks is 85 mm as a mean value, type II and III moulds lead to blocks with a minimum mean height of 72.5 mm and 72.4 mm, respectively, measured in the centre of the block. Water loss due to mould absorbency or non-absorbency may be assumed as the probable cause of the observed behaviour. Due to the minimum dimensions for the edge of the specimens being established at 40 mm, the top portions of type II and III blocks were discarded, hence coming up with 4 pieces of blocks to analyse: (i) type I-b (bottom area of type I); (ii) type I-t (top area of type I); (iii) type II-b (bottom area of type II) and (iv) type III-b (bottom area of type III) (Table 4).

2.3.2. Physico-Chemical Characterisation by SEM/EDX

SEM and EDX observations of specimens were carried out using microscopes, as described in Section 2.1.1, for the identification of the features that can affect mechanical behaviour, namely porosity, cracks, and second phases. Due to the high hydrophilic nature of the blocks, an ambient pressure SEM at low acceleration voltage had to be used, thereby enabling the characterisation of the block specimens after a thin gold conductive layer had been sputtered over the specimen surface.

2.3.3. Physical Characterisation by Computed Tomography

Computed Tomography (hereinafter CT) was carried out on two small specimens that had been diamond-wheel sawn from inside the bottom part of type I blocks (samples I-b, Table 4). Both specimens were analysed using Yxlon Y.COUGAR.SMT equipment, from Comet Yxlon GmbH, Hamburg, Germany. The first sample was scanned as a small cylindrical volume roughly 1 mm high and 1 mm in diameter using a voxel size of 1.1 μm (sample 1). A second sample, a cylindrical volume of 8 mm high and 8 mm diameter, was scanned with a voxel size of 8.86 μm (sample 2). Image analysis was carried out with free software from Dragonfly (2022.2 non-commercial license).

2.4. Specimens

2.4.1. Cut of Specimens

Specimens of a variety of shapes and sizes (Table 5), under different standard specifications (Table 1) were obtained from different areas of the blocks (Table 4). As mentioned in Section 2.3.1, blocks showed several inhomogeneities possibly related to shrinkage and binder decanting. Consequently, certain portions of the blocks had to be discarded (See Table 4 concerning unusable portions). By means of a Wiskehrs M-351-CM cut-off machine, from Wiskehrs Internacional de Elevacion S.A., Zaragoza, Spain, equipped with a 350 mm diameter diamond saw, various types of specimens were obtained (Table 4): (i) type C1 specimens (compression test), which are prisms measuring 40 × 40 × 80 mm, in accordance with EN196-1 [30] and 1015-11 [34], extracted from the bottom and the top of type I blocks (specimens I-C1(b) and I-C1(t)) and from the bottom of type III blocks (III-C1(b)); (ii) type C2 specimens (compression test), which are cubes measuring 40 × 40 × 40 mm. These specimens, following the cubic shape established in IS-4031-4 [44], were cut to the aforementioned dimensions to facilitate comparison with the 40 × 40 × 80 mm specimens described above. Likewise, specimens from the bottom and the top of type I blocks (specimens I-C2(b) and I-C2(t)) and of the bottom of type III blocks (III-C2(b)) were obtained; (iii) type C3 specimens (DPT test), which are prisms measuring 40 × 40 × 16 mm, in accordance with DIN 18555-9 [43]. Furthermore, specimens from the bottom and the top of type I blocks (specimens I-C3(b) and I-C3(t)) and of the bottom of type III blocks (III-C3(b)) were also obtained; and (iv) type F specimens (flexural test): prisms measuring 40 mm × 40 mm × 160 mm, in accordance with EN 196-1 [30]. Specimens were extracted from the bottom and the top of type I blocks (I-F(b) and I-F(t)) and from the bottom of type II blocks (II-F(b)).

Table 5. Quartering of initial sets of blocks in specimens to be tested.

Type of Block	Initial Block	Testing Specimens
Type I (methacrylate)	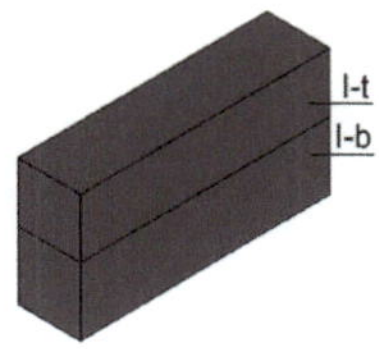	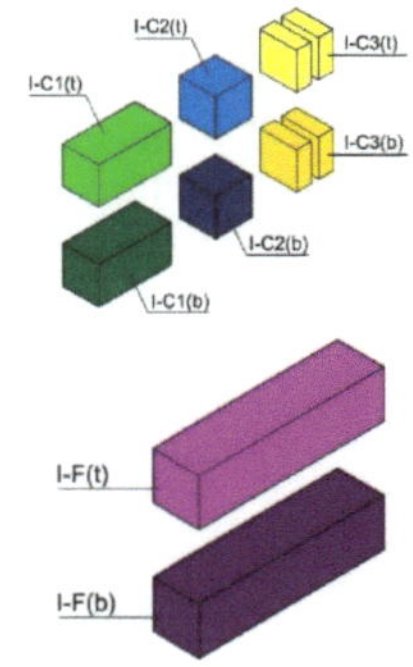
Type II (MDF)	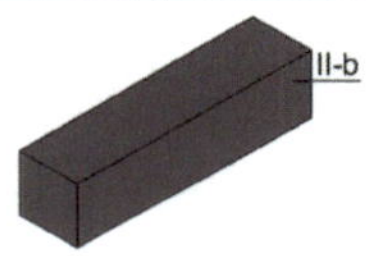	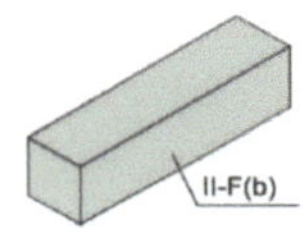
Type III (Methacrylate + MDF)	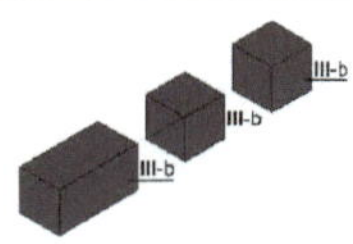	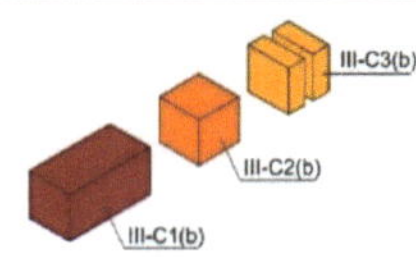

2.4.2. Density

The specimens were weighted in a precision balance in order to measure their density values. Regarding their volume, measurements of dimensions on all the faces of the specimens were carried out and average values of the length of three perpendicular edges were obtained.

2.4.3. Mechanical Testing

The mechanical behaviour of grouts has been successfully characterised by means of compressive tests and flexural tests. According to Section 2.4.1, compressive tests have been carried out on specimens of different shapes and sizes. The variants C1 and C2 are targeted to characterise the compressive behaviour of grouts in standard situations, while C3 is oriented to slender specimens due to the extended use of grouts to seal or to reinforce joints (Table 5). Table 6 shows the design of the experimental campaign in order to characterise the grout under the relevant aforementioned variables.

Table 6. Types and number of tested specimens.

Type of Mould	Specimen	Block Area	Number of Specimens	Dimensions (mm)	Test
Type I	I-C1(b) I-C1(t)	Bottom Top	5	40 × 40 × 80	Compression as per EN 196-1 [30]
	I-C2(b) I-C2(t)	Bottom Top	5	40 × 40 × 40	
	I-C3(b) I-C3(t)	Bottom Top	10	40 × 40 × 16	DPT as per DIN 18555-9 [43]
	I-F(b) I-F(t)	Bottom Top	3	40 × 40 × 160	

Table 6. *Cont.*

Type of Mould	Specimen	Block Area	Number of Specimens	Dimensions (mm)	Test
Type II	II-F(b)	Bottom	3	40 × 40 × 160	Flexural as per EN 196-1 [30]
Type III	III-C1(b)	Bottom	5	40 × 40 × 80	Compression as per EN 196-1 [30]
	III-C2(b)	Bottom	5	40 ×40 × 40	Compression as per EN 196-1 [30]
	III-C3(b)	Bottom	10	40 × 40 × 16	DPT as per DIN 18555-9 [43]

Mechanical tests were carried out by using a digitally controlled Mohr and Federhaff hydraulic framework (class 1). The maximum load is 100 kN and is equipped with position sensors. The complete set of data obtained from the test was reduced by MATLAB software (v. 9.14) by selecting a reduced cloud of points. To this end, points spaced 0.03% of strain for specimens' type C1, C2, and F, and 0.05% of strain for specimens' type C3 were included in the scatter plot.

Types C1 and C2 specimens were tested in accordance with EN 196-1 [30] to ascertain the compressive strength. A constant testing rate of 200 N/s was used. C3 specimens were tested in accordance with DIN 18555-9 [43] (DPT) (Figure 2).

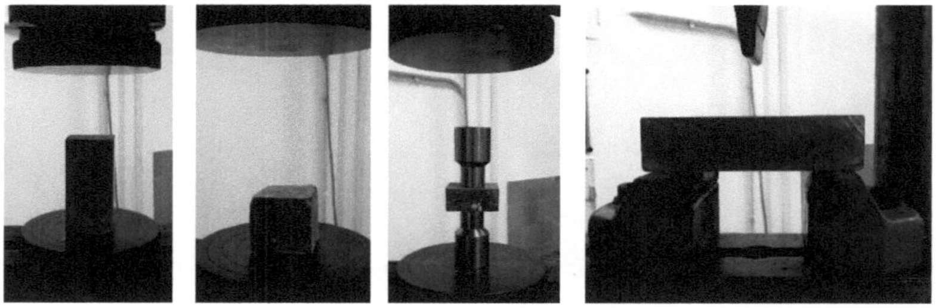

Figure 2. Testing specimens: (**a**) Specimens C1 for the compression test; (**b**) Specimens C2 for the compression test; (**c**) Specimens C3 for the DPT test; (**d**) Specimens F for the flexural test.

Compression strength was obtained in these specimens of reduced thickness by using cylindrical punches of 2.5 cm in diameter and a constant testing rate of 200 N/s. The flexural strength of specimens labelled with the prefix F was determined by the three-point flexural test, in accordance with EN 196-1 [30]. In this test, a load is applied in the central section of a horizontal prism at a low rate (10 N/s). The vertical displacement and the load values were registered by the system until failure. A steel cylinder of 10 mm in diameter was employed to support the specimen and to apply the load, respectively. In total, clouds of points ranging from 3521 to 7341 points have been handled to obtain stress–strain laws for these tests as described in the following section. Slenderness of the specimen during the compressive testing can be deduced from Table 5 to be 2 for type C1 specimens; 1, for type C2 specimens; and 0.4, for type CIII specimens. Due to the impossibility of installing strain gauges in the specimens for DPT, the stiffness of specimens has been evaluated by means of the stress–strain rate of each of the tests that have been performed as the slope of the curve at the origin instead of Young's Modulus.

3. Results

3.1. Binder

3.1.1. Chemical Composition

Elemental analysis of the received binder powder, at several points and areas, shows the mean results as being Ca (34%), Si (13.9%), Al (4.3%), Mg (2.7%). These values are in agreement with the values deduced from the declared oxide composition by the manufacturer (Ca (34.6%), Si (15.9%), Al (5.5%), Mg (4.3%), except for the magnesium content, which lies below the declared value. An increased clinker content in the binder formula may be responsible for the analysed magnesium content. Moreover, decreasing amounts of S, Fe, K, Na, and Cl have also been detected.

As previously discussed, the composition may be consistent with a blend of clinker and blast furnace slag. Moreover, grinding is sufficiently good to render the clinker and slag individual particles hardly distinguishable due to the slight difference expected in their composition, as can be deduced from the uniform grey shades of different particles in Figure 3a.

(**a**)

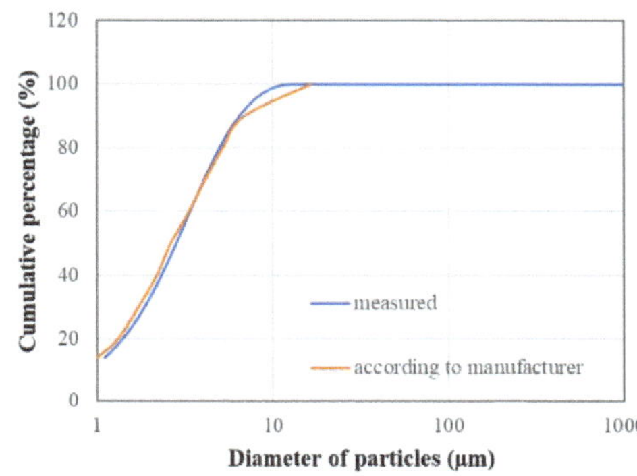

(**b**)

Figure 3. (**a**) SEM back-scattered electrons image of the powdered binder; (**b**) Binder particle size distribution.

Figure 4 corresponds to an elemental chemical EDX mapping of Figure 3a, where brighter areas in each image correspond to a higher content of the represented element. Higher calcium content and lower silica content are expected in clinker particles.

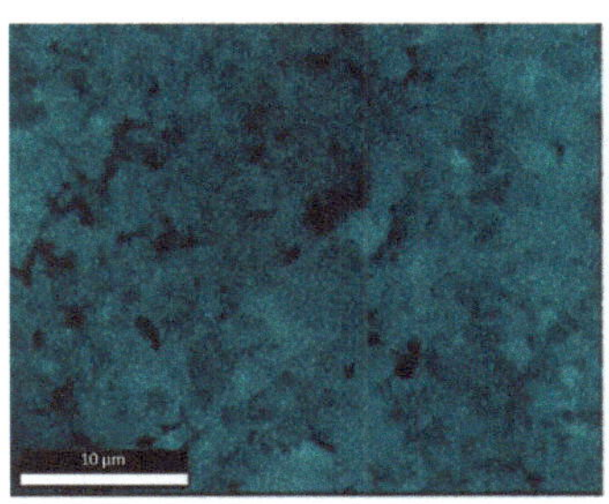

(**a**)

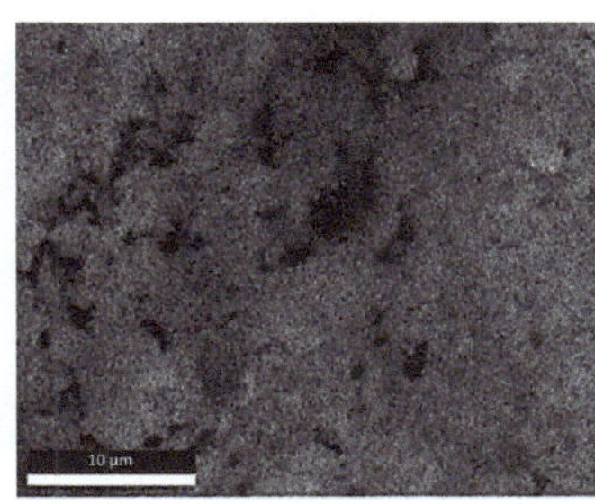

(**b**)

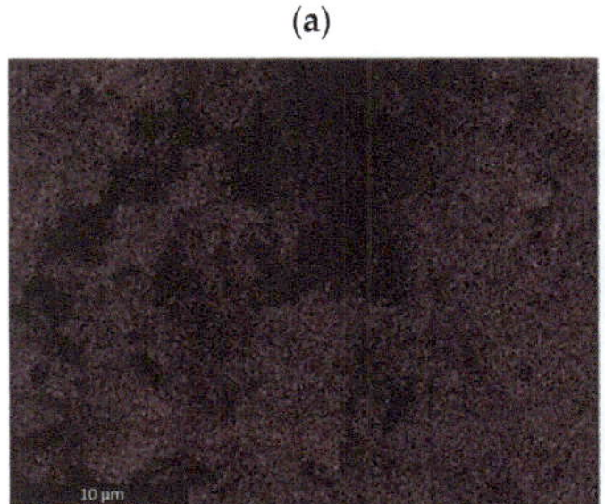

(**c**)

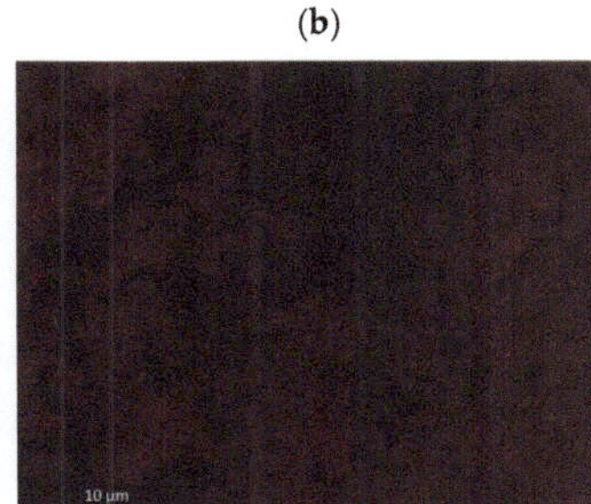

(**d**)

Figure 4. *Cont.*

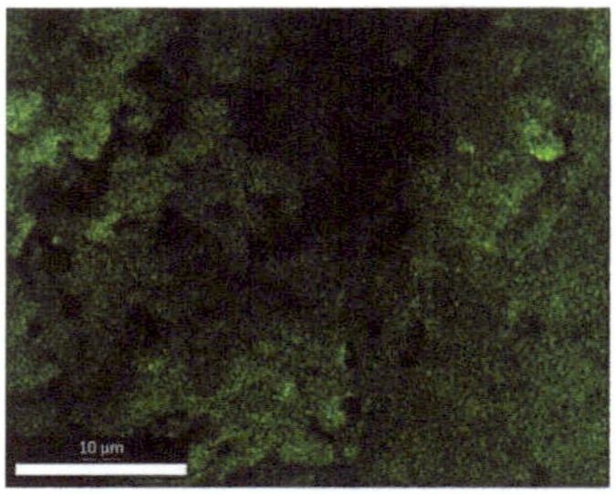

(e)

Figure 4. EDX mapping of several elements for the SEM area of Figure 1: (**a**) Calcium; (**b**) Silicon; (**c**) Aluminium; (**d**) Magnesium; (**e**) Oxygen.

3.1.2. Grain Particle Size

Figure 3b shows binder grain size distribution, which confirms the extremely fine nature of the binder, well suited for use in grout injection projects. According to manufacturer data, the binder grain size is under 12 μm. In fact, actual measurement shows a mere 1.14% of particles over 11.48 μm in diameter while the diameter of 90% of the particles remains under 7.37 μm. By comparing these two distributions, it may be stated that bigger particles are slightly less common than those reported by the manufacturer.

3.2. Cured Blocks

3.2.1. Physic-Chemical Characterisation by SEM/EDX

Figure 5 shows SEM images of the sections taken on several blocks according to the type of mould and the area of the block (top or bottom). Cracking is observed independently of the mould used. Porosity is also observed.

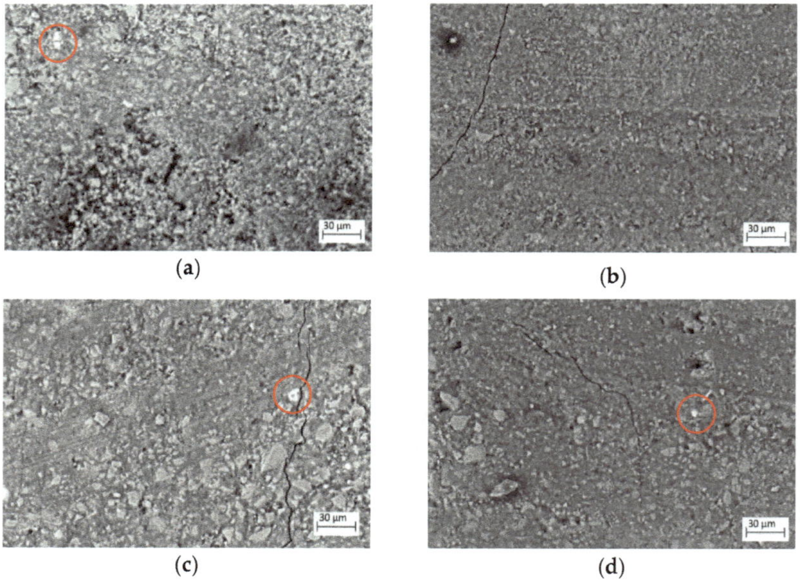

Figure 5. SEM images of sections corresponding to: (**a**) Initial blocks type I-b (methacrylate moulds); (**b**) Initial blocks type I-t (methacrylate moulds); (**c**) Initial blocks type II-b (MDF moulds); (**d**) Initial blocks type III-b (methacrylate-MDF moulds). Inside the red circles, the white phase is associated with an element with a high atomic number (probably iron-rich particles coming from blast-furnace slag).

Several phases are present as expected. The figure corresponding to a type I block at the bottom (Initial block I-b) shows one big particle per field (0.2%), while a type I block at the top (Initial block I-t) shows one or two particles 4 μm in diameter per field (<0.1%). type II blocks at the bottom (Initial block II-b) show one or two particles, 5 to 7 μm in diameter per field (<0.1%). Type III blocks at the bottom (Initial block III-b) show 2 to 4 particles, 6 μm in diameter (0.15%). Given that this phase has a particle size of up to 10 μm, it should be identifiable in the CT images, thanks to its higher density and average atomic number in relation to the matrix.

EDX analysis was carried out on the previously presented SEM images. Two measurements at different areas of the blocks were taken. The identified carbon may be related to the added plasticizer, given that both the clinker and the slag are inorganic compounds. The identified amounts of relative carbon are shown in Figure 6a. Results obtained from the bottom of the blocks show a higher carbon content which indicates a superplasticizer decantation. Furthermore, those blocks partially or totally produced from MDF show a certain level of superplasticizer absorption.

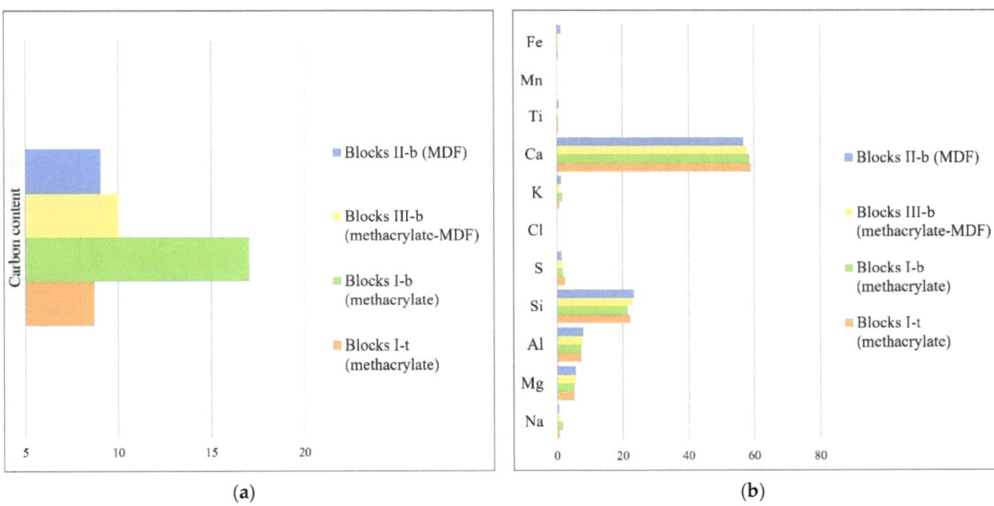

Figure 6. (**a**) Carbon content relative to that of initial blocks type III-b (taken as 10); (**b**) Calcium, Sulphur, Silicon, Aluminium, Magnesium, and Sodium content of specified blocks. Carbon and Oxygen have not been considered in the quantification.

On renormalising the compositions measured for the remaining elements (except Carbon and Oxygen, which may be affected by the decantation and absorption of the superplasticizer), it is observed in Figure 6b that the differences between the various blocks and their areas are minimal. The differences observed do not permit clear conclusions to be drawn between the different samples of the three types of moulds used. In fact, the compositional differences are of the order of those found between the two measurement points of the same type of test specimen.

3.2.2. Physic Characterisation by CT

In the present scan, whose results are shown in Table 7 and Figure 7 for sample 1 and Table 8 for sample 2 (both described in Section 2.3.3), it can be concluded that the mean pore size is of the order of magnitude of binder particle size (Figure 3b).

Regarding sample 1, the total porosity volume was extremely low, at less than 0.1%. White particle content, associated (as previously mentioned in Section 3.2.1) with an element with a high atomic number, probably Fe, is also negligible. These values agree with those found in SEM observations of the samples (Figure 5). While particle distribution is relatively homogeneous (Figure 7a), cracks and pores are clustered in certain areas (grey adjacent areas) (Figure 7b), probably producing mechanically weak zones. Nevertheless, in sample 2, a considerable number of tiny pores detected in sample 1, with a more precise scan, are lost. Consequently, the relative volume of pores and cracks in sample 2 is slightly lower than in sample 1. The same can be stated in connection with the quantification of white particles. The mean diameter of those particles is identified as 61 μm in sample 2 since very small particles with a diameter under 8.86 μm are not detected. Although this method is highly suitable for the detection of tiny pores and particles in the case of the inspected volume is extremely small, in these kinds of intrinsically heterogeneous samples, it may be advisable to scan a larger volume, even if the detection threshold (voxel size) must be increased.

Table 7. Quantitative evaluation of CT observations for sample 1 (voxel size 1.2 μm).

Sample 1 [1]	Number of Features	Mean Diameter (μm)	Sphericity	Total Volume (μm^3)	Relative Volume (%)
Pores + grey adjacent area	4980	8.1 ± 3.1	0.72 ± 0.10	554,456	0.07
Pores	156,561	2.1 ± 1.2	0.95 ± 0.05	684	9×10^{-5}
White particles	9982	2.9 ± 2.0	0.95 ± 0.04	188,453	0.02

[1] Volume tested: 0.76 mm^3.

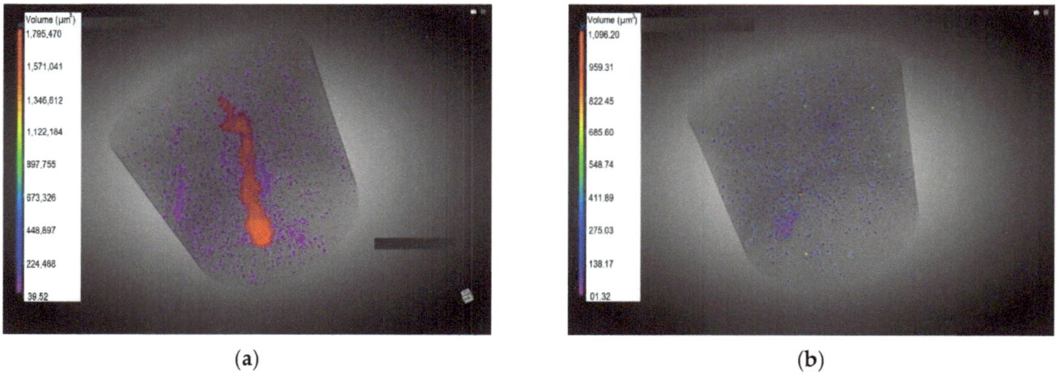

Figure 7. Images obtained from sample 1 CT: (**a**) cracks and pores (colour-coded bar ranges from 39.5 to 1.79×10^6 μm^3); (**b**) white particles (colour-coded bar ranges from 1.3 to 1096.2 μm^3).

Table 8. Quantitative evaluation of CT observations for sample 2 (voxel size 8.86 μm).

Sample 2 [1]	Number of Features	Mean Diameter (μm)	Sphericity	Total Volume ($\mu m^3 \times 10^3$)	Relative Volume (%)
Pores + grey adjacent area	412	215 ± 56	0.32 ± 0.07	213,381	0.05
Pores	40,607	19 ± 16	0.94 ± 0.08	138,777	0.03
White particles	8064	61 ± 20	0.51 ± 0.07	693	2×10^{-4}

[1] Volume tested: 403.8 mm^3.

3.3. Specimens

3.3.1. Density

Table 9 lists the average values of density measured depending on the area of the initial block type (top or bottom) and on the geometry of the specimen extracted from each of them. The differences found between the densities of the specimens confirm the decantation of the binder observed with the naked eye (Table 4).

Despite all the efforts made to produce homogeneous grouts and considering the probable slow setting speed of the particles (due to their small size), a decantation process has occurred. The terminal velocity for a 12 μm spherical particle in pure water can be computed to be 54 cm/h. That value is indeed an upper limit to the real setting velocities, but small density differences are clear between denser specimens obtained from the bottom compared to those lighter specimens obtained from the top area of type I blocks.

Table 9. Average density, maximum force, compressive/flexural strength (with standard deviation) and stress–strain ratio.

Initial Block	Specimen	Average Density (g/cm^3)	Compressive/Flexural Strength (MPa)	Stress–Strain Ratio (MPa)
I-b	I-C1(b)	1.836 ± 0.013	17.6 ± 1.8	3141
	I-C2(b)	1.844 ± 0.014	22.9 ± 1.1	3564
	I-C3(b)	1.831 ± 0.029	27.8 ± 1.4	2346
	I-F(b)	1.837 ± 0.023	6.6 ± 0.6	-
I-t	I-C1(t)	1.730 ± 0.015	12.4 ± 6	2972
	I-C2(t)	1.722 ± 0.023	16.7 ± 0.8	2763
	I-C3(t)	1.726 ± 0.027	16.6 ± 0.9	2360
	I-F(t)	1.731 ± 0.012	3.5 ± 1.6	-
II-b	II-F(b)	1.857 ± 0.029	4 ± 0.7	-
III-b	III-C1(b)	1.866 ± 0.053	13.2 ± 0.6	3557
	III-C2(b)	1.856 ± 0.006	16.2 ± 0.8	2892
	III-C3(b)	1.868 ± 0.071	27.4 ± 1.1	2407

Extracting the specimens from the bottom or the top of type I blocks clearly modifies the results, due to the observed setting of binder in grouts. In this case, the differences in density values are approximately 6% between specimens obtained from the bottom and the top of the block. On the other hand, the density of bottom specimens of type III blocks is higher (1%) than that of type I blocks. Water absorption by MDF will produce a thicker slurry in type II blocks. The same argument is valid in comparing density from bottom specimens from blocks of types I and III.

3.3.2. Mechanical Properties

The results of the mechanical tests are presented in Tables 9 and 10 and Figure 8. Analysing specimens extracted from type I-b initial blocks, it can be assumed that the differences among the three values of compressive strength (I-C1(b), I-C2(b) and I-C3(b)) are over 27%, and from type I-t initial blocks (I-C1(t), I-C2(t) and I-C3(t)), over 40%. For each geometry, a higher strength is obtained for the denser specimens. Nevertheless, the ratio between density and compressive strength is not proportional for all the specimens, mainly due to the effect of the slenderness of the specimens. Moreover, the effect of the slenderness in the value of the compressive resistance is greater than the effect of the specimen density. Specimen C1 shows, for a given density, the lowest compressive strength, while the effect for shapes C2 and C3 is not as clear, as previously mentioned. Two aspects are worth bearing in mind regarding these differences: (i) slender specimens are more likely to present cracks and defects (observed in CT) which diminish the compressive strength; and (ii) the values of compressive strength for specimens C3 are obtained from a lower area (the area of the contact punch) than those of specimens C1 and C2 (the whole cross-section of the specimens).

A similar result can be deduced from the stiffness measurements in specimens obtained from type I blocks. In terms of stiffness, the difference between the specimens extracted from the initial blocks type I-b and type I-t reaches 22%, on comparing results of every specimen shape. Again, for every shape (C1, C2, or C3), a higher specimen density means greater stiffness. However, the effect of the specimen shape is greater than that of the specimen density. In this case, for a given density, specimen shape C3 shows the lowest stiffness whereas the effect of shapes CI and C2 remains unclear. Choosing a type I or type III mould produces no significant differences in the results when solely considering density values. Taking the specimens from the bottom part of the blocks as a reference, the difference in the compressive strength lies in the range of 1 to 25% and is lower in type III block specimens compared to type I block specimens. In contrast, the stress–strain rate presents no clear tendency. Moreover, concerning compression strength, density still plays a role when comparing equal specimen shapes. Specimens with the C1 shape attain the lowest values of compressive strength, mainly because of their higher slenderness. However, specimens with the C3 shape present a lower stress–strain ratio than the C1 shape for the specimens prepared with type 1 and type III moulds. These results cannot be explained solely by the water content of the grouts. Type III moulds absorb water, thereby decreasing the water/binder ratio and producing the well-known increase in strength. However, a reduction in strength (1–25%) is observed even though density increases. Regarding flexural strength and density of specimens obtained from type I and type II blocks, greater strength is shown by the denser specimens obtained from the bottom of the type I blocks, as expected. Due to binder particle sedimentation, the water/binder ratio decreases at the bottom of the block, thereby increasing the density and strength. Specimens obtained from type II blocks show even greater densities, due to water absorption by the mould. Nevertheless, the strength is lower than that shown by specimens from type I blocks.

Table 10. Average density/stress–strain ratio and average density/compressive strength and slenderness for type C specimens.

Initial Block	Specimen	Average Density (g/cm^3)($\times 10^2$)/Stress–Strain Ratio (MPa)	Average Density (g/cm^3)/Compressive Strength (MPa)	Slenderness
I-b	I-C1(b)	0.058	0.10	2
	I-C2(b)	0.052	0.08	1
	I-C3(b)	0.078	0.07	0.4
I-t	I-C1(t)	0.058	0.14	2
	I-C2(t)	0.062	0.10	1
	I-C3(t)	0.073	0.10	0.4
III-b	III-C1(b)	0.052	0.14	2
	III-C2(b)	0.064	0.11	1
	III-C3(b)	0.078	0.07	0.4

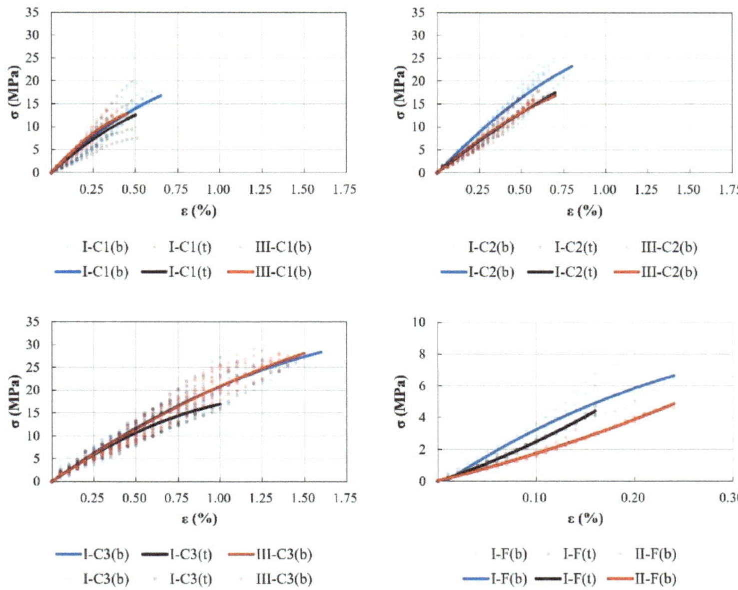

Figure 8. Stress–strain scatter plot and trend lines for the twelve sets of tests, whereby C1, C2, and C3 are tested in compression and F is tested in flexion, according to Table 6.

Regarding the influence of the slenderness of the specimens in compressive strength, a decrease in the strength is evident as the ratio h/L of the specimen increases (Figure 9).

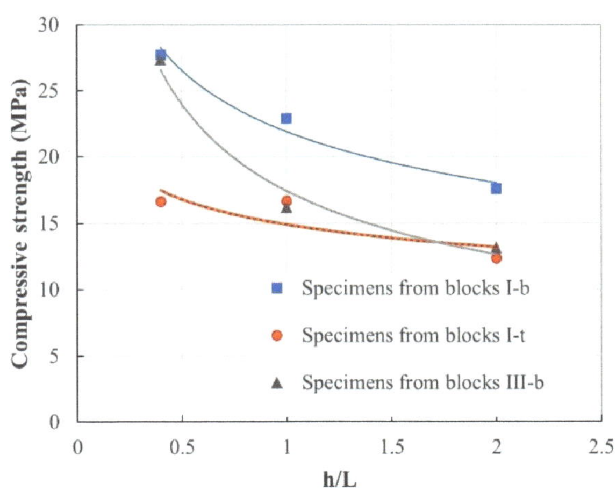

Figure 9. Scatter plot showing the relationship between the compressive strength and the slenderness of specimens obtained from the bottom and the top of initial type I blocks and from the bottom of initial type III blocks.

In this way, as slenderness increases from 0.4 to 2.0, compression strength diminishes by up to 50%. A greater trend to a potential curve can be observed in specimens extracted from the bottom of type I and III blocks (I-b and III-b). Specimens extracted from the top of type I blocks (I-t) show a different performance, with a slightly higher compressive strength for specimens I-C2(t) than that of type III-C2(t).

4. Conclusions

Grout based on SPINOR A12 binder and concentration water/binder equal to 0.75 has been characterised in detail, from the physical, chemical, and mechanical points of view. In the latter case, the different approaches to the material of moulds and the shape of the specimens set out in several international standards regarding grouts have been taken into account. The majority of these aspects adhere to standards about mortars.

Firstly, this research demonstrates that the physical and chemical characteristics of SPINOR® A12, that is, its grain size and composition, are consistent with the values given by the manufacturer.

The material of moulds affects different aspects of the mechanical results:

1. When absorbent material is used for the moulds, grout shrinkage after setting is greater than when using non-absorbent material. This fact underlines the plastic nature of the shrinkage and also demonstrates the water absorption by moulds.
2. Water absorption by moulds also influences the final values of stiffness and compressive and flexural strength of specimens once the grouts have hardened. All these values are lower when using absorbent materials for moulds than in the opposite case. Overall, high values of density lead to better compression and flexural strength values of the specimens. Nevertheless, in certain cases, high density can be a consequence of a lack of water during setting and, consequently, can lead to the appearance of cracking, which unfavourably affects the mechanical properties of specimens.
3. The decanting of the binder provides different characteristics of specimens obtained from the upper and the lower part of the block manufactured with non-absorbent material in moulds, thereby affecting the results of mechanical tests. Due to binder particle fineness, the expected decantation velocity is sufficiently high to produce lower water/binder ratios at the bottom in the initial stages of setting. This is confirmed by density measurements.

The slenderness of specimens constitutes a key factor in the results of the compression tests. An inverse relationship between this parameter and compressive strength has been observed and quantified.

Therefore, there is a great influence of the mould material when preparing the specimens and their geometry characteristics on the mechanical properties of hydraulic ultrafine cement grout. This is why it is important to consider factors such as the porosity of the masonry to be repaired and the relationship between the expected stresses and the injected mixture thickness before designing the experimental campaign and, consequently, knowing the performance of injected grout.

Author Contributions: Conceptualisation, B.H. and J.M.G.; Methodology, B.H.; Validation, B.H. and J.M.G.; Formal Analysis, B.H.; Investigation, B.H.; Resources, B.H.; Data Curation and Mechanical Test Execution, J.A.S.-E.; Writing (Original Draft Preparation), B.H.; Writing (Review and Editing), E.R.-M.; Visualisation, B.H. and E.R.-M.; Supervision, E.R.-M.; Project Administration, F.A. and J.M.G.; Funding Acquisition, F.A., B.H. and E.R.-M. All authors have read and agreed to the published version of the manuscript.

Funding: This research was funded by: PGC2018-098185-A-I100 (FEDER/Ministerio de Ciencia e Innovación-Agencia Estatal de Investigación de España), US-1381350 (FEDER/Consejería de Economía, Conocimiento, Empresas y Universidad de la Junta de Andalucía, Programa Operativo FEDER 2014–2020) and US.20-08 (Consejería de Fomento, Infraestructuras y Ordenación del Territorio de la Junta de Andalucía).

Institutional Review Board Statement: Not applicable.

Informed Consent Statement: Not applicable.

Data Availability Statement: Data are contained within the article.

Acknowledgments: The authors are grateful for the help provided by Juan Manuel Jaramillo and Fernando Tejero for the assistance supplied in the laboratory.

Conflicts of Interest: The authors declare no conflicts of interest.

Nomenclature

DPT	Double Punch Test
GGBS	Ground Granulated Blast Slag
SEM	Scanning Electron Microscopy
EDX	Energy Dispersive X-ray Analysis
CT	Computed Tomography
MDF	Medium-Density Fibreboard

References

1. Warner, J. *Practical Handbook of Grouting: Soil, Rock, and Structures*; John Wiley & Sons: Hoboken, NJ, USA, 2004; 720p.
2. Cambefort, H. The principles and applications of grouting. *Q. J. Eng. Geol. Hydrogeol.* **1977**, *10*, 57–95. [CrossRef]
3. Kutzner, C. *Grouting of Rock and Soil*; Taylor Francis Group: Abingdon, UK, 2020. [CrossRef]
4. Harrison, D.M. *The Grouting Handbook: A Step-by-Step Guide for Foundation Design and Machinery Installation*; Elsevier: Amsterdam, The Netherlands, 2013.
5. Van Rickstal, F. *Grout Injection of Masonry: Scientific Approach and Modeling*; Katholieke Universiteit Leuven: Leuven, Belgium, 2000.
6. Anzani, A.; Cardani, G.; Condoleo, P.; Garavaglia, E.; Saisi, A.; Tedeschi, C.; Tiraboschi, C.; Valluzzi, M.R. Understanding of historical masonry for conservation approaches: The contribution of Prof. *Luigia Binda to Research Advancement. Mater. Struct.* **2018**, *51*, 140. [CrossRef]
7. Binda, L.; Modena, C.; Baronial, G.; Abbaneol, S. *Repair and Investigation Techniques for Stone Masonry Walls*; Elsevier: Amsterdam, The Netherlands, 1997.
8. Modena, C.; Pineschi, F.; Valluzzi, M.R. Valutazione della vulnerabilità sismica di alcune classi di strutture esistenti. In *Sviluppo e Valutazione di Metodi di Rinforzo*; CNR-Gruppo Nazionale per la Difesa dai Terremoti: Roma, Italy, 2000.
9. Luso, E.; Lourenço, P.B. Experimental laboratory design of lime based grouts for masonry consolidation. *Int. J. Archit. Heritage* **2017**, *11*, 1–10. [CrossRef]
10. Toumbakari, E.-E. *Lime-Pozzolan-Cement Grouts and Their Structural Effects on Composite Masonry Walls*; Katholieke Universiteit Leuven: Leuven, Belgium, 2002.
11. Ignoul, S.; van Rickstal, F.; van Gemert, D. Application of mineral grouts. Case study and impact on structural behaviour: Church of St. Catharina at Duisburg (B). In *Structural Analysis of Historical Constructions*; Modena, C., Lourenço, P., Roca, P., Eds.; Springer: Berlin/Heidelberg, Germany, 2005.
12. Sha, F.; Li, S.; Liu, R.; Li, Z.; Zhang, Q. Experimental study on performance of cement-based grouts admixed with fly ash, bentonite, superplasticizer and water glass. *Constr. Build. Mater.* **2018**, *161*, 282–291. [CrossRef]
13. Fonseca, F.S.; Godfrey, R.C.; Siggard, K. Compressive strength of masonry grout containing high amounts of class F fly ash and ground granulated blast furnace slag. *Constr. Build. Mater.* **2015**, *94*, 719–727. [CrossRef]
14. Mirza, J.; Mirza, M.S.; Roy, V.; Saleh, K. Basic rheological and mechanical properties of high-volume fly ash grouts. *Constr. Build. Mater.* **2002**, *16*, 353–363. [CrossRef]
15. Mollamahmuto, M.; Avci, E. Engineering properties of slag-based superfine cement-stabilized clayey soil. *ACI Mater. J.* **2018**, *115*, 541–548. [CrossRef]
16. Mollamahmuto, M.; Avci, E.; Deveci, E.; Gökçe, A. Engineering properties of ultrafine blast furnace slag cement stabilized low plasticity clayey soil. *J. Fac. Eng. Archit. Gazi Univ.* **2021**, *36*, 985–996. [CrossRef]
17. Gutiérrez, J.C.R.; Restrepo, O.; Tobón, J.I. Effects of the addition of metakaolin in Portland cement. *Dyna* **2006**, *73*, 131–141.
18. Nepomuceno, M.C.S.; Bernardo, L.F.A.; Pereira-de-Oliveira, L.A.; Timoteo, R.O. Cement-based grouts for masonry consolidation with high content of limestone filler, metakaolin, glass powder and ceramic waste. *Constr. Build. Mater.* **2021**, *306*, 124947. [CrossRef]
19. Vavričuk, A.; Bokan-Bosiljkov, V.; Kramar, S. The influence of metakaolin on the properties of natural hydraulic lime-based grouts for historic masonry repair. *Constr. Build. Mater.* **2018**, *172*, 706–716. [CrossRef]
20. Teymen, A. Effect of mineral admixture types on the grout strength of fully-grouted rockbolts. *Constr. Build. Mater.* **2017**, *145*, 376–382. [CrossRef]
21. Rodríguez-Mayorga, E.; Yanes, E.; Compán, V.; Sáez-Pérez, A. La restauración del templo parroquial de San Dionisio (Jerez de la Frontera, España). *Inf. Construcción* **2013**, *65*, 5–16. [CrossRef]
22. Mayorga, E.R.; Cobo, A.; Yanes, E.; Saez, A. The Repair of the Structure of Santiago's Church (Jerez De La Frontera, Spain) Using Grout-Injection. *Int. J. Archit. Herit.* **2018**, *13*, 1234–1251. [CrossRef]
23. Rodriguez-Mayorga, E. La Rehabilitación del Templo Parroquial de Santiago Apóstol: Anamnesis, Diagnosis, Terapia y Control. Ph.D. Thesis, University of Sevilla, Sevilla, Spain, 2011. Available online: http://fondosdigitales.us.es/tesis/tesis/2621/rehabilitacion-del-templo-parroquial-de-santiago-apostol-en-jerez-de-la-frontera-anamnesis-diagnosis-terapia-y-control/ (accessed on 3 June 2023).
24. Martinez-Cañete, M.; Rodriguez-Mayorga, E.; Yanes, E.; Saez, A. Intervención en los edificios que gravitan sobre el teatro romano de Cadiz y puesta en valor del mismo. In Proceedings of the 6th Euro-American Congress on Construction Pathology,

Rehabilitation Technology and Heritage Management, REHABEND 2016. Burgos, Spain, 24–26 May 2016; Villegas, L., Blanco, H., Bofill, Y., Lombillo, I., Eds.; University of Cantabria-Building Technology R&D Group: Santander, Spain, 2016; pp. 2131–2138.
25. *ASTM C1019*; Standard Test Method for Sampling and Testing Grout for Masonry. ASTM International: West Conshohocken, PA, USA, 2020.
26. *EN 445:2007*; Grout for Prestressing Tendons-Test Methods. British Standard: Chiswick, London, UK, 2007.
27. BASF Corporation. *Shrinkage of Concrete*; BASF's Master Builders Solutions: Mannheim, Germany, 2016; pp. 1–7.
28. Kosmatka, S.H.; Wilson, M.L. *Design and Control of Concrete Mixture*; Nation Academy: Washington, DC, USA, 2011.
29. Phelan, W.S.; Violetta, B.K.; Eller, R.; Korhonen, C.J.; Nkinamubanzi, P.-C.; Roberts, L.R.; Stokes, D.B.; Strickland, B.R.; Wojakowski, J.B.; Yuers, K.L. *ACI 212.3R-10 Report on Chemical Admixtures for Concrete*; American Concrete Institute: Indianapolis, IN, USA, 2010.
30. *EN 196-1:2016*; Methods of Testing Cement-Part 1: Determination of Strength. European Committee for Standarization: Brussels, Belgium, 2016.
31. Baltazar, L.G.; Henriques, F.M.A.; Cidade, M.T. Grouts with improved durability for masonry consolidation: An experimental study with non-standard specimens. *Key Eng. Mater.* **2017**, *74*, 480–487. [CrossRef]
32. Zhang, B.; Zhu, H.; Wu, G.; Wang, Q.; Li, T. Improvement of bond performance between concrete and CFRP bars with optimized additional aluminum ribs anchorage. *Constr. Build Mater.* **2020**, *241*, 118012. [CrossRef]
33. Luso, E.; Lourenço, P.B. Experimental characterization of commercial lime based grouts for stone masonry consolidation. *Constr. Build. Mater.* **2016**, *102*, 216–225. [CrossRef]
34. *EN 1015-11:2019*; Methods of Test for Mortar for Masonry, Part 11: Determination of Flexural and Compressive Strength of Hardened Mortar. European Committee for Standarization: Brussels, Belgium, 2019.
35. *ASTM C942-15*; Standard Test Method for Compressive Strength of Grouts for Preplaced-Aggregate Concrete in the Laboratory. ASTM International: West Conshohocken, PA, USA, 2021.
36. *IS 4031-4*; Methods of Physical Tests for Hydraulic Cement, Part 4: Determination of Consistency of Standard Cement Paste. Bureau of Indian Standards: New Dheli, India, 2005.
37. Tokyay, M.; Özdemir, M. Specimen shape and size effect on the compressive strength of higher strength concrete. *Cem. Concr. Res.* **1997**, *27*, 1281–1289. [CrossRef]
38. Tuncan, M.; Arioz, O.; Ramyar, K.; Karasu, B. Assessing concrete strength by means of small diameter cores. *Constr. Build Mater.* **2008**, *22*, 981–988. [CrossRef]
39. Nikbin, I.M.; Eslami, M.; Rezvani, D.S.M. An experimental comparative survey on the interpretation of concrete core strength results. *Eur. J. Res.* **2009**, *37*, 445–456.
40. Drdácký, M. Non-Standard Testing on Mechanical Characteristics of Historic Mortars. *Int. J. Archit. Heritage* **2011**, *5*, 383–394. [CrossRef]
41. Drdácký, M.; Slížková, Z. Mechanical characteristics of historical mortars from tests on small-sample non-standard specimens. *Mater. Sci. Appl. Chem.* **2008**, *17*, 20–29.
42. Soares, A.; Flores-Colen, I.; de Brito, J. Influence of slenderness on the compressive strength evaluation of cores of renders. *Mater. Struct.* **2015**, *48*, 1449–1460. [CrossRef]
43. *DIN 18555-9*; Testing of Mortars Containing Mineral Binders-Part 9: Hardened Mortars-Determination of the Mortar Compressive Strength in the Bed Joint. German Institute for Standardisation: Berlin, Germany, 1999.
44. *IS 10086: 2018*; Indian Standard Specification for Moulds for Use in Tests of Cement and Concrete. Bureau of Indian Standards: New Dheli, India, 2018.
45. *GB/T 17671-2021*; Test Method of Cement Mortar Strength. The Standardization Administration of the People's Republic of China: Beijing, China, 2021.
46. *JC/T 726-2005*; Mould for Cement Mortars; China Institute of Building Materials Science and Technology: Beijing, China, 2005.
47. Zhang, G.; Qiu, D.; Wang, S.; Wang, P. Effects of plastic expansive agent on the fluidity, mechanical strength, dimensional stability and hydration of high performance cementitious grouts. *Constr. Build. Mater.* **2020**, *243*, 118204. [CrossRef]
48. Mahmood, W.; Mohammed, A.S.; Asteris, P.G.; Kurda, R.; Armaghani, D.J. Modeling Flexural and Compressive Strengths Behaviour of Cement-Grouted Sands Modified with Water Reducer Polymer. *Appl. Sci* **2022**, *12*, 1016. [CrossRef]
49. Albusaisi, K.M.; Al-Busaltan, S.F.; Kadhim, M.A. Characterizing the Mechanical Properties of Sustainable Modified Cementitious Grout for Semi-Flexible Mixture. *IOP Conf. Ser. Earth Environ. Sci.* **2021**, *856*, 12047. [CrossRef]
50. *ASTM C48-22*; Standard Test Method for Flexural Strength of Concrete (Using Simple Beam with Third-Point Loading). ASTM International: West Conshohocken, PA, USA, 2022.
51. Suministros y Servicios Tecnológicos, Microcementos. Available online: https://microcementos.com/descargas/fichas-tecnicas-y-seguridad/spinor-A12/ficha-tecnica-spinor-A12.pdf (accessed on 13 November 2023).
52. *EN 197-1:2011*; Cement Part 1: Composition, Specifications and Conformity Criteria for Common Cements. BSI Standards Publication: London, UK, 2011.
53. Celik, F.; Canakci, H. An investigation of rheological properties of cement-based grout mixed with rice husk ash (RHA). *Constr. Build. Mater.* **2015**, *91*, 187–194. [CrossRef]

applied sciences

MDPI

Article

Effect of the Silanization of Aerosil OX50 in the Properties of Light-Cured Dental Composites

Selena de la Caridad Díaz-Rodríguez [1], Oridayma Tarano-Artigas [1], Wilberth Herrera-Kao [2], Juan Valerio Cauich-Rodríguez [2], José Manuel Cervantes-Uc [2], Ana Rosa-Sainz [3,*], Amisel Almirall La Serna [1] and Yaymarilis Veranes-Pantoja [1,*]

[1] Centro de Biomateriales, Universidad de La Habana, Ave. Universidad, e/Ronda y G, Plaza de la Revolución, La Habana C.P. 10400, Cuba; sele20diaz@gmail.com (S.d.l.C.D.-R.); oridayma@gmail.com (O.T.-A.); amiselalmirall@gmail.com (A.A.L.S.)

[2] Centro de Investigación Científica de Yucatán (CICY), Calle 43, No. 130 x 32 y 34, Col. Chuburná de Hidalgo, Mérida C.P. 97205, Yucatán, Mexico; gywahkao@cicy.mx (W.H.-K.); jvcr@cicy.mx (J.V.C.-R.); manceruc@cicy.mx (J.M.C.-U.)

[3] Escuela Politécnica Superior, Departamento de Ingeniería y Ciencia de los Materiales y del Transporte, Universidad de Sevilla, Calle Virgen de África, 7, 41011 Sevilla, Spain

* Correspondence: arosa@us.es (A.R.-S.); yaymarilis@gmail.com (Y.V.-P.); Tel.: +53-52866259 (Y.V.-P.)

Abstract: In this study, the surface of aerosil OX50 was treated with γ-methacryloxypropyltrimethoxysilane in order to evaluate its effect on light-cured dental composites. Four composites were prepared with Bis-GMA/tetraethylene glycol dimethacrylate/γ-methacryloxypropyltrimethoxysilane, Bis-GMA/tetraethylene glycol dimethacrylate, and a 45% treated or untreated aerosil OX50, using the camphorquinone/N,N-dimethylaminoethyl methacrylate pair as initiator. Evidence of filler silanization was provided by FTIR as a low-intensity absorption at 1707 cm^{-1} (carbonyl functional group) while thermogravimetric analysis showed a mass loss of approximately 2% associated with the decomposition of γ-methacryloxypropyltrimethoxysilane. The experimental composites studied meet the requirements of the ISO 4049:2019 standard for depth of cure, water sorption, and solubility. The composites are shown to be thermally stable and presented a degree of conversion higher than 70%, being higher than that reported for many commercial composites. Based on the observed properties, the best formulations were those in which the silane is incorporated into the matrix and the filler was previously treated.

Keywords: aerosil OX50; light-cured dental composites; silanization

Citation: Díaz-Rodríguez, S.d.l.C.; Tarano-Artigas, O.; Herrera-Kao, W.; Cauich-Rodríguez, J.V.; Cervantes-Uc, J.M.; Rosa-Sainz, A.; La Serna, A.A.; Veranes-Pantoja, Y. Effect of the Silanization of Aerosil OX50 in the Properties of Light-Cured Dental Composites. *Appl. Sci.* **2024**, *14*, 2453. https://doi.org/10.3390/app14062453

Academic Editor: Vittorio Checchi

Received: 28 January 2024
Revised: 7 March 2024
Accepted: 12 March 2024
Published: 14 March 2024

1. Introduction

Dental caries is one of the most widespread diseases in humanity, and it is defined as an infectious, multifactorial, chronic, and localized disease that affects the hard tissues of the tooth. It is produced by the action of organic acids from microbial fermentation of carbohydrates in the diet, the characteristics of the dental tissue, and the hygiene of each individual [1]. For a long time, the most widely used restorative material in the repair of dental tissue has been silver amalgam. Its adequate physical–mechanical properties and ease of handling have made it the ideal material for the restoration of posterior teeth. However, it presents drawbacks such as its unnatural appearance, the lack of adhesion to dental tissue that generates more sacrifice of healthy dental tissue, and the contamination produced by mercury [2]. For this reason, some countries prohibit the use of amalgam, and work is underway worldwide on the gradual reduction of the use of compounds that contain mercury, which is declared in the Minamata Convention, and its total elimination is proposed for the year 2030 [3,4].

The appearance in restorative dentistry of light-cured composite resins has been one of the most significant contributions as it considers the preservation of healthy tissue, reduction of microleakage, and prevention of postoperative sensitivity. In general, composite resins or dental composites are made up of a mixture of acrylic monomers made up of an organic matrix, filler particles, a coupling agent, whose function is to achieve an optimal bond between both components, and an initiator system [5]. The organic matrix of dental composite resins is basically made up of two types of monomers that are divided according to their viscosity into base monomers and diluents. 2,2 [p-(2-hydroxy-3-methacryloxypropoxy) phenyl] propane (Bis-GMA) is the most commonly used base monomer for this purpose due to its properties such as low shrinkage upon polymerization, rigidity, and high resistance [6]. However, it has a high viscosity that negatively affects the final properties of the composite, requiring the use of diluent monomers such as tri- and tetraethylene glycol dimethacrylates (TEGDMA and TEEGDMA, respectively) [7]. Quartz, glass, titanium oxide, and silicon oxide are generally used as inorganic fillers, providing an increase in the resistance to traction, compression, and abrasion of composites. Hydroxyapatite could be a promising bioactive filler for dental resin composites, since it is the main biological component of dentin and enamel [8,9]. In addition to reducing shrinkage in composites it is desirable that mechanical properties are like that of the dental structure in order to guarantee a restoration that meets physical, chemical, and mechanical needs, i.e., an aesthetic and biocompatible dental composite. Currently, there is a trend to use nanofillers to obtain composite resins with dental applications, since these improve the wear properties of the composites, such is the case of aerosil OX50. This is a fumed silica made up of spherical particles of 40 nm in diameter. This silica is obtained by flame hydrolysis of silicon tetrachloride, $SiCl_4$ [10,11].

Among the most important properties of dental resins for clinical use are the degree of conversion, water sorption and solubility, and the mechanical properties. The values required for these properties are well known and have been established in various international standards [12]. The degree of conversion of the composites expresses the number of double bonds of the methacrylate monomeric groups that react. It can be determined by infrared spectroscopy, considering the decrease in the intensity of the absorption band of the double bonds corresponding to the aliphatic carbons [9]. The depth of cure is also a property that indicates the effectiveness of the polymerization process and is carried out according to the stablished parameters in the ISO 4049 standard [12,13]. Other properties that are evaluated and regulated (ISO 4049) are water sorption and solubility. The latter is of great importance since these materials will be placed in a humid environment where unreacted monomers may dissolve or the filler particles can be leached into the oral environment, affecting its clinical behavior and its biocompatibility [14].

Obtaining good properties in composite resins depends largely on the bond between the matrix and the filler. With a good bond, the detachment of inorganic particles is avoided, as well as the penetration of water in the interface, and then the deterioration of the restoration. To ensure the formation of a proper interphase, the filler is treated with coupling agents, generally silanes. The most widely used silane is γ-methacryloxypropyltrimethoxysilane (MPS), capable of chemically interacting with both the matrix and the surface of the filler [15,16].

In the 1970s, "two-step" silanization methods were developed, either by dry contact or in solution, which involve a pretreatment of the filler with the silane before its incorporation into the monomeric mixture for the formulation of composite resins. Other methods that have been used include the "one-step" treatment, which consists of incorporating the silane directly into the organic matrix [17]. Several types of aerosil have been used as previously reported and MPS silanized aerosil with smaller particle size rendered better adhesion with a Bis-GMA matrix but the amount of sorbed water increased. However, the use of larger particle sizes (40 nm) like aerosil OX50 did not severely affect their mechanical performance as their bending strength is slightly lowered without sacrificing flexural modulus [17].

Recently, changes have been introduced in the composition of dental resins with the aim of improving their properties and in order to reduce the use of bisphenol A, where the main concern is the possible release of monomers [5]. In addition to these alternative monomers, the use of monomers with antibacterial activity has also been pursued but the most significant enhancements and changes have been in fillers such as size reduction and shape enhancement (plate-like, rod-like, nanoparticles) [18].

In this research, MPS is introduced into organic matrices not only as a partial diluent of Bis-GMA (Bis-GMA/TEEGDMA/MPS) but also as a coupling agent. Therefore, the aim of this study was to evaluate the influence of the surface treatment of aerosil OX50, used as an inorganic filler, on various properties of dental composites including degree of conversion, depth of cure, water sorption, and solubility.

2. Materials and Methods

In this section, the materials and methods used are presented. The section is subdivided into: silanization of aerosil OX50, preparation of dental composite formulations, and characterization of the filler and experimental composites.

2.1. Silanization of Aerosil OX50

To improve the surface adhesion between the organic matrix and inorganic particles, the filler was treated with MPS (Sigma Aldrich, Burlington, MA, USA). The silanizing solution was prepared with 80 mL of acetone, 0.5 mL of H_2O, 1 mL of MPS, stirred for 20 min, and then 1 g of aerosil OX50 (DEGUSSA, Tokyo, Japan) was added. Silanization in acetone/water mixtures was conducted as in previous studies that demonstrated a better incorporation of silicon on hydroxipatatite surfaces in comparison to the use of methanol/water mixtures [19]. Magnetic stirring was continued for 2 h. Subsequently, it was placed in an ultrasonic bath for 20 min and stirred again for 2 h with a magnetic stirrer. It was decanted and washed with sufficient acetone. The treated aerosil OX50 was dried for 12 h in an oven at 60 °C. Figure 1 shows the schematic diagram of the procedure performed to obtain the silanized aerosil OX50.

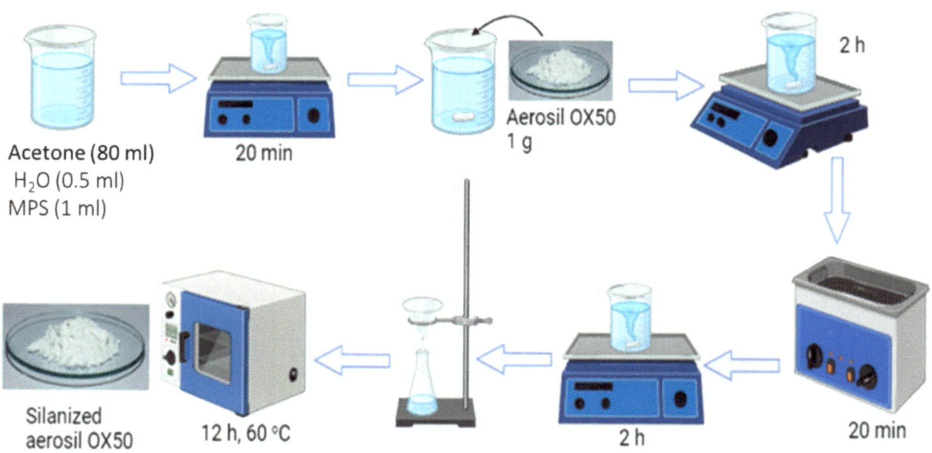

Figure 1. Scheme of the silanization process.

2.2. Preparation of Dental Composite Formulations

Four experimental light-cured dental composites were prepared with two different monomeric matrices and 45% aerosil OX50 as an inorganic filler (Table 1). The M1 matrix is made up of Bis-GMA/TEEGDMA/MPS (60:30:10) and the M4 matrix is made up of Bis-GMA/TEEGDMA (60:40). All composites contain Bis-GMA from Sigma Aldrich, USA, in their organic matrix while tetraethylene glycol dimethacrylate (TEEGDMA) was purchased from Fluka. The C145 composite has MPS incorporated into its organic matrix as part of the diluent, while C445 does not. In addition, the C145s and C445s composites contain filler previously silanized with MPS. The camphorquinone (CQ, Aldrich, USA)/N,N-dimethylaminoethyl methacrylate (MDMAE, Riedel de Haen, Seelze, Germany) pair was used as the initiator system. As reference material, the commercial Tetric Evo Ceram® (TT) was used. All the reagents used were 98% pure.

Table 1. Composition of experimental light-cured dental composites.

Composites	Organic Matrix	Inorganic Filler
C145 C445	Bis-GMA/TEEGDMA/MPS (M1) Bis-GMA/TEEGDMA (M4)	Aerosil OX50 (45%)
C145s C445s	Bis-GMA/TEEGDMA/MPS (M1) Bis-GMA/TEEGDMA (M4)	Aerosil OX50 silanized (45%)
TT	Bis-GMA/TEGDMA	Mixtures of oxides and copolymers (82%)

The light source employed to cure the materials was a light-emitting diode (LED) from Woodpecker (Guilin, China) with a wavelength range from 420 to 480 nm and a light intensity of 1000 mW/cm^2.

2.3. Characterization of the Filler and Experimental Composites

To verify the occurrence of filler silanization and to analyze the thermal behavior of dental composites, a thermogravimetric analysis (TGA) of 50 to 800 °C was performed in a Perkin Elmer TGA-7 (Norwalk, CT, USA), with a heating rate of 10 °C/min and nitrogen flow of 50 mL/min. For the study, 3 mm diameter and 1 mm thickness samples were prepared after polymerization for 40 s.

The morphology of silanized and unsilanized aerosil OX50 was observed on a TESCAN VEGA3 scanning electron microscope, and it was coated using an ion sputter coater SBC-12 (KYKY TECHNOLOGY CO., Beijing, China). The study was carried out at an acceleration voltage of 25 kV with a magnification of 50,000 times. From the micrographs obtained, 100 measurements of aerosil OX50 particles were made using the ImageJ program, version 1.54d. These measurements allowed us to obtain particle size distribution of the filler and thus estimate their average size.

Fourier transform infrared spectroscopy (FTIR) was used for the characterization of the filler and determination of the degree of conversion in the composites. The aerosil OX50 was analyzed in a Thermo Nicolet 8700 spectrometer (Madison, WI, USA), with a resolution of 4 cm^{-1}, 100 scans, and in the spectral interval of 4000–500 cm^{-1}. The samples were dispersed in KBr and the degree of conversion (DC) determined by recording transmission spectra in the 4000–600 cm^{-1} range, with the same resolution and 50 scans. Three replicates of each sample were run at 0, 1, 2, 3, 4, 5, 10, 15, 20, 25, 30, 35, 40, and 80 s after irradiation. The DC of every dental resin composite was calculated from the absorbance intensities (R) of the aliphatic C=C bond (1638 cm^{-1}) against the internal standard of the aromatic C=C bond (1608 cm^{-1}), determined before and after polymerization. The degree of conversion (%) was calculated according to Equation (1) [20–22].

$$\text{DC (\%)} = 100 \times [1 - (R_{cured}/R_{not\ cured})] \quad (1)$$

The depth of cure, water sorption (A), and solubility (S) tests were evaluated according to the specifications described in the ISO 4049:2019 standard [12]. For the depth of cure, three cylindrical specimens (6 mm in height and 4 mm in diameter) were irradiated for 40 s on one side. After the polymerization process, the specimen was removed from the mold and any soft material was scraped using a plastic spatula. The height was measured with a micrometer (0–25 mm, 0.01 mm precision), the obtained value was divided by two, and the measured magnitude was reported as the depth of cure of the material. Water sorption and solubility tests were performed using five disc-shaped specimens (1 mm in height and 15 mm in diameter). The samples were cured with nine overlapped irradiations on one side (40 s for each irradiation; 360 s in total). Diameter and thickness are measured with a micrometer (0–25 mm) to calculate the volume (V). The samples are weighed and stored in a desiccator until constant weight, the final mass is taken as m_1. Then, they are immersed in distilled water (m_2) for seven days at 37 °C and placed again in a desiccator until a constant mass (m_3). For the calculations, Equations (2) and (3) were used.

$$A = (m_2 - m_3)/V \tag{2}$$

$$S = (m_1 - m_3)/V \tag{3}$$

The Origin8 computer program was used for data processing. Statistical analysis was performed by comparing means and variances with the Statgraphics Plus 5.5 program, using Student's "t" statistic and Fisher's "F" statistic with 95% confidence. The mean and standard deviation (Sd), as an interval (±), were reported for all cases.

3. Results

This section presents the obtained results, which are divided into: characterization of the filler, physical appearance of the experimental composites, and characterization of the experimental composites.

3.1. Characterization of the Filler

FTIR spectra, mass loss (TGA), and the first derivative (DTGA) of the silanized and untreated aerosil OX50 are shown in Figure 2a,b. The FTIR spectrum of non-silanized aerosil OX50 showed two typical bands of silicon oxide at 1100 and 801 cm^{-1}. In the case of silanized aerosil OX50, an almost imperceptible third band can be observed around 1707 cm^{-1}. In the TGA of the untreated aerosil OX50 no mass loss was observed, which can be explained by the high stability of SiO_2. In contrast, silanized aerosil OX50 showed a decomposition temperature at 437 °C with a mass loss of approximately 2%.

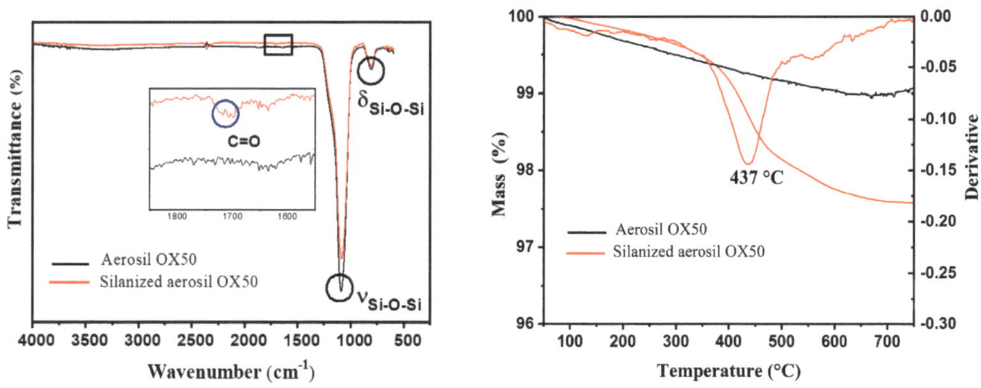

Figure 2. Characterization of the filler: (**a**) FTIR spectrum; (**b**) TGA thermograms of untreated and silanized aerosil OX50.

Several micrographs of aerosil OX50 were taken before and after silanization and representative ones are shown in Figure 3. In general, aerosil OX50 particles are spherical in shape and easily agglomerate, which is typical of nanoparticles. The agglomerates in the untreated aerosil OX50 (Figure 3a) are round and 150 nm in size. In the case of the silanized aerosil OX50 (Figure 3b), there is an increase in the cohesion between the agglomerated particles.

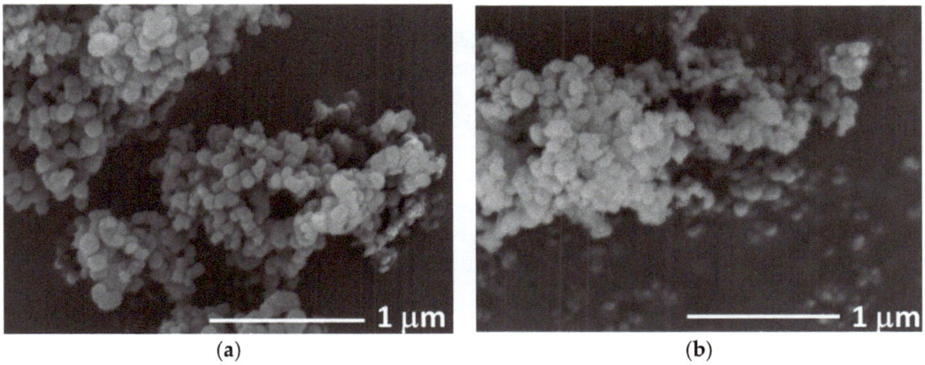

Figure 3. Characterization of the filler: (**a**) micrograph of unsilanized aerosil OX50; (**b**) micrograph of silanized aerosil OX50.

The particle size distribution of aerosil OX50 (Figure 4), obtained from micrograph measurements, shows a normal distribution. Most of the particles are in the 40–50 nm range in agreement with the manufacturer's description. Furthermore, no polydispersity is observed, and the average particle size was 44 ± 1 nm.

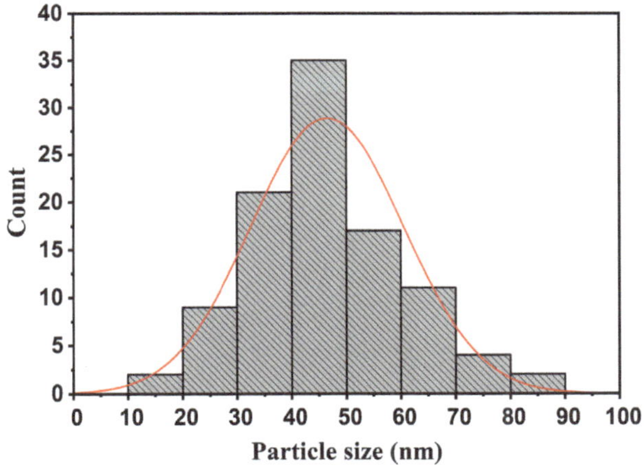

Figure 4. Particle size distribution of aerosil OX50.

3.2. Physical Appearance of the Experimental Composites

The four experimental composites, prepared with the monomeric matrices M1 (Bis-GMA/TEEGDMA/MPS) and M4 (Bis-GMA/TEEGDMA) and 45% aerosil OX50, are homogeneous pastes, easy to handle, and of similar consistency and color to commercial resins. The most viscous composites, as visually appraised and the most difficult to manipulate, are those that contain the M4 matrix, because the diluent monomer (TEEGDMA) is more viscous than the TEEGDMA/MPS mixture. In addition, in composites prepared with the Bis-GMA/TEEGDMA/MPS matrix, the filler is better incorporated. The commercial resin is a homogeneous, cream-colored paste, slightly more viscous than that prepared with the Bis-GMA/TEEGDMA matrix. The monomeric mixtures have a similar color (yellow) due to the presence of camphorquinone, with a color change after polymerization.

3.3. Characterization of the Experimental Composites

The measurement of the degree of conversion of the unsilanized and silanized experimental composites and the monomeric mixtures was carried out at 0, 1, 2, 3, 4, 5, 10, 15,

20, 25, 30, 35, 40, and 80 s of irradiation. Figure 5a shows the FTIR spectrum of the C145 composite at the different irradiation times mentioned above. The curves corresponding to each time are indicated in different colors from blue for time 0 to red for 80 s. The two bands under study in this research can be observed in the spectrum. The band corresponding to the aliphatic C=C bond appears around 1637 cm^{-1}, while the one associated with the vibrations of the C=C aromatic bond can be seen around 1612 cm^{-1}. Figure 5b shows the results obtained from the determination of the degree of conversion with time. Generally, only the degree of conversion at the final irradiation time, which in normally 40 s, is reported. In this study, in addition, the results of the degree of conversion at several times up to 80 s are reported and analyzed.

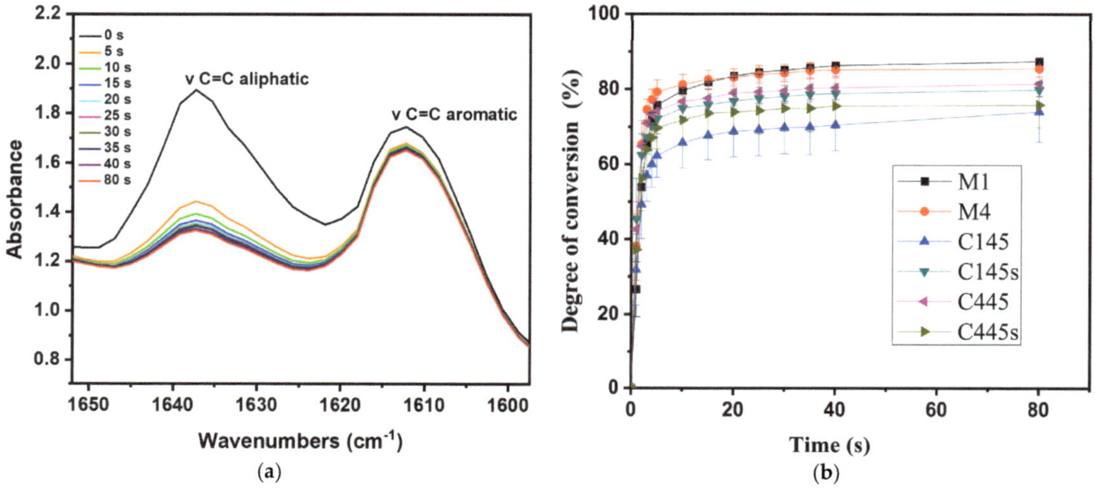

Figure 5. (**a**) FTIR spectrum of the C145 composite in the region from 1680–1580 cm^{-1} at different irradiation times; (**b**) Degree of conversion vs. irradiation time of the matrices and composites.

The results of the depth of cure and degree of conversion at 40 s for the matrices and composites are shown in Table 2. Composite materials meet the requirements of the ISO 4049:2019 standard [12] for cure depth, showing values greater than 2 mm. The highest values are exhibited by the monomeric matrices M1 and M4, (2.95 $\pm$ 0.01) mm and (2.94 $\pm$ 0.01) mm, respectively. In addition, there are no statistically significant differences between the monomeric matrices M1 and M4, nor between the composites prepared with them, so no effect of the silanization process was observed for this property. The experimental composites showed a greater value of curing depth than the commercial TT composite used as a reference.

Table 2. Degree of conversion at 40 s and depth of cure.

Samples	DC (%)	Depth of Cure (mm)
Property Value	**55–80**	**$\geq$2**
M1	86.3 (0.1) [a]	2.95 (0.01) [a]
M4	85 (2) [a,b]	2.94 (0.01) [a,b]
C145	71 (5) [c]	2.93 (0.02) [b]
C145s	79 (2) [d]	2.93 (0.01) [b]
C445	80 (3) [b,d]	2.93 (0.01) [a,b]
C445s	76 (6) [c,d]	2.93 (0.02) [a,b]
TT	63 (4) [e]	2.67 (0.09) [c]

[a,b,c,d,e] Statistically homogeneous groups (analysis by properties).

In the case of the degree of conversion for the matrices and composites at 40 s of irradiation, values higher than 67% and, in several cases, more than 75% (C145s and C445) were obtained. As shown in Table 2, all experimental composites present DC values higher than TT, except for composite C145, which does not have significant differences from it. Furthermore, the highest values of this property correspond to matrices M1 and M4, (86.3 ± 0.1)% and (85 ± 2)%, respectively.

The results of the water sorption and solubility test are shown in Figure 6. The highest values of this property evaluated for the different formulations correspond to the pristine monomeric matrices M1 and M4, which slightly exceed the limit value in the ISO 4049:2019 standard [12]. In the case of absorption, the best values correspond to the C145s, C445 and C445s composites, and no significant differences between them are observed. Furthermore, the commercial TT composite showed the lowest water sorption values and significant differences with respect to the experimental composites. Regarding water solubility, no significant differences were observed between C145 and C145s, nor between C445 and C445s.

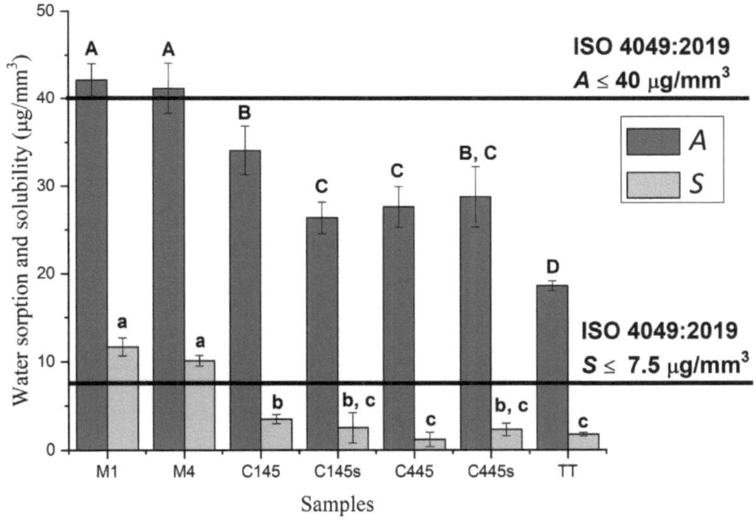

Figure 6. Water sorption and solubility for the matrices and composites. A, B, C, D and a, b, c are the statistically homogeneous groups for absorption and solubility, respectively.

The thermal degradation of dental composites and matrices can be visualized from the thermograms shown in Figure 7. As observed in Figure 7a, M1 and M4 have highest mass loss (90% at 700 °C and 89% at 500 °C, respectively) with respect to the other samples. In the case of composites, the greatest mass loss corresponds to the commercial TT (24.40%), followed by C145 (47.08%), C145s (49.00%), C445 (49.36%), and C445s (55.68%). The analysis of the first derivative (Figure 7b) indicates that the degradation of the tested materials occurs in an interval of 250–650 °C. In addition, the matrices present two decomposition temperatures (M1: 400, 424 °C; M4: 408, 432 °C), and the C145, C145s, and C445 composites experience a single decomposition at 430, 424, and 420 °C, respectively. The commercial composite experiences two endothermic effects, one at 368 °C and the other at 451 °C. On the other hand, the C445s composite experiences three endothermic effects (372, 420, and 600 °C).

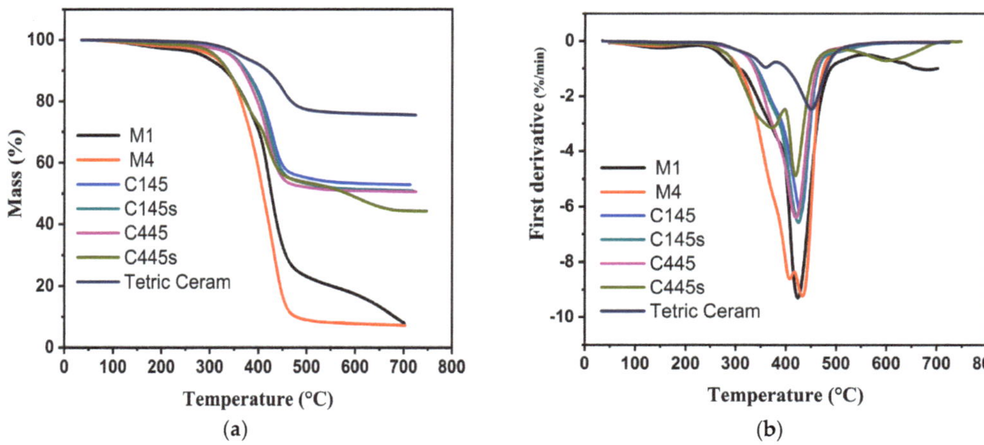

Figure 7. (**a**) Thermogravimetric curves; (**b**) First derivative of the thermogravimetric curves of matrices and composites.

4. Discussion

Silicon dioxide presents three characteristic absorption bands at 1080, 800, and 450 cm^{-1}, which correspond to the vibration modes of stretching, bending, and rocking of the Si-O-Si bonds, respectively. In the IR spectra of the untreated and silanized aerosil OX50 (Figure 2a) only two of the characteristic bands of silicon dioxide are observed, since the spectrum was recorded between 4500 and 650 cm^{-1}. The highest-intensity band appears around 1100 cm^{-1} and the least intense one at approximately 801 cm^{-1}, corresponding to the stretching and bending vibrations of the Si-O-Si bonds, respectively. The authors have verified the occurrence of silanization by the presence of a band of low intensity around 1707 cm^{-1} in the spectrum of silanized aerosil OX50 (Figure 2a), corresponding to the vibrations of the carbonyl group present in the MPS coupling agent [23]. The low intensity of this band suggests that a very small percentage of silane is incorporated, making it almost imperceptible for this characterization technique.

When the TGA curves of the unsilanized aerosil OX50 (Figure 2b) are analyzed, it is observed that the mass of the sample remains practically constant, which is due to the thermal stability of silicon dioxide. In the case of the treated aerosil OX50, the loss of 2% mass is associated with the decomposition of the MPS. In this sense, it has been reported that the physical decomposition of silane physically absorbed on the surface occurs between 50 and 150 °C, while the silane chemically bonded to the filler decomposes between 200 and 600 °C [24]. These results demonstrate the appearance of the silanization process of the filler.

The authors report that nanoparticle resins are highly aesthetic and their physical and mechanical properties are superior to others, presenting less contraction during polymerization and high resistance to fracture, compression, and traction [25]. With a smaller particle size, a better finish of the restoration is achieved, which is observed in its surface texture, reducing the possibilities of biodegradation of the material. In addition, this technology has allowed the mechanical properties of the resin to be suitable for both the anterior and posterior sectors [26]. In the micrographs of aerosil OX50 (Figure 3) it can be seen that it is formed by spherical particles in agreement with Aminoroaya et al. [8], suggesting that dental composites containing smaller spherical silica particles result in more uniform stress distribution and a reduction of stress at the filler–matrix interface. An increase in filler size is expected to increase the stress concentration at the filler–matrix interface, causing more susceptibility to crack nucleation in this area due to the different rigidities of organic matrix and inorganic filler, leading to lower mechanical performance, such as lower strength of the dental composites.

According to the manufacturer's technical indications [27], aerosil OX50 is a hydrophilic fumed silica and its average particle size is 40 nm. The slight increases in the quantity and size of the agglomerates observed for the silanized aerosil OX50 (Figure 3b), as well as the small irregularities generated, have no influence on the morphology of the particles. Despite these differences in the agglomerates, it is considered that the silanization process does not alter the particle morphology. These results coincide with those obtained by other authors who suggest that the presence of the silane molecule covalently bound to the silicon oxide implies new interactions between the nanoparticles and, therefore, a new arrangement of the aggregates. One way to explain this change is that the appropriate modification of the surface of a nanoparticle by a silane coupling agent involves the removal of the hydroxyl groups present on its surface together with the formation of new siloxane bonds (-Si-O-Si) between them. This new interaction causes the alteration in the agglomerates of the nanoparticles before and after the silanization process. In addition, during the development of this process, the first reaction that occurs is the hydrolysis of the alkoxy groups of the silane molecule that become silanol groups (–Si-OH) and, when this occurs, the silane molecules can react with each other and form dimers or oligomers.

Light-cured composite resins have the advantage of being systems with unlimited working time before the material cures. However, the efficiency in polymerization is of prime importance in their behavior. The degree of conversion indicates the percentage of double bonds that have been converted into single bonds, that is, the number of monomers in the resin that have reacted to form polymers and can be determined by FTIR. In Figure 5a, it can be seen how the band corresponding to the vibrations of the aliphatic C=C double bonds decreases in intensity as the irradiation time increases, which demonstrates the progress of the polymerization process. Properties such as compressive and diametric compression strength, elastic modulus, as well as color stability, solubility, and cytotoxicity, are closely related to the conversion. A low degree of conversion favors the presence of residual monomer inside the cavity, which can be toxic to the pulp, demonstrating that this property is closely related to the biocompatibility of composite resins. For these materials, the conversion from monomer to polymer occurs from 55–75% [28]. The nature of the monomers that make up the monomeric matrix, the filler content, the intensity of the irradiated light, and the irradiation time are aspects of great influence on this property. As can be seen in Figure 5b, in all cases there is a sudden increase in the degree of conversion early in irradiation. Then, this increase becomes less pronounced until 5 s of polymerization, from which time until 80 s the value of the property does not change appreciably, a factor that is due to the decrease in active centers (free radicals) and to their low mobility when the viscosity increases with the occurrence of the polymerization reaction.

The results obtained for the degree of conversion (Table 2) are totally satisfactory, as they give a measure of the lower toxicity that they can present if we compare them with the commercial composites in the literature. The higher DC values in matrices M1 and M4 are because the monomeric mixtures do not contain filler. It is known that the inorganic filler influences this property, since it increases the viscosity of the composite and hinders the diffusion of light emitted during irradiation and in some cases also the mobility of the organic molecules. The composites obtained have a degree of conversion between 65% and 85%, which is higher than those reported in other studies, as in the case of Nor Ain Fatihah Azlisham et al. in 2023 [29]. They evaluated the degree of conversion of different dental resins containing nanohybrid silica derived from rice husk as inorganic filler (40, 50, and 60% by weight). For these composites, they obtained conversion values from 60–65%.

The result of the statistical test for the degree of conversion shows that there are significant differences between the formulated materials. To determine which composite materials present significant differences among themselves, a multiple comparison test was performed. This showed that there are no significant differences between the degree of conversion of the Bis-GMA/TEEGDMA/MPS (M1) and Bis-GMA/TEEGDMA (M4) monomeric matrices and between the composites prepared with M4 (C445 and C445s). In the case of composites prepared with the monomeric matrix containing MPS (M1), there are significant differences between C145 and C145s and the highest value of the property corresponds to the C145s composite.

The depth of cure is one of the standard properties that should be evaluated in light-cured composite resins and is an indirect measure of the polymerization efficiency, which is decisive in the behavior of composites. Uncured material can migrate into the oral environment and cause allergic reactions in some patients, as well as stimulate the growth of bacteria around the restoration [30]. The ISO 4049:2019 standard [12] requires that these materials have a curing depth greater than 1.5 mm and that the loss of uncured material must be less than 0.5 mm, that is, that the optimal curing depth is equal to or greater than 2 mm. The maximum value recommended to dentists to layer composites when performing restorations in deep cavities is 2 mm. As can be seen in Table 2, the highest values correspond to the monomeric matrices M1 and M4, (2.95 $\pm$ 0.01) and (2.94 $\pm$ 0.01) mm, respectively. One of the factors that explains the high depth of cure values for the monomeric mixtures is the structure of the monomers used in the preparation of the mixtures (Bis-GMA, TEEGDMA), which polymerize by the two existing methacrylate groups at their ends. Thus, a higher degree of cross-linking is achieved with a lower amount of residual double bonds, which produces a higher efficiency in polymerization.

It has been reported that with thiourethane-functionalized fillers, it is possible to improve the properties of dental resins, such as the curing depth. This was attributed to the increase in refractive index provided by the presence of thiol, thiocarbamate, and aromatic groups, which minimizes the mismatch with the refractive index of the filler during polymerization and improves light penetration at depth [21].

Water sorption and solubility are parameters that must be controlled in dental materials. Both are related to the amounts of unreacted monomers and other non-polymerizable molecules, as well as to some small polymer chains that can migrate from the cure matrix into the medium. The sorption in the oral cavity can reduce the physical and mechanical properties of composites. This reduction is mainly attributed to the hydrolytic rupture of the bond between the filler and silane, which created a fissure. The sorption is then considered as a measure of the structural changes in the polymeric matrix, and the solubility is considered as an indirect measure of the toxicity of the material. The ISO 4049:2019 standard [12] refers to adequate sorption values as those that are less than or equal to 40 $\mu g/mm^3$ and solubility values that are less than or equal to 7.5 $\mu g/mm^3$. The highest values of these properties correspond to the monomeric matrices and slightly exceed the limit reported in the standard. In this sense, it is known that the Bis-GMA, TEEGDMA, and MPS monomers present in the organic matrix have oxygen atoms that, when interacting with the protic solvent, form hydrogen bonds with it, favoring water sorption. The statistical analysis and the subsequent multiple comparison test developed with the water sorption values showed significant differences between the C145 and C145s composites, the latter being the one with the lowest value. It should be remembered that the C145s composite is the one that contains MPS as part of the diluent and the silanized filler, a factor that contributed to the result, demonstrating the effectiveness of the treatment by both methods to improve the surface bond. No statistically significant differences were found between the monomeric matrices M1 and M4, nor between the C445 and C445s composites, so in this case there are no effects of the particle silanization on their solubility. These results are not in complete correspondence with those obtained for the depth of cure. Regarding the water solubility of the composites, the statistical analysis showed that there are no statistically significant differences between the composites prepared with the same matrix, so the surface treatment of the filler had no impact on this property. The results obtained for the C445 and C445s composites coincide with the results obtained for the degree of conversion.

Thermogravimetric tests allow the characterization of the thermal degradation processes that occur in different materials. In the case of the composites evaluated, the mass losses correspond to the relative composition of each of them and are associated with the degradation of the polymeric matrix that is part of the composites. Their degradation occurs from 250 to 650 °C, a temperature that coincides with those reported in the literature for commercial composites [31]. The C445s composite presents three endothermic effects at 372, 420, and 600 °C corresponding to the gradual decomposition of the organic matrix. In this sense, it is known that the monomers that make up these organic matrices decompose, resulting in products such as methacrylic acid, 2-hydroxyethyl methacrylate, and phenolic compounds [23]. In the case of the commercial TT composite, the thermogravimetric test corroborated that the percentage of inorganic filler is around 76.60% and, according to what is reported in the literature, this value should be 80–82%. The difference between these values is due to the fact that the filler also contains copolymers that degrade during heat treatment.

It is important to remember that the maximum temperature that the oral environment supports when eating hot food is 50–60 °C [32,33]. Bearing this in mind, dental materials must have decomposition temperatures higher than those mentioned above, otherwise they would soften, causing restoration failure. We can then affirm that the composites obtained are thermally stable since their decomposition begins at 250 °C.

5. Conclusions

It was possible to demonstrate the appearance of the silanization process by FTIR and TGA, where in the case of the silanized aerosil OX50 a mass loss of around 2% is observed and associated with the decomposition of the MPS. On the other hand, the composites presented a degree of conversion higher than 70%, which in many cases exceeds those reported in the literature. The characterization of the experimental composites showed that they meet the requirements of the ISO 4049:2019 standard for depth of cure and water sorption and solubility. Thermal analysis showed that the studied composites are thermally stable, since the degradation of the matrix begins at 250 °C. The composites obtained have adequate properties that make them suitable for use as dental filler and, in addition, the prior silanization of the filler, together with the introduction of MPS as a diluent in the polymeric matrix, allowed greater compatibility between the organic matrix and the filler, which is manifested in the final properties of the C145s composite.

Author Contributions: Conceptualization and methodology, Y.V.-P., S.d.l.C.D.-R. and O.T.-A.; formal analysis, Y.V.-P., S.d.l.C.D.-R., O.T.-A., A.A.L.S., A.R.-S., J.V.C.-R., J.M.C.-U. and W.H.-K.; investigation, Y.V.-P., S.d.l.C.D.-R., O.T.-A., A.A.L.S., A.R.-S., J.V.C.-R., J.M.C.-U. and W.H.-K.; data curation, S.d.l.C.D.-R., O.T.-A. and Y.V.-P.; writing—original draft preparation, S.d.l.C.D.-R., O.T.-A. and Y.V.-P.; writing—review and editing, S.d.l.C.D.-R., O.T.-A., A.A.L.S., A.R.-S., J.V.C.-R., J.M.C.-U., W.H.-K. and Y.V.-P.; supervision, Y.V.-P. All authors have read and agreed to the published version of the manuscript.

Funding: This research received no external funding.

Institutional Review Board Statement: Not applicable.

Informed Consent Statement: Not applicable.

Data Availability Statement: The original contributions presented in the study are included in the article.

Acknowledgments: Thanks to the Secretariat of Foreign Affairs (SRE) of the Mexican Embassy and the Mexican Agency for International Development Cooperation (AMEXCID) for the scholarship granted to one of the authors that allowed the characterization of the composite resins and the filler. The authors are grateful to J. García-Couce for her assistance in the review of the document and Nicolás Alberto Sirgado Pérez for his support in the morphological characterization of aerosil OX50.

Conflicts of Interest: The authors declare no conflicts of interest.

References

1. Duarte de Oliveira, F.J.; Ferreira da Silva Filho, P.S.; Fernandes Costa, M.J.; Rabelo Caldas, M.R.G.; Dutra Borges, B.C.; Gadelha de Araújo, D.F. A comprehensive review of the antibacterial activity of dimethylaminohexadecyl methacrylate (DMAHDM) and its influence on mechanical properties of resin-based dental materials. *Jpn. Dent. Sci. Rev.* **2021**, *57*, 60–70. [CrossRef] [PubMed]
2. Makvandi, P.; Jamaledin, R.; Jabbari, M.; Nikfarjam, N.; Borzacchiello, A. Antibacterial quaternary ammonium compounds in dental materials: A systematic review. *Dent. Mater.* **2018**, *34*, 851–867. [CrossRef] [PubMed]
3. *Convenio de Minamata Sobre el Mercurio—Texto y Anexos*; Programa de Naciones Unidas para el Medio Ambiente: Nairobi, Kenya, 2019; pp. 3–15.
4. Gomez, G.J. Perspectivas del Uso de la Amalgama Dental bajo el Convenio de Minamata: Tendencias Nacionales e Internacionales. *CES Odontol.* **2020**, *33*, 53–63. [CrossRef]
5. Pérez-Mondragón, A.A.; Cuevas-Suárez, C.E.; González-López, J.A.; Trejo-Carbajal, N.; Meléndez-Rodríguez, M.; Herrera-González, A.M. Preparation and evaluation of a BisGMA-free dental composite resin based on a novel trimethacrylate monomer. *Dent. Mater. Off. Publ. Acad. Dent. Mater.* **2020**, *36*, 542–550. [CrossRef] [PubMed]
6. Pratap, B.; Gupta, R.K.; Bhardwaj, B.; Nag, M. Resin based restorative dental materials: Characteristics and future perspectives. *Jpn. Dent. Sci. Rev.* **2019**, *55*, 126–138. [CrossRef] [PubMed]
7. Martim, G.C.; Kupfer, V.L.; Moisés, M.P.; dos Santos, A.; Buzzetti, P.H.M.; Rinaldi, A.W.; Rubira, A.F.; Girotto, E.M. Physical-chemical properties of dental composites and adhesives containing silane-modified SBA-15. *J. Mech. Behav. Biomed. Mater.* **2018**, *80*, 277–284. [CrossRef] [PubMed]
8. Aminoroaya, A.; Neisiany, R.E.; Khorasani, S.N.; Panahi, P.; Das, O.; Madry, H.; Cucchiarini, M.; Ramakrishna, S. A review of dental composites: Challenges, chemistry aspects, filler influences, and future insights. *Compos. Part B Eng.* **2021**, *216*, 108852. [CrossRef]
9. Barot, T.; Rawtani, D.; Kulkarni, P. Physicochemical and biological assessment of silver nanoparticles immobilized Halloysite nanotubes-based resin composite for dental applications. *Heliyon* **2020**, *6*, e03601. [CrossRef]
10. Moragues, A.; Ruiz Soto, A.; De Miguel, M.; Reyes, E.; Cristina, A. Nano-scale aluminium interaction in synthetic hydrated calcium silicate gel studied by 29Si MAS-NMR. *Bol. Soc. Esp. Ceram. Vidr.* **2022**, *62*, 388–401. [CrossRef]
11. Nader, R.; Hamieh, T.; Villieras, F.; Angelina.Razafitianamaharav; Toufaily, J.; McHeik, A.S.; Thomas, F. Study of Gas Adsorption on Biphasic Nanostructured Surfaces. *Phys. Procedia* **2014**, *55*, 373–382. [CrossRef]
12. *ISO 4049 E*; Dentistry—Polymer-Based Restorative Materials. ISO: Geneva, Switzerland, 2019.
13. Haji, Z.; Ghafoor, R. Effect of Curing Modes on Depth-of-Cure in Bulk-Fill Composite. *Int. Dent. J.* **2021**, *71*, S40. [CrossRef]
14. Alkhouri, N.; Xia, W.; Ashley, P.; Young, A. The effect of varying monocalcium phosphate and polylysine levels on dental composite properties. *J. Mech. Behav. Biomed. Mater.* **2023**, *145*, 106039. [CrossRef]
15. Thadathil Varghese, J.; Cho, K.; Raju; Farrar, P.; Prentice, L.; Prusty, B.G. Influence of silane coupling agent on the mechanical performance of flowable fibre-reinforced dental composites. *Dent. Mater.* **2022**, *38*, 1173–1183. [CrossRef] [PubMed]
16. Cavalcante, L.M.; Ferraz, L.G.; Antunes, K.B.; Garcia, I.M.; Schneider, L.F.J.; Collares, F.M. Silane content influences physicochemical properties in nanostructured model composites. *Dent. Mater. Off. Publ. Acad. Dent. Mater.* **2021**, *37*, e85–e93. [CrossRef] [PubMed]
17. Karabela, M.M.; Sideridou, I.D. Synthesis and study of properties of dental resin composites with different nanosilica particles size. *Dent. Mater. Off. Publ. Acad. Dent. Mater.* **2011**, *27*, 825–835. [CrossRef] [PubMed]
18. Elfakhri, F.; Alkahtani, R.; Li, C.; Khaliq, J. Influence of filler characteristics on the performance of dental composites: A comprehensive review. *Ceram. Int.* **2022**, *48*, 27280–27294. [CrossRef]
19. Cisneros-Pineda, O.G.; Herrera Kao, W.; Loría-Bastarrachea, M.I.; Veranes-Pantoja, Y.; Cauich-Rodríguez, J.V.; Cervantes-Uc, J.M. Towards optimization of the silanization process of hydroxyapatite for its use in bone cement formulations. *Mater. Sci. Eng. C* **2014**, *40*, 157–163. [CrossRef]
20. Vidal, M.L.; Rego, G.F.; Viana, G.M.; Cabral, L.M.; Souza, J.P.B.; Silikas, N.; Schneider, L.F.; Cavalcante, L.M. Physical and chemical properties of model composites containing quaternary ammonium methacrylates. *Dent. Mater. Off. Publ. Acad. Dent. Mater.* **2018**, *34*, 143–151. [CrossRef] [PubMed]
21. Goulart, M.; Fugolin, A.P.; Lewis, S.H.; Rodrigues, J.A.; Erhardt, M.C.; Pfeifer, C.S. Thiourethane filler functionalization for dental resin composites: Concentration-dependent effects on toughening, stress reduction and depth of cure. *Mater. Sci. Eng. C* **2021**, *118*, 111535. [CrossRef]
22. Abid Althaqafi, K.; Alshabib, A.; Satterthwaite, J.; Silikas, N. Properties of A Model Self-Healing Microcapsule-Based Dental Composite Reinforced with Silica Nanoparticles. *J. Funct. Biomater.* **2022**, *13*, 19. [CrossRef]
23. Alayola, J.J. Efecto de la Incorporación de Nanoarcillas Sobre las Propiedades de Materiales Compuestos Dentales. Master's Thesis, Centro de Investigaciones Científicas de Yucatán, A.C., Mérida, Mexico, 2017.
24. Zha, C.; Wang, W.; Lu, Y.; Zhang, L. Constructing covalent interface in rubber/clay nanocomposite by combining structural modification and interlamellar silylation of montmorillonite. *ACS Appl. Mater. Interfaces* **2014**, *6*, 18769–18779. [CrossRef]
25. Zubrzycki, J.; Klepka, T.; Marchewka, M.; Zubrzycki, R. Tests of Dental Properties of Composite Materials Containing Nanohybrid Filler. *Materials* **2023**, *16*, 348. [CrossRef]
26. Althaqafi, K.A.; Satterthwaite, J.; Silikas, N. A review and current state of autonomic self-healing microcapsules-based dental resin composites. *Dent. Mater. Off. Publ. Acad. Dent. Mater.* **2020**, *36*, 329–342. [CrossRef]

27. Susila Anand, V.; Balasubramanian, V. Effect of resin chemistry on depth of cure and cytotoxicity of dental resin composites. *Mater. Sci. Eng. B* **2014**, *181*, 33–38. [CrossRef]
28. Cherchali, F.Z.; Attik, N.; Mouzali, M.; Tommasino, J.B.; Abouelleil, H.; Decoret, D.; Seux, D.; Grosgogeat, B. Structural stability of DHMAI antibacterial dental composite following in vitro biological aging. *Dent. Mater. Off. Publ. Acad. Dent. Mater.* **2020**, *36*, 1161–1169. [CrossRef] [PubMed]
29. Azlisham, N.A.F.; Johari, Y.; Mohamad, D.; Yhaya, M.F.; Mahmood, Z. Degree of conversion and physicomechanical properties of newly developed flowable composite derived from rice husk using urethane dimethacrylate monomer. *J. Eng. Med.* **2023**, *237*, 1339–1347. [CrossRef] [PubMed]
30. Encalada-Alayola, J.J.; Veranes-Pantoja, Y.; Uribe-Calderón, J.A.; Cauich-Rodríguez, J.V.; Cervantes-Uc, J.M. Effect of Type and Concentration of Nanoclay on the Mechanical and Physicochemical Properties of Bis-GMA/TTEGDMA Dental Resins. *Polymers* **2020**, *12*, 601. [CrossRef] [PubMed]
31. Kamalak, H.; Aksu Canbay, C.; Yigit, O.; Altin, S. Physico-mechanical and thermal characteristics of commercially available and newly developed dental flowable composites. *J. Mol. Struct.* **2017**, *1156*, 314–319. [CrossRef]
32. Al Ayyan, W.; Al Halabi, M.; Hussein, I.; Khamis, A.H.; Kowash, M. A systematic review and meta-analysis of primary teeth caries studies in Gulf Cooperation Council States. *Saudi Dent. J.* **2018**, *30*, 175–182. [CrossRef] [PubMed]
33. Tarano-Artigas, O.; Cauich-Rodriguez, J.V.; Cervantes-Uc, J.M.; Veleva-Muleshkova, L.; Veranes-Pantoja, Y. Evaluación termo-mecánica de composites dentales fotopolimerizables. *Rev. Cuba. Química* **2022**, *34*, 194–210.

polymers

MDPI

Article

Feasibility of Invasive Brown Seaweed *Rugulopteryx okamurae* as Source of Alginate: Characterization of Products and Evaluation of Derived Gels

Ismael Santana *, Manuel Felix and Carlos Bengoechea

Escuela Politécnica Superior, Universidad de Sevilla, Calle Virgen de África, 7, 41011 Sevilla, Spain; mfelix@us.es (M.F.); cbengoechea@us.es (C.B.)
* Correspondence: isantana@us.es; Tel.: +34-954-557-179

Abstract: *Rugulopteryx okamurae* (RO) is an invasive brown seaweed that causes severe environmental problems in the Mediterranean Sea. This work proposed an extraction method that enables their use as a raw material for producing sodium alginate. Alginate was successfully extracted from this invasive seaweed, with its gelling performance in the presence of Ca^{2+} ions comparable to existing commercial alginates. The mannuronic acid (M)-to-guluronic (G) acid ratio in the ^{1}H-NMR profile indicated a higher percentage of G in the RO-extracted alginate, which implies a greater formation of so-called egg box structures. These differences resulted in their different rheological behaviour, as sodium alginate aqueous solutions exhibited a greater viscosity (η at 1 s^{-1} = 3.8 ± 0.052 Pa·s) than commercial alginate (2.8 ± 0.024 Pa·s), which is related to the egg box structure developed. When gelled in the presence of calcium, an increase in the value of the elastic modulus was observed. However, the value of the tan δ for the extracted alginate was lower than that of commercial alginate gels, confirming a structure more densely packed, which implies a different restructuring of the alginate chain when gelling. These results confirm the suitability of using invasive *Rugulopteryx okamurae* as a source of calcium alginate gels. In this way, sustainable bio-based materials may be produced from undesired biomass that currently poses a threat to the ecosystem.

Keywords: alginate extraction; gelation; *Rugulopteryx okamurae*; SAOS

Citation: Santana, I.; Felix, M.; Bengoechea, C. Feasibility of Invasive Brown Seaweed *Rugulopteryx okamurae* as Source of Alginate: Characterization of Products and Evaluation of Derived Gels. *Polymers* **2024**, *16*, 702. https://doi.org/10.3390/polym16050702

Academic Editor: Bruno Medronho

Received: 9 January 2024
Revised: 25 February 2024
Accepted: 28 February 2024
Published: 5 March 2024

1. Introduction

Seaweeds are a natural resource of a wide variety of natural compounds, of which carbohydrates are the most numerous and can reach up to 71.2% of their dry weight, depending on the type of seaweed [1,2]. Most of these carbohydrates are generally found in the form of polysaccharides, whose composition will vary depending on the species of seaweed. Green seaweeds, for example, have a polysaccharide composition similar to that of land plants compared to other algae species [3]. In some species of green seaweeds, cellulose is one of the main polysaccharides that make up their cell wall, apart from other sulphated polysaccharides, such as ulvan or galactan, among others [4]. Regarding red seaweeds, within the wide spectrum of polysaccharides it contains, it is worth highlighting agar, carrageenans and cellulose, among others [5]. These polysaccharides are raw materials usually useful in the industry for film production, where some of the algae used for this purpose are *Eucheuma sp.* or *Palmaria sp.*, among many more [6]. Finally, the polysaccharide composition of brown algae consists mainly of alginate, laminarian, or fucoidan, which have a wide variety of uses such as in the food industry, pharmaceutical, films, etc. Some brown seaweed used for the extraction of these polysaccharides are *Sargassum sp.*, *Laminaria sp.*, or *Undaria sp.* [7,8]. Thus, the extraction of sodium alginate from algae has already been reported; however, the relationship between their structure and functional properties is limited, especially in the case of the invasive species *Rugulopteryx okamurae*.

Alginate is a linear polysaccharide found in the matrix and cell wall of brown seaweeds and is formed by blocks of mannuronic (M) acid and guluronic (G) acids linked by glycosidic bonds. These blocks are found in heteropolymeric sequences along the chain they form, varying their length, order, and position between different alginates [9]. Commercially used alginate is usually found in the form of soluble sodium alginate, for which an extraction process is carried out from the insoluble salts of calcium, potassium, or magnesium alginate that form the cell wall [7,10]. In its soluble salt form, sodium alginate has several uses in the pharmaceutical and food industries thanks to its gelling, coagulant, and encapsulating properties [11]. It is also interesting to highlight the ability of alginate to form gels in the presence of calcium, which has great interest in industry [12].

Although commercial sodium alginate is already available, there are many advantages of using seaweed as a source of this compound. Firstly, they are greatly available and diverse, and they do not compete with other crops. Thus, they do not require land or irrigation for their cultivation, and in many cases, their growth is not uncontrolled. More specifically, the seaweed used in this work was *Rugulopteryx okamurae* (RO). It is an invasive brown seaweed. It is a species of algae that has been considered invasive in Spain since 2020 [13], found mainly in the Mediterranean Sea and western European coasts, currently affecting other countries such as Morocco and France but with prospects of continuing to proliferate [14]. This seaweed, native to the northwest Pacific Ocean, was introduced via the Strait of Gibraltar, where, thanks to the warm waters, a favourable environment is produced for its proliferation [15]. Nowadays, it represents a huge problem since tons of arrivals of RO reach the Mediterranean coasts every year, threatening their use not only for tourism but also the lives of fishers since it pushes away native species, multiplying rapidly and quickly colonizing the Mediterranean coasts [13]. Since the problem cannot be addressed from a biological point of view, finding applications for this seaweed could serve as a path to control their uncontrolled dissemination.

In the present work, the extraction and characterization of sodium alginate from the invasive brown algae RO was carried out. ^{1}H-NMR, DSC, TGA, and FTIR analysis were carried out in order to obtain structural information, including its molecular weight, which can be later used to properly justify its properties. Subsequently, aqueous solutions were prepared, and their rheological properties were evaluated and eventually compared with that obtained with commercial sodium alginate by oscillatory rheological tests. Finally, the controlled gelation of calcium alginate was carried out by ion exchange via a dialysis membrane. The mechanical spectra of the calcium alginate gels were obtained. The water absorption and retention properties of freeze-dried gels were assessed by their water uptake capacity and water holding capacity (WUC and WHC, respectively). The results obtained permitted the correlation of the molecular structure of the extracted sodium alginate with the functional properties of the developed gels, in agreement with the egg box structure proposed by other authors.

2. Materials and Methods

2.1. Alginate Extraction Process

The sodium alginate was extracted according to the scheme represented in Figure 1. This extraction was carried out following the procedure suggested by Calumpong et al. [16] with some modifications. First, 30 mg/mL of milled and freeze-dried seaweed was soaked overnight in 2% formaldehyde to eliminate pigments. The presence of phenolic compounds in the seaweed pigment during the alginate extraction with Na_2CO_3 may contribute to the degradation of the polysaccharide. Hence, formaldehyde was employed to generate an insoluble polymer solution, subsequently extracted during the acid treatment [17]. The extract obtained was rinsed with Milli-Q grade water and soaked in a 0.2 M HCl solution for 24 h to protonate the alginic acid. The resulting solid was washed with Milli-Q grade water and stirred for 5 h at 65 °C in a 2% Na_2CO_3 solution. The supernatant was collected by centrifugation at 5000× *g* for 30 min, and it was precipitated with ethanol 95% (3 volumes).

The solid generated was washed with acetone and dried at 50 °C overnight. Finally, the sample was freeze-dried and stored at −20 °C [16,18,19].

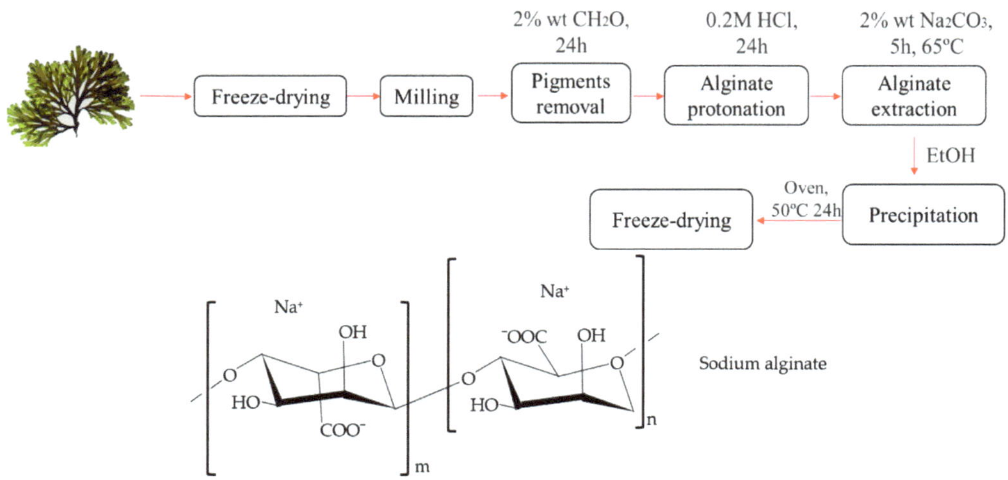

Figure 1. Flow chart of the sodium alginate extraction from *Rugulopteryx okamurae* and structure of sodium alginate.

2.2. Characterization of Commercial and RO Extracted Alginate

2.2.1. Differential Scanning Calorimetry (DSC)

DSC analysis was carried out by a DSC Q20 (TA Instruments, New Castle, DE, USA) equipped with an autosampler unit and was used to perform DSC. Overall, 5–10 mg samples were weighed; they were placed into hermetic-sealed aluminum pans. Samples were initially equilibrated at −50 °C (10 min), and then, they were heated at 10 °C min^{-1} up to 325 °C with a continuous nitrogen flow (50 mL min^{-1}).

2.2.2. Thermogravimetric Analysis (TGA)

TGA analysis was carried out in an SDT Q600 V20.9 Build 20 (TA Instruments, New Castle, DE, USA). The test was performed from 25 to 600 °C at a heating rate of 10 °C/min, with a mass flow of 100 mL/min under N_2 atmosphere. The weight of the sample was always approximately 10 mg.

2.2.3. Fourier Transform Infrared Spectroscopy (ATR-FTIR)

Extracted and commercial alginate were both measured using an FTIR Invenio X spectrometer (Bruker, Billerica, MA, USA) with an ATR objective. Tests were carried out for values of wavenumber from 4000 to 400 cm^{-1} with a resolution of 4 cm^{-1}.

2.2.4. Proton Nuclear Magnetic Resonance (^{1}H-NMR)

Solutions containing commercial and extracted alginates (20.2 mg/mL) were prepared in deuterated water (D_2O 99.9%) and then lyophilized. Subsequently, 100 µL of a solution of TSP-d4 (sodium 2,2,3,3-tetradeutero-3-trimethylsilylpropionate) and 900 µL of D_2O (99.9%) were added to each lyophilizate, dissolved, and 600 µL was taken for analysis. ^{1}H-NMR study was carried out on a Bruker Avance NEO 500 MHz spectrometer (Bruker, Billerica, MA, USA) at 80 °C. A 90° pulse and 32 sweeps were used, with a waiting time between sweeps of 20 s. Overall, 64k points were used to register the free induction decay (FID). ^{1}H-NMR spectra were recorded both with and without presaturation of the residual solvent signal (semi-deuterated water, DHO).

2.2.5. Determination of Intrinsic Viscosity and Molecular Mass of the Extracted Alginate

Intrinsic viscosity, [η], is a measure of the hydrodynamic volume occupied by a macromolecule, which depends on the molecular mass and the solvent [20]. Intrinsic viscosity measurements were carried out from sodium alginate solutions in 0.1 M NaCl at concentrations of 0.2, 0.4, 0.6, 0.8, and 1% wt. using a Ubbelhode viscosimeter. NaCl was chosen as the solvent since the constants k and a used in Equation (2) for the determination of the molecular mass depend on the solvent, and this was the one used by Mark Houwink due to its affinity with the polymer and its reduction in electrostatic repulsions [20,21]. The temperature was kept constant at 25 °C using a Tectron Bio thermostatic bath (J.P. Selecta, Barcelona, Spain). The efflux time of the solvent at the different concentrations prepared was measured in triplicate [21]. Fedors' Equation (1) was used to relate the concentrations and efflux times with the intrinsic viscosity of the RO-extracted alginate [22]:

$$\frac{1}{2\left(\eta_r{}^{\frac{1}{2}}-1\right)}=\frac{1}{C[\eta]}-\frac{1}{C_{max}[\eta]} \quad (1)$$

where η_r is the reduced viscosity, which is the ratio of the viscosity of the samples to that of the solvent; C is the concentration of each sample; and C_{max} is a polymer concentration parameter.

Once the intrinsic viscosity of the alginate was determined, the Mark Houwink–Sakurada Equation (2) was used to obtain the molecular mass of the extracted alginate [23].

$$[\eta]=k(M)^a \quad (2)$$

where $[\eta]$ is the intrinsic viscosity, M is the molecular mass of the polymer, and k and a are empirical coefficients dependent on the type of polymer. These coefficients (k and a, 0.023 and 0.984, respectively) were taken from an empirical regression of several measurements of different alginates from seaweeds (*Fucus vesicularus*, *Laminaria hyperborean* [24], *Laminaria digitata* [25], *Laminaria cloustoni* [26], and *Macrocystis pyrifera* [27]) collected by Clementi et al. (1998) [20].

2.3. *Gelation Process of Extracted and Commercial Alginate*

Alginate gelation was carried out by dialysis using dialysis tubing cellulose membrane 33 mm flat D9652-100FT (Sigma-Aldrich, St. Louis, MO, USA) to obtain a broad and homogeneous structure of approximately 10 cm high and 20 mm in diameter. First, sodium alginate was dissolved in distilled water. Then, this solution was placed in the dialysis tubing cellulose membranes, which were immersed in a $CaCl_2$ solution and left to sit overnight. An equivalent amount of glucono-δ-lactone was added to acidify the medium and facilitate the release of Ca^{2+} ions [28–31].

2.4. *Characterization of Calcium Alginate Gels*

2.4.1. Viscosity of Alginate Solutions

The AR-2000 rheometer (TA Instruments, New Castle, DE, USA) was used to determine the viscosity of the different sodium alginate solutions (1 and 2% wt.) when dissolved in water before gelation. Frequency sweep tests (from 0.1 to 100 Hz) were carried out with low inertia aluminum plates to increase sensibility at 25 °C. All the measurements were carried out within the linear viscoelastic range (LVR), previously determined by strain sweep tests. The temperature was controlled by a Peltier system.

2.4.2. Rheological Properties of Gels

Gelled calcium alginate was cut into smaller pieces 25 mm high and 20 mm in diameter using a scalpel. Frequency sweep tests were carried out with a Haake MARS 40 Rheometer (Thermo Fisher Scientific, Waltham, MA, USA) using parallel rough plates of 20 mm in diameter. Tests were performed from 0.1 to 10 Hz for gels at an alginate concentration of 0.5, 1, and 2% wt. at 20 °C just after obtaining gels. The initial water content of the gels was assessed by measuring the weight difference between the gels obtained immediately after gelation and the lyophilized counterparts, yielding a water content result of 42.16 ± 3.34%.

2.4.3. Water Uptake Capacity (WUC), Soluble Matter Loss (SML) and Water Holding Capacity (WHC)

Starting from freeze-dried gels stored in a desiccator, the first step consisted of drying samples in an oven at 50 °C overnight (20 h) (w_1). Then, they were immersed in distilled water for 24 h and weighed (w_2). Right after immersion, samples were centrifuged at 2000 rpm for 10 min (w_3). Finally, gels were freeze-dried and weighed (w_4). Thus, WUC, SML, and WHC were determined by weight differences following Equations (3)–(5), respectively:

$$WUC\ (\%) = \frac{(w_2 - w_4)}{w_4} \cdot 100 \tag{3}$$

$$SML\ (\%) = \frac{(w_1 - w_3)}{w_1} \cdot 100 \tag{4}$$

$$WHC\ (\%) = \frac{(w_2 - w_3)}{w_2} \cdot 100 \tag{5}$$

where w indicates the weight, and the subscript number is the step of the process.

2.4.4. Scanning Electron Microscopy (SEM)

Freeze-dried gels were examined by SEM using a ZEISS EVO microscope (ZEISS, Oberkochen, Germany). Samples were coated with an Au-Pd layer. Samples were Au-Pd coated and observed at a working distance of 6 mm, with a bean current of 18 pA and with an acceleration voltage of 10 kV. Images obtained were analyzed by software ImageJ version v1.54i (Bethesda, MD, USA).

2.5. Statistical Analysis

Significant differences are indicated with superscript letters ($p < 0.05$). The measurements were carried out at least in triplicate.

3. Results and Discussion

3.1. Alginate Characterization

The extracted alginate from RO was subjected to different tests in order to (i) ensure its suitability for gelling properties and (ii) compare its performance with respect to commercial samples.

3.1.1. ATR-FTIR and ^{1}H-NMR

Figure 2A shows the FTIR profile of the alginate obtained from RO seaweed and commercial alginate. ATR-FTIR spectroscopy can be used to analyze a wide range of organic and inorganic materials, being their FTIR spectrum characteristics since it is relative to their covalent bonding pairs and functional groups ([32]). Alginic acid has a characteristic structure of β-D-mannuronic acid and α-L-guluronic acid that swap the homopolymeric and alternating regions. Although this is the general structure, the concentrations of guluronic and mannuronic acids depend not only on the species but also on ambient conditions such as the season, temperature, and seaweed [33].

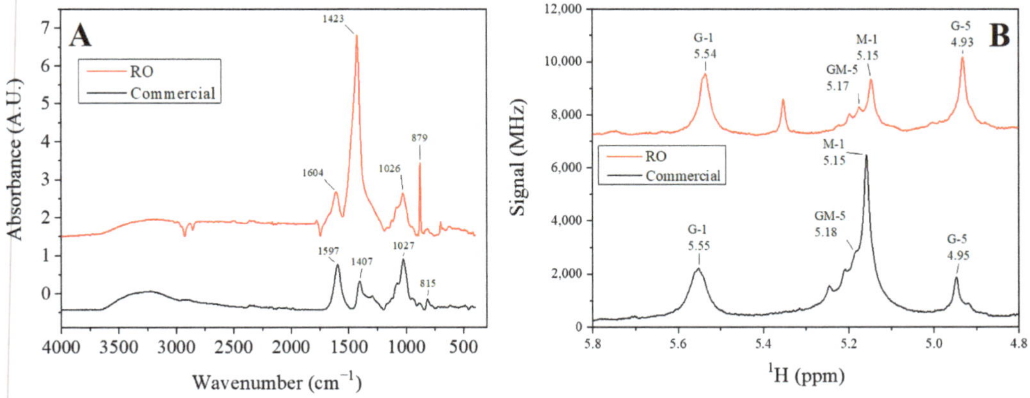

Figure 2. Fourier Transform Infrared analysis (**A**) and ^{1}H-NMR analysis (**B**) of commercial and *Rugulopteryx okamurae* (RO) extracted alginate.

Thus, the coincidence in the position of the different peaks observed confirms that the powder obtained following the experimental procedure shown in Figure 1 is sodium alginate. The first peak rose at ~1600 cm^{-1} [34]. This peak, together with the peak observed at ~1400 cm^{-1}, corresponds to the vibrations of the O-C-O bond of the carboxylate group, where the former signal can be attributed to asymmetric stretching and the latter symmetric stretching vibration, with a contribution of C-OH deformation vibration [19]. The difference between the heights and the offset may indicate the interference of some impurity from the seaweed. Moreover, both alginates (extracted and commercial) exhibited a peak at ~1026 cm^{-1}, which corresponds with the C-O stretching vibrations [35]. It can be mentioned that more intense peaks were obtained for the extracted alginate compared to the ones obtained for commercial alginate at 1600 and 880 cm^{-1}. These peaks can be attributed to the presence of sodium carbonate, which has those characteristics peaks too. The small peak found on commercial alginate around 815 cm^{-1} is attributed to mannuronic acid residues [18]. Despite variations in peak heights and phase shifts, the obtained powder is predominantly composed of sodium alginate, as evidenced by the presence of its characteristic peaks. However, as stated above, its performance depends on both structure and ambient condition growth. Further analyses were conducted to assess the performance of the extracted powder.

Figure 2B shows the ^{1}H-NMR analysis of both RO extracted and commercial alginates. This method is the most appropriate to determine the composition of the structure of alginate, obtaining the fractions of guluronic (G) and mannuronic (M) acid blocks that make up the alginate by comparison of the area of the spectrum signals corresponding to each M or G group, following Grasdalen et al. [36] Equations (6)–(9). A different peak is observed in the profile referring to the RO-extracted alginate at approximately 5.35 ppm, possibly due to impurities remaining from the algae during the alginate extraction process.

$$F_G = \frac{I_A}{I_B + I_C} \tag{6}$$

$$F_M = 1 - F_G \tag{7}$$

$$M/G = \frac{1 - F_G}{F_G} \tag{8}$$

$$F_{GG} = \frac{I_C}{I_B + I_C} \tag{9}$$

where F_M is the fraction of M blocks in the alginate structure; F_G is the fraction of G blocks; F_{GG} is the dyad sequence of G blocks; and I_A, I_B, and I_C are the areas under the curve of the signals corresponding to G-1, (GM-5 + M-1) and G-5, respectively.

Thus, higher values of F_G and F_{GG} were obtained by the RO-extracted alginate (F_G = 0.53 and F_{GG} = 0.53) compared to the commercial alginate (F_G =0.47 and F_{GG} = 0.23) and, therefore, lower values for the M/G in the RO extracted alginate (0.88) than in the commercial one (1.11). Similar results as the RO-extracted alginate were obtained by Larsen et al. (2003) [37] for alginate extracted from brown seaweeds *Sargassum latifolium* and *Sargassum asperiforium*. This structural difference implies a greater possibility of the formation of the structure known as the "egg box" produced by the cross-linking of Ca^{2+} ions with G-block fractions when calcium alginate gels are produced, thus improving its mechanical properties [38]. Also, authors such as Li et al. 2007, after studies carried out by X-ray diffraction, confirmed the presence of these structures, mainly when gelation occurs slowly in large pieces of gel, as is the case [39]. This structural difference between the alginate extracted from RO and the commercial one implies the possibility of obtaining materials with greater resistance using the same processing methods.

3.1.2. Determination of Intrinsic Viscosity and Molecular Weight

Using Equation (1) [40], $\frac{1}{2\left(\eta_r^{\frac{1}{2}}-1\right)}$ was plotted against $\frac{1}{C}$, observing in a linear dependence (R_2 = 0.9994). Supplementary Figure S1 shows the correlation line between the viscosity and the concentration of the alginate solutions.

From the slope of the fitted straight line (0.4929), an intrinsic viscosity of 2.03 dL/g was obtained for the RO extracted alginate, and a value of 1.93 dL/g for the medium viscosity commercial alginate. Similar values were obtained for other brown seaweeds, such as *Laminaria digitata*, by Fertah et al. [18].

Once the intrinsic viscosity was known, the Mark Houwink Equation (2) was applied to determine the molecular mass of sodium alginate. A molecular mass of 94.874 kDa was obtained for the RO-extracted alginate and 90.241 kDa for the commercial medium viscosity alginate. These values are within the orders of magnitude obtained for other alginates extracted from seaweed and collected by Kaidi et al. [41]. These results, along with the ones obtained in Section 3.1.1, suggest that the sodium alginate extracted from the RO seaweed is suitable for some applications, as is the case of binding agents in tablets or thickening agents in gels and creams [42].

3.1.3. DSC and TGA

Figure 3A shows the TGA profile of both the sodium alginate powder and the calcium alginate gel before and after the gelation process, respectively, as well as the first derivate for the sample extracted from RO. Figure 3B shows the DSC profile for these systems. This figure evidences the presence of four characteristic peaks in the first derivate signal when the samples are heated in an inert atmosphere. The first peak was observed at ~85 °C (it implies a loss of 10% of the initial weight), corresponding to the dehydration of the alginic salt.

Thus, comparing this result with the first peak observed in the TGA measurements, which corresponds to an endothermic event at ~75–95 °C, was previously attributed to hydrated salts [43].

Moreover, the DSC plot also shows an exothermic event at ~250 °C, which was also observed in the TGA profile (Figure 2A). This thermal event corresponded to the thermal decomposition of the biopolymer (found to be around 240–260 °C), which agrees with the results obtained by other authors such as Soares et al. [44] for commercial sodium alginate. The cross-linking of calcium with alginate on gels produced resistance to thermal degradation, which was reflected in a shift of this peak towards ~285 °C in TGA experiments when alginate is gelled (Figure 3A), and also a widening and delaying of the thermal event observed in the DSC profile (Figure 3B). This thermal decomposition produces the loss of about 20% of the initial mass of the sample of alginate and almost 30% in the case of the calcium alginate gel. From these values onwards, the mass loss responds to the formation of carbonate due to the decomposition of the carbonaceous materials. Similar results were observed by several authors [43,45] for calcium alginate nanoparticles used for drug delivery and biodegradable poly-(vinyl caprolactam) grafted onto alginate microgel, respectively.

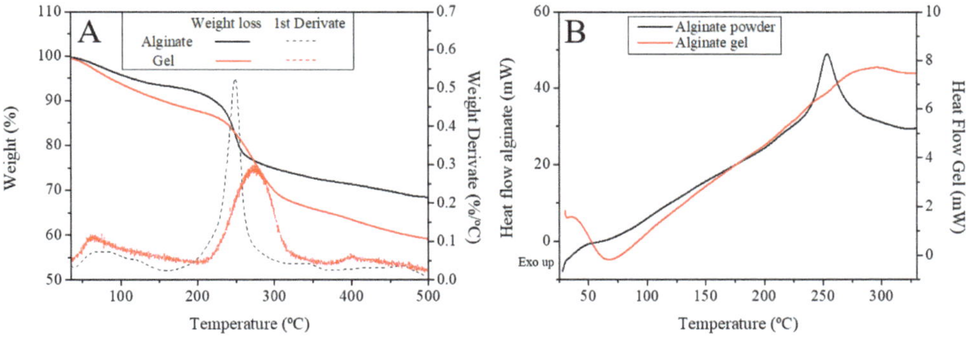

Figure 3. Thermogravimetric Analysis (TGA) (**A**) and Differential Scanning Calorimetry (DSC) (**B**) of the *Rugulopteryx okamurae* (RO) extracted alginate made under N_2 atmosphere.

3.2. Characterization of the Commercial and RO-Extracted Alginate Gels

3.2.1. Rheological Characterization of Sodium Alginate Solution and Calcium Alginate Gels

Figure 4 displays the flow curves acquired for sodium alginate solutions at concentrations of 1% and 2% by weight (Figure 4A). Additionally, it illustrates the relationship between normalized viscoelastic moduli ($G'/G'_{0.1}$ and $G''/G'_{0.1}$, where G' corresponds to the elastic modulus, G'' to the viscous modulus, and $G'_{0.1}$ corresponds to the elastic modulus at 0.1 Hz) and frequency (Figure 4B). The figure also presents the variation in G' and tan δ at 1 Hz (G'_1, tan δ_1, respectively) with the percentage of alginate at concentrations of 0.5%, 1%, and 2% by weight (Figure 4C). This analysis encompasses both commercial alginate variants (low and medium viscosity) and RO-extracted alginate gels. The flow curves obtained (Figure 4A) show an increase in viscosity values as the percentage of alginate in the solution increases. This result is common in other polymers and biopolymer solutions since the interaction between molecules increases when more molecules are in the solution [46,47]. Moreover, comparing the viscosity of the different alginates, the medium viscosity of commercial alginate is very similar to that of the extracted alginate, both being higher than the low viscosity commercial alginate (0.031 $\pm$ 0.0074, 3.8 $\pm$ 0.052, 2.8 $\pm$ 0.024 Pa·s for commercial low viscosity, medium viscosity, and RO-extracted alginate 2% at 1 s^{-1}, respectively). The similarity of these values between commercial and RO-extracted alginate suggested that the number of G groups does not influence the rheology of sodium alginate solutions, as it does in calcium alginate gels due to the formation of "egg box" structure when the ions are replaced [29]. Moreover, this plot also evidenced that when there are limited interactions among biopolymer chains (curves exhibiting the lower viscosity) the solutions exhibited a Newtonian behaviour. However, when these interactions are greater, a shear thinning behaviour was observed since the values obtained decreased with the shear rate [48].

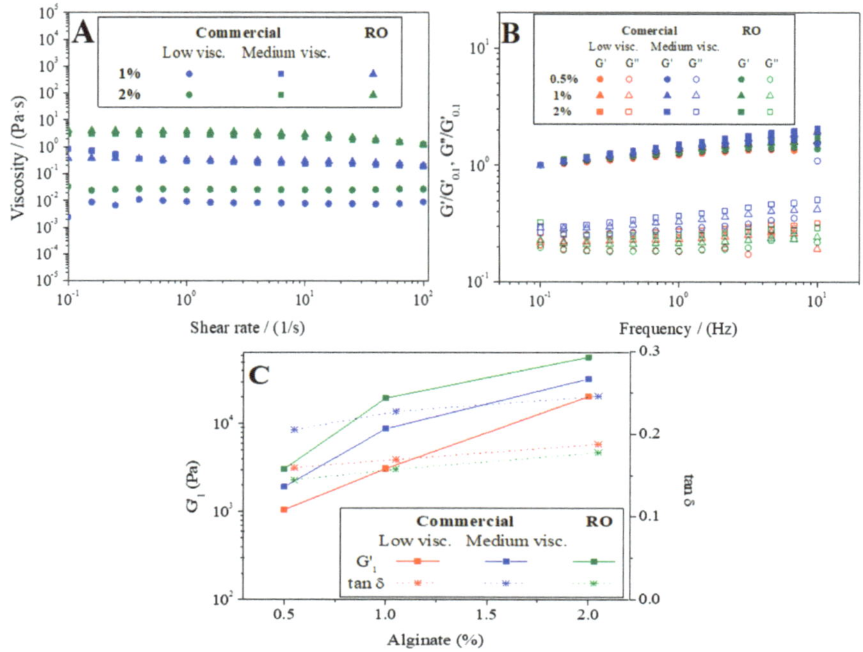

Figure 4. (**A**) Evolution of viscosity with share rate for different systems before gelation prepared at 1 and 2% wt.; (**B**) evolution of G'/G'_1 and G''/G'_1 modulus with the frequency of commercial low and medium viscosity alginate gels and extracted alginate gel from *Rugulopteryx Okamurae* (RO), prepared at 0.5, 1, and 2% wt; and (**C**) evolution of G'_1 and tan d with the percentage of alginate forming gels.

Figure 4B shows the G' and G'' moduli normalized (i.e., $G'/G'_{0.1}$ and $G''/G'_{0.1}$) against frequency for the calcium alginate gels. According to the results obtained, all systems displayed an elastic behaviour as the storage modulus is always above the loss modulus, confirming the gelation of sodium alginate as calcium alginate following the procedure above-mentioned, regardless of the alginate nature analyzed, which agreed with the results obtained by LeRoux et al. [49] for calcium alginate gels. Furthermore, when comparing gels with the same percentage of polymer, the frequency dependence of commercial alginates was slightly higher than that of RO-extracted alginate. This distinction becomes clearer when examining the slope values for the normalized elastic modulus lines, these being 0.124 ± 0.005, 0.156 ± 0.003, and 0.110 ± 0.006 for gels prepared from commercial alginate (with low and medium viscosities) and RO-extracted alginate, respectively. This finding highlights a diminished frequency dependence of the viscoelastic moduli in gels prepared with RO in contrast to their commercial counterparts. Thus, the structure of the gel obtained from the RO-extracted alginate is expected to be less dependent on the relaxation time than the one formed by commercial alginate.

For a better understanding of the rheological behaviour of the calcium alginate gels obtained, the evolution of the elastic modulus (G') and the tan δ at 1 Hz (G'_1 and tan δ_1, respectively) as a function of the alginate concentration (Figure 4C). First of all, when comparing the evolution of G'_1 of the gels with the results of Figure 4A, it is observed for commercial medium viscosity alginate and RO-extracted alginate that, despite having identical viscosity values in solution, significant differences emerge in their elastic moduli when gelation occurs. Moreover, the gelation is confirmed by increase in viscosity, which can be estimated approximately by comparing the initial shear viscosity obtained solutions (e.g., around 0.36 and 3.91 Pa·s for the 1 and 2% extracted sodium alginate solutions, respectively) and the value of the complex viscosity (η^*) obtained for those systems after gelation (i.e., around 19,800 and 58,300 Pa·s for the 1 and 2% extracted sodium alginate solutions, respectively).

Thus, this figure evidences that an increase in the percentage of alginate also involved an increase in the elastic modulus value (20,380, 32,400, and 57,400 Pa for commercial low viscosity, medium viscosity, and RO-extracted alginate 2% at 1 Hz, respectively). This figure confirms that the value of the elastic moduli was higher for gels made from RO-extracted alginate than those obtained for the medium-viscosity commercial alginate, which results in more marked solid behaviour for the RO-extracted calcium alginate gels. This result can be related to the higher number of "egg box" structures formed by the higher guluronic acid fractions in the structure of the extracted alginate chain compared to the commercial one, as previously reported in the ^{1}H-NMR analysis.

When the behaviour of tan δ_1 is observed, a slight increase occurred when increasing the amount of alginate forming gels until, from 1% onwards, no significant differences are observed. This implies that there is no direct relationship between tan δ_1 with greater chain cross-linking with increasing Ca^{2+} anchoring sites when the percentage of alginate in the gels is increased [49]. In summary, increasing the polymer concentration in the presence of Ca^{2+} produces an increase in G' and stiffer gels with more solid behaviour, while at the same time, the tan δ_1 value increases to a certain extent, due to a dissipation of intermolecular interactions and confirming an independence of the tan δ_1 value with the Ca^{2+} concentration [49]. It is also interesting to note that the RO-extracted alginate gel has lower tan δ_1 values than the commercial ones, which confirms a greater solid character than commercial calcium alginate gels at the same concentration. The results obtained by these sodium alginate gels agree with previous results reported by LeRoux et al. [49] for calcium alginate gels when they were analyzed by compressive and shear stresses. Thus, these results confirm the suitability of the alginate extracted from the invasive algae RO, which eventually led to higher mechanical properties than conventional alginates nowadays available for the industry.

3.2.2. Water Uptake Capacity (WUC), Soluble Matter Loss (SML) and Water Holding Capacity (WHC)

Figure 5A shows the WUC and the SML of the three systems studied (RO-extracted alginate as well as low and medium viscosity commercial alginate) at different alginate concentrations (0.5, 1.0, and 2.0). It can be observed that WUC and SML values decreased when increasing the content of alginate in gels. When commercial alginates are analyzed, a significant decrease is observed with increasing viscosity. Nevertheless, although the extracted alginate has similar viscosity values to medium viscosity commercial alginate (Figure 4A), WUC values higher than these are observed, obtaining intermediate values between the commercial alginate of low and medium viscosity (1220 $\pm$ 163.3, 909.7 $\pm$ 21.6 and 1114.3 $\pm$ 106.4% for commercial low viscosity, medium viscosity and RO-extracted alginate gels at 1%, respectively). When the evolution of WUC was compared with the viscoelastic properties of the gels obtained (Figure 4C) (G' in increasing order: low viscosity commercial, medium viscosity commercial and RO-extracted alginate), there was a general decrease in the WUC value as the amount of alginate in gels increased, but RO-extracted alginate gels display higher values of WUC than medium viscosity commercial alginate gels, even when G' is higher for RO-extracted alginate gels. It is important to highlight that below 2% wt. alginate, hydrogels with superabsorbent capacity were obtained, with more than 1000% water uptake capacity [50]. Thus, these results confirm not only the increase in the rheological properties of the RO-extracted alginate but also the functional properties, as in the case of WUC.

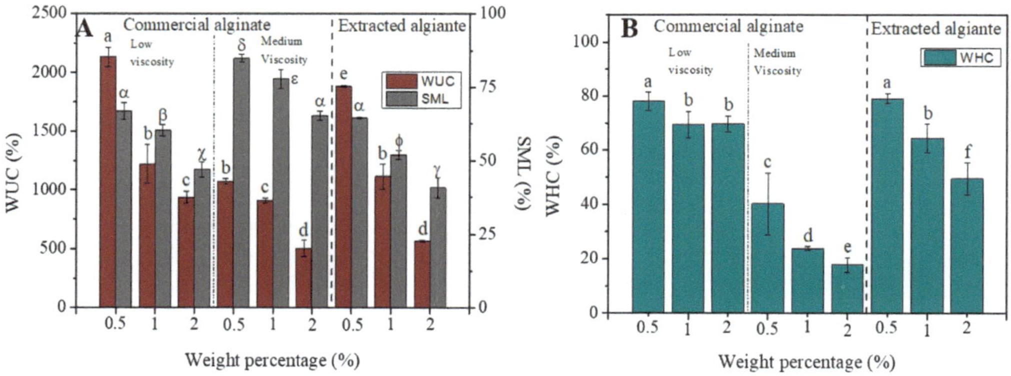

Figure 5. Water uptake capacity (WUC) and soluble matter loss (SML) (**A**), and water holding capacity (WHC) (**B**) of commercial low and medium viscosity alginate gels and extracted alginate gel from *Rugulopteryx Okamurae* (RO). Different letters within a column indicate significant differences ($p < 0.05$).

SML values decrease as the alginate content in the gels increases (0.5% > 1% > 2%), as happened with WUC values. However, SML values are lower for RO-extracted alginate gels than for commercial ones, which implies more stable and less soluble structures than those. This is related to the higher G′ values (Figure 4C) and to the egg box structure that is formed in the case of RO-extracted alginate gels.

Figure 5B shows the evolution of WHC as a function of alginate content (0.5, 1 and 2 wt.%) for gels from both RO-extracted and commercial alginates. The same pattern is observed for the WUC values, decreasing as the percentage of alginate in the gels increases and showing higher values for RO-extracted alginate gels. These gels, despite having a higher G′ than the commercial alginate gel of medium viscosity, present WHC values more similar to that of low viscosity. Thus, the developed alginate structure had more free polar hydrophilic hydroxyl groups and carboxyl groups than the commercial medium viscosity one [51]. This structure developed not only allowed a suitable performance when mechanical properties were tested, but also these polar groups led to a high affinity for water, which determined the WUC and the WHC of the gels obtained.

3.2.3. Scanning Electron Microscopy (SEM) of Calcium Alginate Gels

Figure 6 shows the SEM images obtained for the RO-extracted and commercial calcium alginate gels after freeze-drying. Figure 6A shows the gel obtained from the RO-extracted alginate, which was characterized by a slightly rough shape, with some pores in its structure. However, the surface commercial alginate observed in Figure 6B evidenced a far smoother surface with no visible pores on the surface. Alginate polymer chains contain hydrophilic groups (-OH or -COOH) in their structure, contributing to water absorption [52]. The porous structure in RO gel (Figure 6A) enhances water absorption due to increased surface area and favourable hydrophilic interactions, surpassing the commercial gel (Figure 6B). This greater porosity in Figure 6A than in Figure 6B can also explain the increase in water holding capacity in the RO gels, where significant differences were observed between the two gels (Figure 5B).

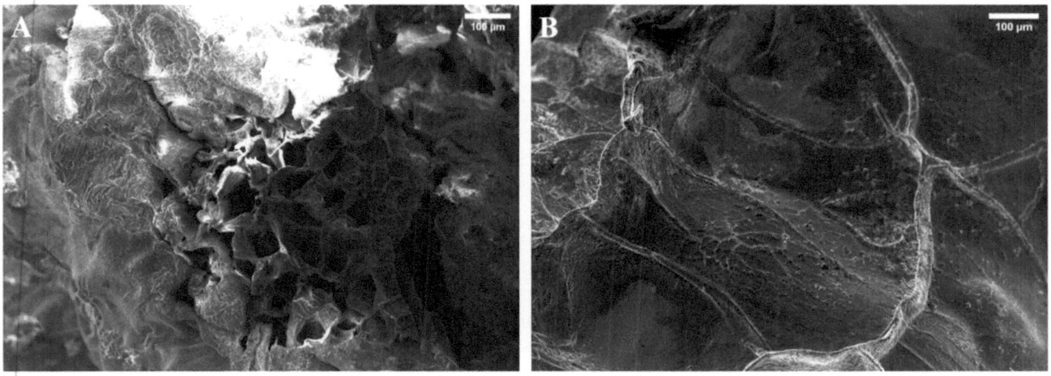

Figure 6. SEM images of calcium alginate gels from alginate extracted from *Rugulopteryx Okamurae* (RO) (**A**) and medium viscosity commercial alginate (**B**) after freeze-drying at 100× magnification.

4. Conclusions

This study successfully extracted alginate from the invasive brown alga *Rugulopteryx okamurae* (RO), with a yield of approximately 74%. Characterization techniques employed confirmed the identity of the extracted alginate when compared with commercial brown seaweed alginate. The results obtained by RMN conducted to assume that the structure of the alginate obtained has a higher proportion of guluronic acid (G) than mannuronic acid (M) (0.53 and 0.47 for extracted and commercial medium viscosity alginate, respectively), which favoured the formation of "egg box" structures by replacing sodium with calcium during gelation. The mechanical spectra obtained showed a prevailing elastic behaviour in the frequency range studied for all the samples analyzed. By increasing the weight percentage of alginate in each gel, an increase in the elastic modulus was observed. Thus, elastic modulus at 1 Hz increased from 3.69 ± 0.56 to 58.20 ± 1.12 kPa for RO-extracted alginate, from 1.91 ± 0.15 to 32.40 ± 2.90 to for medium viscosity commercial alginate, and from 1.12 ± 0.013 to 17.37 ± 4.26 kPa for low viscosity commercial alginate when alginate increased from 0.5 to 2%, respectively. It also showed a higher frequency dependence with increasing concentration, except for the RO-extracted alginate gel, which showed an even lower frequency dependence. These results were consistent with the tan δ values, which increased with alginate content in the gels, obtaining lower tan δ values at 1 Hz for the RO-extracted alginate (0.17 ± 0.0056, 0.23 ± 0.011 and 0.16 ± 0.0035 for the reference low viscosity and medium viscosity commercial alginate and RO-extracted commercial alginate at 1%, respectively). These results indicate that the gelation of the extracted alginate produces gels that are stiffer and have a more solid-like character than the two commercial alginates used as references in this work. With increasing elastic modulus and alginate content, there was a general decrease in water absorption and retention values. The extracted alginate showed higher water absorption values compared to the medium viscosity commercial alginate, with values of 1114.3 ± 106.4% at 1% RO-extracted alginate, suggesting potential for obtaining superabsorbent gels. While preliminary findings suggest superior structural and mechanical properties of alginate extracted from Rugulopteryx okamurae compared to two commercially available alginates, it is essential to acknowledge the need for further extensive research to substantiate these claims. Initial results provide promising insights, warranting continued investigation into the potentially enhanced qualities of *Rugulopteryx okamurae*-derived alginate for various applications.

Supplementary Materials: The following supporting information can be downloaded at https://www.mdpi.com/article/10.3390/polym16050702/s1. Figure S1: Correlation between viscosity and the inverse of concentration for RO-extracted alginate solutions at different concentrations following Fedors' Equation (1).

Author Contributions: Conceptualization, M.F. and C.B.; methodology, M.F. and C.B.; software, I.S.; validation, M.F. and C.B.; formal analysis, M.F. and C.B.; investigation, I.S.; resources, C.B.; data curation, I.S.; writing—original draft preparation, I.S.; writing—review and editing, M.F. and C.B.; visualization, C.B.; supervision, M.F. and C.B.; project administration, C.B.; funding acquisition, C.B. All authors have read and agreed to the published version of the manuscript.

Funding: The authors acknowledge the project PID2021-124294OB-C21 funded by Ministerio de Ciencia, Innovación y Universidades (MCIN/AEI/10.13039/501100011033/) and by FEDER Una manera de hacer Europa.

Data Availability Statement: Data are contained within the article.

Acknowledgments: The authors thank CITIUS for granting access to and their assistance with the Functional characterization and Microscopy service. Finally, the authors also kindly thank Ismael Hachero (from IFAPA El Toruño) for providing the raw material used in this study.

Conflicts of Interest: The authors declare no conflicts of interest.

References

1. Salehi, B.; Sharifi-Rad, J.; Seca, A.M.L.; Pinto, D.C.G.A.; Michalak, I.; Trincone, A.; Mishra, A.P.; Nigam, M.; Zam, W.; Martins, N. Current Trends on Seaweeds: Looking at Chemical Composition, Phytopharmacology, and Cosmetic Applications. *Molecules* **2019**, *24*, 4182. [CrossRef]
2. Holdt, S.L.; Kraan, S. Bioactive Compounds in Seaweed: Functional Food Applications and Legislation. *J. Appl. Phycol.* **2011**, *23*, 543–597. [CrossRef]
3. Moral, A.; Aguado, R.; Castelló, R.; Tijero, A.; Ballesteros, M. Potential Use of Green Alga *Ulva* sp. for Papermaking. *Bioresources* **2019**, *14*, 6851–6862. [CrossRef]
4. Zollmann, M.; Robin, A.; Prabhu, M.; Polikovsky, M.; Gillis, A.; Greiserman, S.; Golberg, A. Green Technology in Green Macroalgal Biorefineries. *Phycologia* **2019**, *58*, 516–534. [CrossRef]
5. Goyanes, S.N.; D'Accorso, N.B. *Industrial Applications of Renewable Biomass Products: Past, Present and Future*; Springer: Cham, Switzerland, 2017; pp. 1–332.
6. Freile-Pelegrín, Y.; Madera-Santana, T.J. Biodegradable Polymer Blends and Composites from Seaweeds. In *Handbook of Composites from Renewable Materials*; John Wiley & Sons: Hoboken, NJ, USA, 2017; Volume 1–8, pp. 419–438. [CrossRef]
7. Lim, C.; Yusoff, S.; Ng, C.G.; Lim, P.E.; Ching, Y.C. Bioplastic Made from Seaweed Polysaccharides with Green Production Methods. *J. Environ. Chem. Eng.* **2021**, *9*, 105895. [CrossRef]
8. Rioux, L.E.; Turgeon, S.L.; Beaulieu, M. Characterization of Polysaccharides Extracted from Brown Seaweeds. *Carbohydr. Polym.* **2007**, *69*, 530–537. [CrossRef]
9. Cao, L.; Lu, W.; Mata, A.; Nishinari, K.; Fang, Y. Egg-Box Model-Based Gelation of Alginate and Pectin: A Review. *Carbohydr. Polym.* **2020**, *242*, 116389. [CrossRef]
10. Sachan, N.K.; Pushkar, S.; Jha, A.; Bhattcharya, A. Sodium Alginate: The Wonder Polymer for Controlled Drug Delivery. *J. Pharm. Res.* **2009**, *2*, 1191–1199.
11. Hu, C.; Lu, W.; Mata, A.; Nishinari, K.; Fang, Y. Ions-Induced Gelation of Alginate: Mechanisms and Applications. *Int. J. Biol. Macromol.* **2021**, *177*, 578–588. [CrossRef]
12. Hu, C.; Lu, W.; Sun, C.; Zhao, Y.; Zhang, Y.; Fang, Y. Gelation Behavior and Mechanism of Alginate with Calcium: Dependence on Monovalent Counterions. *Carbohydr. Polym.* **2022**, *294*, 119788. [CrossRef]
13. Altamirano Jeschke, M.; Zanolla, M. *Análisis de Riesgos de La Macroalga Exótica Rugulopteryx Okamurae*; Ministerio Para La Transición Ecológica y Reto Demográfico: Madrid, Spain, 2019; p. 69.
14. MITECO. *Estrategia De Control Del Alga Rugulopteryx Okamurae En España*; MITECO: Madrid, Spain, 2022; pp. 1–93.
15. Santana, I.; Félix, M.; Guerrero, A.; Bengoechea, C. Processing and Characterization of Bioplastics from the Invasive Seaweed Rugulopteryx Okamurae. *Polymers* **2022**, *14*, 355. [CrossRef]
16. Calumpong, H.P.; Maypa, A.P.; Magbanua, M. Population and Alginate Yield and Quality Assessment of Four Sargassum Species in Negros Island, Central Philippines. *Hydrobiologia* **1999**, *398–399*, 211–215. [CrossRef]
17. Saji, S.; Hebden, A.; Goswami, P.; Du, C. A Brief Review on the Development of Alginate Extraction Process and Its Sustainability. *Sustainability* **2022**, *14*, 5181. [CrossRef]
18. Fertah, M.; Belfkira, A.; Dahmane, E.M.; Taourirte, M.; Brouillette, F. Extraction and Characterization of Sodium Alginate from Moroccan Laminaria Digitata Brown Seaweed. *Arab. J. Chem.* **2017**, *10*, S3707–S3714. [CrossRef]
19. Fenoradosoa, T.A.; Ali, G.; Delattre, C.; Laroche, C.; Petit, E.; Wadouachi, A.; Michaud, P. Extraction and Characterization of an Alginate from the Brown Seaweed Sargassum Turbinarioides Grunow. *J. Appl. Phycol.* **2010**, *22*, 131–137. [CrossRef]
20. Clementi, F.; Mancini, M.; Moresi, M. Rheology of Alginate from Azotobacter Vinelandii in Aqueous Dispersions. *J. Food Eng.* **1998**, *36*, 51–62. [CrossRef]

21. da Costa, M.P.M.; Delpech, M.C.; de Mello Ferreira, I.L.; de Macedo Cruz, M.T.; Castanharo, J.A.; Cruz, M.D. Evaluation of Single-Point Equations to Determine Intrinsic Viscosity of Sodium Alginate and Chitosan with High Deacetylation Degree. *Polym. Test.* **2017**, *63*, 427–433. [CrossRef]
22. Fedors, R.F. An Equation Suitable for Describing the Viscosity of Dilute to Moderately Concentrated Polymer Solutions. *Polymer* **1979**, *20*, 225–228. [CrossRef]
23. Liggat, J. *Physical Chemistry of Macromolecules: Basic Principles and Issues*, 2nd ed.; John Wiley & Sons: Hoboken, NJ, USA, 1995; Volume 19, ISBN 9786468600.
24. Mackie, W.; Noy, R.; Sellen, D.B. Solution Properties of Sodium Alginate. *Biopolymers* **1980**, *19*, 1839–1860. [CrossRef]
25. Smidsrød, O. Solution Properties of Alginate. *October* **1970**, *13*, 359–372. [CrossRef]
26. Donnan, F.G.; Rose, R.C. Osmotic pressure, molecular weight, and viscosity of sodium alginate. *Can. J. Res.* **1950**, *28b*, 105–113. [CrossRef]
27. Martinsen, A.; Skjåk-Bræk, G.; Smidsrød, O.; Zanetti, F.; Paoletti, S. Comparison of Different Methods for Determination of Molecular Weight and Molecular Weight Distribution of Alginates. *Carbohydr. Polym.* **1991**, *15*, 171–193. [CrossRef]
28. Mancini, M.; Moresi, M.; Rancini, R. Mechanical Properties of Alginate Gels: Empirical Characterisation. *J. Food Eng.* **1999**, *39*, 369–378. [CrossRef]
29. Fu, S.; Thacker, A.; Sperger, D.M.; Boni, R.L.; Buckner, I.S.; Velankar, S.; Munson, E.J.; Block, L.H. Relevance of Rheological Properties of Sodium Alginate in Solution to Calcium Alginate Gel Properties. *AAPS PharmSciTech* **2011**, *12*, 453–460. [CrossRef]
30. Mancini, M.; Moresi, M.; Rancini, R. Uniaxial Compression and Stress Relaxation Tests on Alginate Gels. *J. Texture Stud.* **1999**, *30*, 639–657. [CrossRef]
31. Antonio, S.; Presb, G.; Peña-par, L.; Iv, E.; Patricia, K.; Galv, V.; Vop, M.; Kumpov, I.; Elizalde-herrera, L.E. Characterization of Sodium Alginate Hydrogels Reinforced with Nanoparticles of Hydroxyapatite for Biomedical Applications. *Polymers* **2021**, *13*, 2927.
32. Shabanian, M.; Hajibeygi, M.; Raeisi, A. FTIR Characterization of Layered Double Hydroxides and Modified Layered Double Hydroxides. In *Layered Double Hydroxide Polymer Nanocomposites*; Woodhead Publishing: Sawston, UK, 2020; pp. 77–101. [CrossRef]
33. Sakugawa, K.; Ikeda, A.; Takemura, A.; Ono, H. Simplified Method for Estimation of Composition of Alginates by FTIR. *J. Appl. Polym. Sci.* **2004**, *93*, 1372–1377. [CrossRef]
34. Flórez-Fernández, N.; Domínguez, H.; Torres, M.D. A Green Approach for Alginate Extraction from Sargassum Muticum Brown Seaweed Using Ultrasound-Assisted Technique. *Int. J. Biol. Macromol.* **2019**, *124*, 451–459. [CrossRef]
35. Bajpai, M.; Shukla, P.; Bajpai, S.K. Enhancement in the Stability of Alginate Gels Prepared with Mixed Solution of Divalent Ions Using a Diffusion through Dialysis Tube (DTDT) Approach. *J. Macromol. Sci. Part A Pure Appl. Chem.* **2017**, *54*, 301–310. [CrossRef]
36. Grasdalen, H. Study of the Composition and Sequence of Uronate Residues in Alginate. *J. Chem. Inf. Model.* **1979**, *68*, 23–31. [CrossRef]
37. Larsen, B.; Salem, D.M.S.A.; Sallam, M.A.E.; Mishrikey, M.M.; Beltagy, A.I. Characterization of the Alginates from Algae Harvested at the Egyptian Red Sea Coast. *Carbohydr. Res.* **2003**, *338*, 2325–2336. [CrossRef]
38. Smidsr0d, O.; Skak-Braek, G. Alginate as Immobilization Matrix for Cells. *Trends Biotechnol.* **1990**, *8*, 71–78. [CrossRef] [PubMed]
39. Li, L.; Fang, Y.; Vreeker, R.; Appelqvist, I.; Mendes, E. Reexamining the Egg-Box Model in Calcium—Alginate Gels with X-Ray Diffraction. *Biomacromolecules* **2007**, *8*, 464–468. [CrossRef] [PubMed]
40. Alberto Masuelli, M.; Omar Illanes, C. Review of the Characterization of Sodium Alginate by Intrinsic Viscosity Measurements. Comparative Analysis between Conventional and Single Point Methods. *Int. J. BioMater. Sci. Eng.* **2014**, *1*, 1–11.
41. Kaidi, S.; Bentiss, F.; Jama, C.; Khaya, K.; Belattmania, Z.; Reani, A.; Sabour, B. Isolation and Structural Characterization of Alginates from the Kelp Species Laminaria Ochroleuca and Saccorhiza Polyschides from the Atlantic Coast of Morocco. *Colloids Interfaces* **2022**, *6*, 51. [CrossRef]
42. Tønnesen, H.H.; Karlsen, J. Alginate in Drug Delivery Systems. *Drug Dev. Ind. Pharm.* **2002**, *28*, 621–630. [CrossRef]
43. Abulateefeh, S.R.; Taha, M.O. Enhanced Drug Encapsulation and Extended Release Profiles of Calcium-Alginate Nanoparticles by Using Tannic Acid as a Bridging Cross-Linking Agent. *J. Microencapsul.* **2015**, *32*, 96–105. [CrossRef] [PubMed]
44. Soares, J.P.; Santos, J.E.; Chierice, G.O.; Cavalheiro, E.T.G. Thermal Behavior of Alginic Acid and Its Sodium Salt. *Eclet. Quim.* **2004**, *29*, 57–63. [CrossRef]
45. Madhusudana Rao, K.; Krishna Rao, K.S.V.; Sudhakar, P.; Chowdoji Rao, K.; Subha, M.C.S. Synthesis and Characterization of Biodegradable Poly (Vinyl Caprolactam) Grafted on to Sodium Alginate and Its Microgels for Controlled Release Studies of an Anticancer Drug. *J. Appl. Pharm. Sci.* **2013**, *3*, 61–69.
46. Freed, K.F.; Edwards, S.F. Polymer Viscosity in Concentrated Solutions. *J. Chem. Phys.* **1974**, *61*, 3626–3633. [CrossRef]
47. Telis, V.R.N.; Telis-Romero, J.; Mazzotti, H.B.; Gabas, A.L. Viscosity of Aqueous Carbohydrate Solutions at Different Temperatures and Concentrations. *Int. J. Food Prop.* **2007**, *10*, 185–195. [CrossRef]
48. Gómez-Díaz, D.; Navaza, J.M. Rheology of Aqueous Solutions of Food Additives. *J. Food Eng.* **2003**, *56*, 387–392. [CrossRef]
49. LeRoux, M.A.; Guilak, F.; Setton, L.A. Compressive and Shear Properties of Alginate Gel: Effects of Sodium Ions and Alginate Concentration. *J. Biomed. Mater. Res.* **1999**, *47*, 46–53. [CrossRef]
50. Buchholz, F.L.; Graham, A.T. (Eds.) *Modern Superabsorbent Polymer Technology*; Wiley-VCH: Weinheim, Germany, 1998; p. 304. ISBN 978-0-471-19411-8.

51. Zhang, X.; Wang, X.; Fan, W.; Liu, Y.; Wang, Q.; Weng, L. Fabrication, Property and Application of Calcium Alginate Fiber: A Review. *Polymers* **2022**, *14*, 3227. [CrossRef]
52. Milivojević, M.; Popović, A.; Pajić-Lijaković, I.; Šoštarić, I.; Kolašinac, S.; Stevanović, Z.D. Alginate Gel-Based Carriers for Encapsulation of Carotenoids: On Challenges and Applications. *Gels* **2023**, *9*, 620. [CrossRef]

sensors

Article

Protecting FPGA-Based Cryptohardware Implementations from Fault Attacks Using ADCs

Francisco Eugenio Potestad-Ordóñez [1,2,*,†], Alejandro Casado-Galán [2] and Erica Tena-Sánchez [1,2,†]

1 Escuela Politécnica Superior, Universidad de Sevilla, 41011 Seville, Spain; etena@us.es
2 Instituto de Microelectrónica de Sevilla, IMSE-CNM (CSIC; Universidad de Sevilla), 41092 Seville, Spain; casado@imse-cnm.csic.es
* Correspondence: fpotestad@us.es
† These authors contributed equally to this work.

Abstract: The majority of data exchanged between connected devices are confidential and must be protected against unauthorized access. To ensure data protection, so-called cryptographic algorithms are used. These algorithms have proven to be mathematically secure against brute force due to the key length, but their physical implementations are vulnerable against physical attacks. The physical implementation of these algorithms can result in the disclosure of information that can be used to access confidential data. Some of the most powerful hardware attacks presented in the literature are called fault injection attacks. These attacks involve introducing a malfunction into the normal operation of the device and then analyzing the data obtained by comparing them with the expected behavior. Some of the most common methods for injecting faults are the variation of the supply voltage and temperature or the injection of electromagnetic pulses. In this paper, a hardware design methodology using analog-to-digital converters (ADCs) is presented to detect attacks on cryptocircuits and prevent information leakage during fault injection attacks. To assess the effectiveness of the proposed design approach, FPGA-based ADC modules were designed that detect changes in temperature and supply voltage. Two setups were implemented to test the scheme against voltage and temperature variations and injections of electromagnetic pulses. The results obtained demonstrate that, in 100% of the cases, when the correct operating voltage and temperature range were established, the detectors could activate an alarm signal when the cryptographic module was attacked, thus avoiding confidential information leakage and protecting data from being exploited.

Keywords: hardware security; voltage attack; temperature attack; electromagnetic attack; countermeasures; FPGA

Citation: Potestad-Ordóñez, F.E.; Casado-Galán, A.; Tena-Sánchez, E. Protecting FPGA-Based Cryptohardware Implementations from Fault Attacks Using ADCs. *Sensors* **2024**, *24*, 1598. https://doi.org/10.3390/s24051598

Academic Editor: Antonio J. Jara

Received: 30 January 2024
Revised: 23 February 2024
Accepted: 28 February 2024
Published: 29 February 2024

1. Introduction

The proliferation of the Internet of Things (IoT) has caused a rapid and significant increase in the number of interconnected devices. These interconnected devices process a large amount of data, most of which are sensitive data from users. Due to the methods that hackers are constantly developing to gain access to users' secret information, protecting sensitive data and countering attacks by third parties has become a constant challenge. Recent research has highlighted the importance of analyzing the security of IoT solutions in their various design stages [1].

Mahmoud et al. [2] and Xu et al. [3] pointed out the importance of and challenges related to security in the IoT, highlighting secure implementations under strong constraints of power consumption and area, among other considerations. The creation of new cryptographic algorithms that can safeguard data and meet the stringent requirements of applications is an ongoing process. One of the solutions proposed in [4] was the use of so-called lightweight cryptography as a solution to security problems and the use of solutions whose cost has the lowest possible impact on implementation. Finally, Ref. [5] warned of the potential dangers of not considering the security of embedded applications used in this field.

A cryptographic module is deemed secure in practice if there is no known attack that can successfully break it in a reasonable amount of time and with a reasonable amount of computing power. Although algorithms are mathematically secure, their physical implementation on hardware can be exploited by third parties to reveal confidential information through side-channel attacks (SCAs) or fault injection attacks (FIAs) [6,7].

There are various attack techniques that vary greatly in terms of the execution cost and time, equipment needed, and expertise required. These two characteristics can be used to classify them as active or passive and invasive or non-invasive [7,8]. Figure 1 summarizes the attack classification. Invasive attacks involve manipulating the device, typically by unpacking the chip and gaining access to its internal layers. Non-invasive attacks, on the other hand, collect data from the device (such as accessible I/O, power consumption, and execution time) without making any changes. In a passive attack, confidential data are exposed while the cryptographic system is functioning correctly by examining side-channel information such as power usage, timing, or electromagnetic radiation. However, an active attack changes the functionality of the device during operation, manipulating its input, power supply, or environmental conditions such as temperature. The malfunctioning of the process and the results it produces can be used to uncover the secret key. Invasive attacks, whether passive or active, are costly in terms of time and money and can often cause irreparable damage to the cryptographic device. Non-invasive attacks are a major concern for the cryptographic community due to their low cost, minimal equipment requirements, and high success rate.

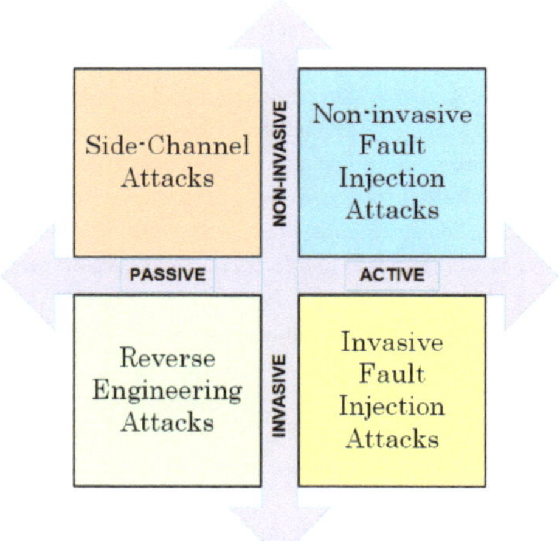

Figure 1. Attack classification.

With this in mind, great effort has been put into implementing cryptographic algorithms securely and efficiently on hardware platforms. It is important to point out that this paper concentrates on private-key ciphers, yet there are numerous studies that have concentrated on the implementation of public-key algorithms and their ability to withstand attacks from third parties. As evidence of this, there have been studies on effective executions of public-key algorithms such as the elliptic curve and RSA [9–12], as well as theoretical attacks [13–15]. Consequently, this study concentrates on the implementation of a security system on private-key encryption systems, while public-key encryption systems are not included. In this paper, we focus on non-invasive active attacks, specifically those that aim to manipulate the supply voltages and the operating temperature of the circuit that implements a cryptographic algorithm, and electromagnetic attacks. The proposed methodology enables one to implement a system that can identify these kinds of attacks and prevent the potential retrieval of confidential data stored in the circuit. The Advanced Encryption Standard (AES) is employed as a test vehicle. This countermeasure is not specifically designed for the AES encryption algorithm; it is used just as an example. It is a system that works in parallel and can be applied to any other cryptographic circuit or hardware-based cryptographic element, since it detects attacks on the hardware rather than the algorithm that is implemented in it. Therefore, this scheme can be applied in parallel to protect any other cipher implemented on the hardware.

The main objective of this work is the presentation and evaluation of a detection scheme methodology based on analog-to-digital converters to protect cryptohardware implementations. The contributions of the paper are the following:

- The classification of the most relevant types of attacks on physical implementations, focusing on attacks by the manipulation of supply voltage and temperature and the injection of electromagnetic pulses.
- The proposal of a methodology for the design of the detection scheme based on analog-to-digital converters, serving as a countermeasure to prevent information leakage.
- The development of two experimental case studies to evaluate the proposed design methodology. One focused on voltage and temperature manipulation. The other pertained to the injection of electromagnetic pulses.
- The comparison of the results obtained from the proposed system with the schemes reported in the literature.

The remainder of the paper is organized as follows. Section 2 presents an overview of the AES used as a test vehicle and the physical implementation attacks reported in the literature, focusing on voltage, temperature, and electromagnetic (EM) radiation. Section 3 describes the proposed design methodology to detect attacks based on temperature, voltage, and EM radiation. Section 4 presents the setups used to test the design methodology, the results obtained, and a comparison with previous schemes reported in the literature. Finally, Section 5 presents the conclusions derived from this work.

2. Physical Fault Injection Implementation Attacks

In the case of non-invasive active attacks based on fault injection, the attacker maliciously aims to cause transitory faults in the operations of cryptographic algorithms. Faults must not be permanent, as this would render the circuit unusable and eliminate the possibility of a differential study. With this in mind, the attacker will try to determine the erroneous behavior under different types of fault, which depends on the algorithm under attack, and compare it with the correct behavior of the circuit. This mathematical comparison is known as Differential Fault Analysis (DFA) [16] and allows one to establish the relationship between the produced faults and the internal information of the cipher. Depending on the encryption algorithm, the desired faults will be of different types: single-bit, multi-bit, single-byte, multi-byte, faults in the same byte, faults in different bytes, or even random faults. In the case of the AES cipher, analyses can be found in the following references [17–20]. These differ as to the place where the fault must be inserted, such as the state matrix, S-box operation, and KeySchedule. In addition, the differential analysis will differ depending on the encryption or decryption round in which the fault is inserted. Table 1 shows the classification of DFA attacks on the AES, considering if the attack is performed on the state, the S-box, or KeySchedule; the type of fault; and the round where the fault must be injected.

In short, the AES encryption scheme [21] is the official standard set by the National Institute of Standards and Technology (NIST) to replace the Data Encryption Standard (DES). It employs the Rijndael algorithm with input blocks of 128 bits. The AES conducts the input transformation across several rounds, with the number of rounds varying based on the key size selected (10 rounds for 128 bits, 12 rounds for 192 bits, and 14 rounds for 256 bits). The rounds involve manipulating the input state S, which is 16 bytes in size, using the SubBytes(), ShiftRows(), MixColumn(), and AddRoundKey() operations. In our experiments, we focused on the 128-bit key setup. This cipher is the most widely used and implemented in the vast majority of technologies because it is the most widely used standard for the symmetric block ciphering of information.

Regarding attacks on the AES cipher, the two main works are those presented by Giraud [17] and Dussart [18]. The first one established the assumptions that an attacker must follow in order to recover the secret key contained in the cryptocircuit when a single bit-level fault is injected into the state matrix. These attacks are carried out on the eighth round of encryption. On the other hand, the second paper presented a differential analysis based on the injection of multi-bit faults in the same byte of the state matrix. In this case, the attacks are performed from the seventh round and before the ninth round, prior to the MixColumn() operation. These were the first DFA attacks on the AES, but after them many more appeared continuously, modifying the assumptions of the attacks and the type, round, and location of the fault necessary to compromise the security of the cipher [19,20].

Table 1. Classification of DFA attacks on AES.

Reference	Attack on State	Attack on S-Box	Attack on KeySchedule	Type of Attack	Attack in Round
[17]	Yes	No	Yes	S. bit | S. byte	8 and 9
[18]	Yes	~	No	M. bit (same byte)	7, 8, and 9
[19]	Yes	Yes	No	S. byte	8
[20]	Yes	~	No	S. byte— random	During encryption

S. bit and M. bit = single-bit and multi-bit, respectively. S. byte and M. byte = single-byte and multi-byte, respectively.

2.1. Voltage and Temperature Attacks

In the case of the supply voltage, an increase or decrease in the voltage supply to the chip above the tolerance level of the devices (typically 10%) can cause faults in the combinational operations or in the bits stored in the flip-flops (FFs). These faults can affect part of the circuit or cause widespread faults [22]. Additionally, tampering with the power supply can lead more easily to fault injection. In [23], the authors showed how by manipulating the supply voltage, fault injection through clock manipulation is made simpler, without the need to drive the circuit to extremes ar which it is more difficult for the attacker to achieve effective faults in cryptocircuits. A representative scheme of an attack modifying the circuit supply voltage can be seen in Figure 2.

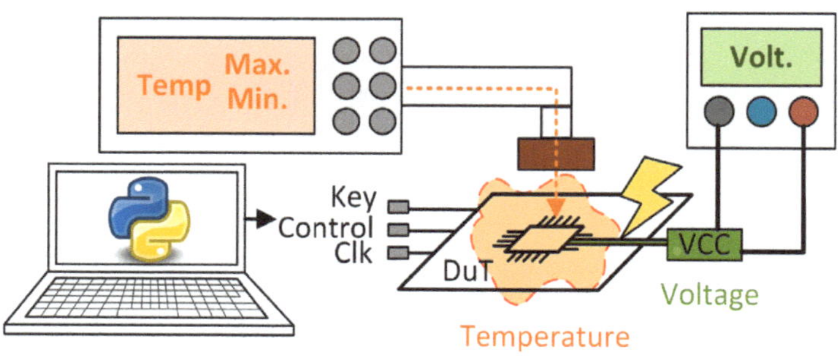

Figure 2. Voltage and temperature attack representation.

When considering temperature attacks, exceeding the temperature range specified by chip manufacturers for proper operation can deliberately induce faults in the chip. By configuring the chip temperature to a level at which write operations are functional while reads are not, or vice versa, multiple attacks can be launched. Various kinds of faults can occur based on the susceptibility of the components to temperature. Typically, the attacker lacks specific control over the exact type of fault that will occur [24,25]. An illustrative example of an attack involving the manipulation of the circuit's ambient temperature is depicted in Figure 2.

2.2. Electromagnetic Attacks

Electromagnetic Fault Injection (EMFI) attacks are based on the introduction of errors in an integrated circuit using an electromagnetic pulse (EMP). When the electromagnetic field of the EMP penetrates the device, it produces anomalous voltage differences and currents within the components of the circuit. A representative scheme of an attack using an electromagnetic pulse injector can be seen in Figure 3. Inducing Foucault currents on the chip surface can cause a fault of up to a single bit [26]. In addition to the security aspect, there is also the characterization and study of the resilience of integrated circuits to fault insertions produced by the environment itself, such as electromagnetic interference or cosmic rays, when circuits are used in space applications [27]. However, the use of passive protections, such as passive shields provided by metal layers that cover sensitive parts of the chip, makes the injection of EM faults more difficult because a very high level of precision is required [6].

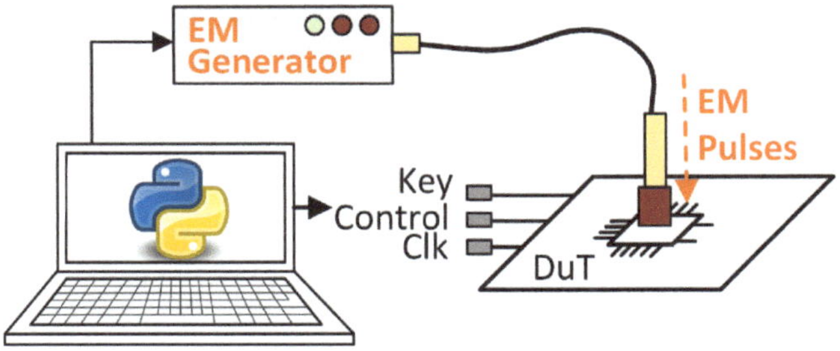

Figure 3. Electromagnetic attack representation.

3. Design Methodology to Protect against Attacks

Taking into account the attacks mentioned above, the proposed solution uses the Xilinx Analog-to-Digital Converters (XADCs) provided by Xilinx in its Field-Programmable Gate Array (FPGA) devices. The XADC is a basic building block that enables analog mixed-signal (AMS) functionality. By combining analog blocks with programmable logic, it is possible to create customized analog interfaces for a wide range of applications.

The possible applications that can be developed with these components include the reading and monitoring of the analog values of the operating voltages of the device. Therefore, in this work, this component was used to establish operating ranges outside of which the system will understand that the device is being maliciously manipulated and will therefore activate an error signal that will allow the system to detect that it is being attacked.

With the aim of designing a secure protection against these types of attacks, it is necessary to follow a design methodology that allows for the most efficient scheme. A schematic representation of the design methodology is presented in Figure 4. Firstly, it is necessary to characterize the cryptocircuit that is to be protected. Therefore, the maximum frequency, voltage supply, and temperature characteristics must be established, following the manufacturer's data sheets. Once these parameters have been established, it is necessary to determine the possible vulnerabilities of the physical implementation, namely, the layout, power supply lines, etc. Thirdly, the designer must configure the XADC to cover the voltage and temperature ranges that are considered correct. At this point, out-of-range values are the aim of attackers when a fault is desired. Finally, a response must be selected against the manipulation of the circuit. This response could be different depending on the attack scenario considered. For example, responses to Ineffective Fault Attacks (IFAs) will be different from the responses to DFA attacks.

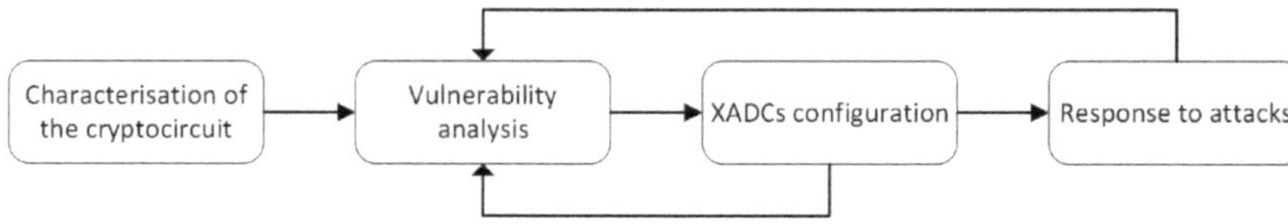

Figure 4. Design methodology process.

In response to the attack, a scheme is proposed, as shown in Figure 5. Note that the cryptocircuit identifier is used since the system can be implemented to protect different cryptocircuits. In this scheme, the encryption algorithm and the XADC, which monitors the voltages and temperatures within the set range, are working in parallel. As long as the protection does not detect any anomalous state, the cryptocircuit can operate normally and output the encrypted data without any interference. However, if the protection system detects that the system is being attacked by any of the techniques described above, it generates an alarm signal that allows for the activation of the response to the attack and alerts the system or user of this cipher so that they are aware of the situation. This alarm would allow a root of trust that works independently, encrypting the information, to know the scenario in which it finds itself and that its security is compromised.

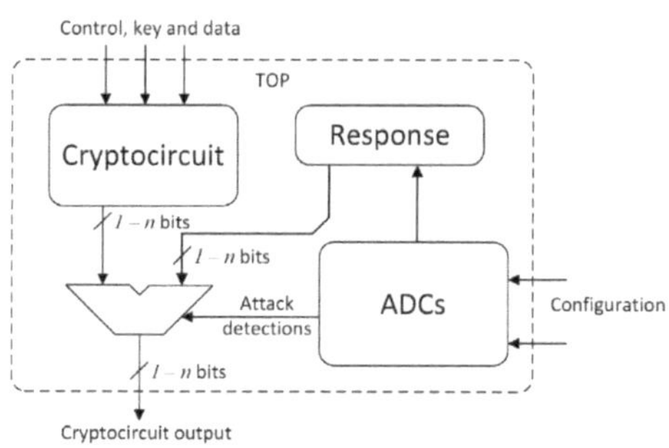

Figure 5. Schematic representation of the top module of protection.

When the system detects that it is under attack, the alarm signal is triggered, and the response is activated. In this scheme, the response consists of writing a zero value in the output. With this response, an attacker cannot know whether the attack was effective or not due to the fact that any data related to the data stored or processed during encryption or decryption are given in the output, completely blocking DFA or IFA attacks.

It should be noted that FPGA ADCs have both internal and external inputs. In our case, we used internal signals that monitor the voltage and temperature ranges of the device. These inputs are not accessible from the outside, and, therefore, if an attacker tries to manipulate them, in addition to the enormous complexity that this would entail, any manipulation would cause the device to be out of its operating ranges, and thus the alarms would be triggered. Note that the inputs shown in Figure 5 denoted as *Configuration* are only accessible during the design process.

4. Setups and Results

To test the performance of this protection, a Xilinx Nexys 4 board with an Artix 7 100T FPGA (Xilinx Inc. from San Jose, CA, USA) was used. In addition to the board, different tools were used: a Thermonics ATS-505-S-2 (Temptronic Corporation, Mansfield, MA, USA) temperature control system, a Keysight e36312A (Keysight Technologies, Santa Rosa, CA, USA) power supply, an Agilent InfiniiVision DSO7054A oscilloscope (Agilent, Santa Clara, CA, USA) with 4 G/samples and a bandwidth of 500 MHz, and a NewAE EM pulse generator called ChipSHOUTER (NewAE Technology Inc., Dartmouth, Canada). Both setups can be observed in Figures 6 and 7.

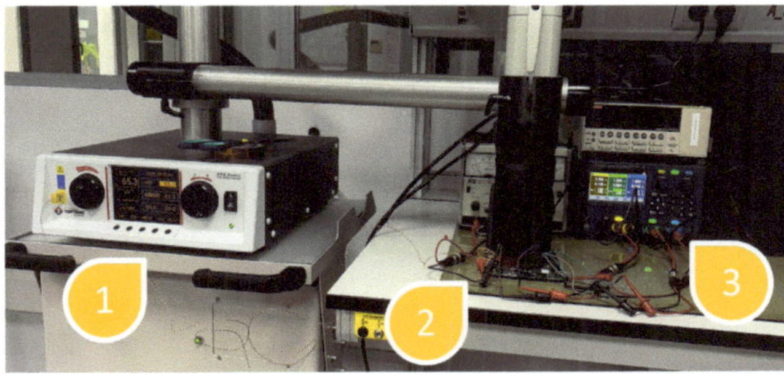

Figure 6. Experimental setup for testing voltage and temperature schemes.

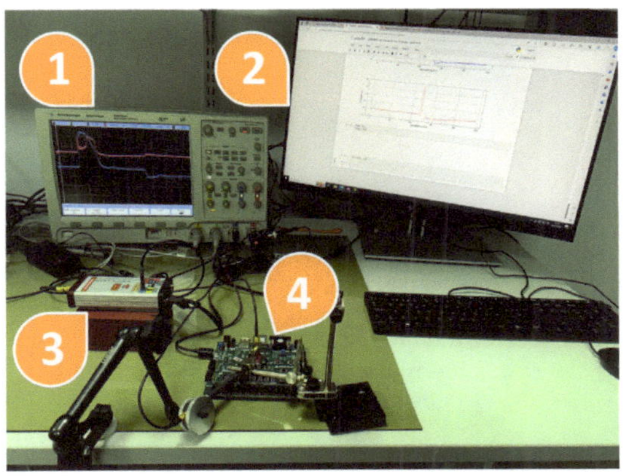

Figure 7. Experimental setup for testing electromagnetic scheme: (1) oscilloscope, (2) PC, (3) ChipSHOUTER, (4) Nexys A7 board.

Different devices can be used to induce an EMP. In this work, we used the ChipSHOUTER (CW520) from NewAE Technology. This device charges a bank of capacitors that, when fired, are discharged through a probe tip, producing a short-duration high-intensity EMP in the vicinity of the tip. These probe tips are made of a conductor coil with a ferrite core inside that creates a (mainly) magnetic field. The voltage at which the capacitors can be charged and the approximate duration of the EMP are parameters that can be adjusted via a digital interface with a PC. The voltage is directly proportional to the intensity of the magnetic field produced, and the duration of the pulse also corresponds to the duration of the electromagnetic perturbation. In order to characterize the EMP, we placed a probe from the Rohde&Schwarz HZ-15 near-field electromagnetic probes kit (Rohde & Schwar, Munich, Germany) near the coil of the ChipSHOUTER. This probe measured the magnetic field inside its own coil, and the variation over time was recorded with an oscilloscope.

In Figure 8, we can see how the magnetic field induced near the probe tip corresponds to the variation in the voltage inside the ChipSHOUTER. We can see that the aspects of both traces are similar. The main differences are at the start and the end of the pulse generation. At first, since the voltage scales very fast, we can see a spike in the magnetic field produced by the quick variation. Then, the magnetic field is more or less constant during the pulse and relaxes to another constant value slightly different from zero. This could be because a current remained in the coil of the probe tip, and it produced a less intense magnetic field until it disappeared. This final relaxation to the initial state was much longer than the pulse, but it barely produced an alteration in the attacked signals.

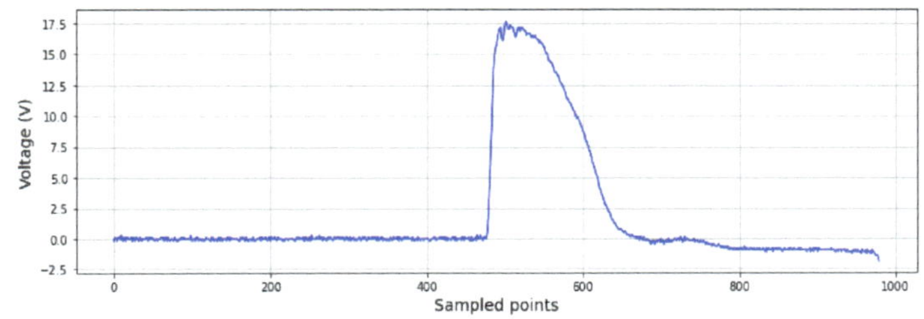

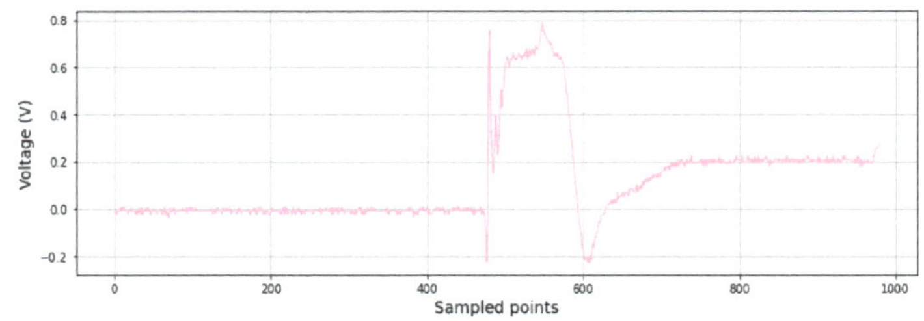

Figure 8. Comparison between the voltage levels of the ChipSHOUTER (**top**) and the magnetic field measured at the probe tip (**bottom**).

We can infer then that the current and voltage variation induced by the EMP were higher the faster the intensity of the magnetic field changed. This makes sense if we observe Faraday's law of induction (Equation (1)), which states that the electromotive force produced along a closed path is proportional to the rate of change of the magnetic flux enclosed by the path:

$$\varepsilon = -\frac{d\varphi_B}{dt}, \quad (1)$$

where ε is the electromotive force, and φ_B is the magnetic flux.

4.1. Voltage and Temperature Setup

Figure 6 presents the configuration for the manipulation of temperature and voltage ((1) Thermonics, (2) Nexys board, and (3) supply voltage). In the case of the power supply voltage, the main power supply voltage, V_{CCINT}; the memory voltages of Random-Access Memory (RAM), V_{CCBRAM}; and the internal auxiliary power supply voltage of the FPGA itself, V_{CCAUX}, are protected against malicious variations. The values established as performance limits were those established following the manufacturer's characterization tests. Specifically, for fault detection, the established ranges within which operation is considered correct are those presented in Table 2. Outside of these ranges, the protection scheme considers the device to be under attack.

Table 2. Operating ranges for supply voltage.

V_{CCINT} (V)		V_{CCBRAM} (V)		V_{CCAUX} (V)	
Min.	Max.	Min.	Max.	Min.	Max.
0.88	1.05	0.88	1.05	0.88	1.8

For the voltage case, it is necessary to consider tools (2) and (3) in Figure 6. Figure 9 shows the result of the monitoring of voltage variation using the Xilinx Vivado (Xilinx Inc. from San Jose, CA, USA) tool. The supply voltage was modified to obtain a voltage outside the ranges (purple line), with a minimum of 0.87 V and a maximum of 1.06 V. Notice that the green line is the temperature variation of the circuit under these conditions.

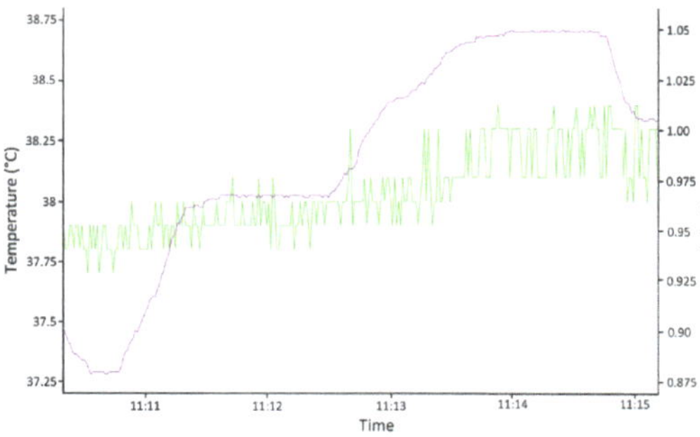

Figure 9. Screenshot of the voltage test outside the established ranges. Purple: Voltage; Green: Temperature.

For the case of temperature variations, the same scheme was used as for voltage variations. In this case, the XADC was configured to obtain the operating temperature values of the device and to check if the device was outside the normal operating ranges. For this purpose, a temperature range was defined for which the device should operate correctly and outside of which the system is considered to be under attack. The temperature range considered as normal operation is between 60 °C and 0 °C. This range was determined by the operating ranges given by the manufacturer, considering that our experimental application's maximum temperature was fixed at 60 °C for the protection of the laboratory equipment and the device under test. This range was established according to the characterization data, which were considered normal behavior for a circuit in a typical work process.

4.2. Voltage and Temperature Results

For this test, the Xilinx Nexys 4 board, Figure 6 (2); a Thermonics ATS-505-S-2 temperature control system, Figure 6 (1); and a computer with Xilinx Vivado software were used to monitor the results.

In this case, only the temperature above the normal operating range could be tested, since subjecting the system to temperatures that were too low could produce small frozen water spots and irreversibly damage the device. Figure 10 shows the result of monitoring the temperature variation using the Xilinx Vivado tool. The temperature was modified to obtain a temperature outside the range (green line), which was the maximum value at 65.5 °C. The test again corroborated that the error signal was correctly triggered by turning on an LED on the board when the temperature was out of range. Note that the purple line is the voltage supply of the circuit in this test.

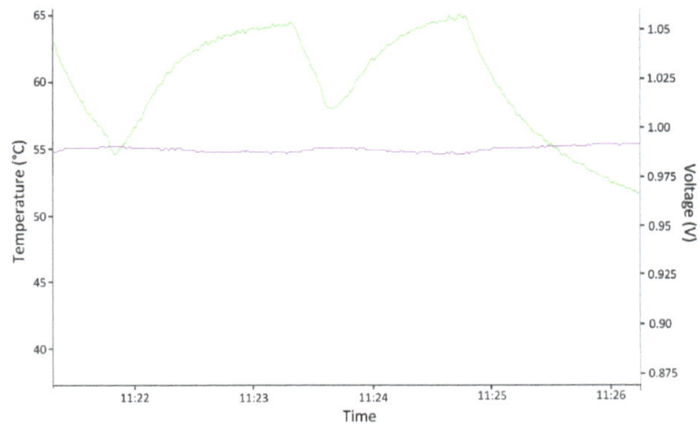

Figure 10. Screenshot of the temperature test outside the established range. Purple: Voltage; Green: Temperature.

4.3. Electromagnetic Setup

In Figure 7, the EM fault attack setup can be observed. Figure 11 presents in more detail the same setup for EM attacks. In Figure 7, (1) is the Agilent InfiniiVision DSO7054A oscilloscope, (2) is a personal computer, (3) is the ChipSHOUTER, and (4) is the Nexys board. In Figure 11, (1) is a Rohde&Schwarz HZ-15, (2) is the probe tip, and (3) is the Nexys board. The main objective was to test whether inserting an EMP into the FPGA would detect a voltage rise (or fall) that could be detected by the XADC. For this, a 4 mm ChipSHOUTER probe tip (diameter of the ferrite core) was used directly above the FPGA encapsulation (approximately 1 mm). The magnetic field probe surrounding the probe tip of the ChipSHOUTER was used to measure the injected magnetic field. More specifically, this probe measured the component of the magnetic field vector that was perpendicular to the plane where the probe coil was located.

Figure 11. Details of experimental setup for testing electromagnetic scheme: (1) near-field magnetic probe, (2) probe tip of the ChipSHOUTER directly pointing at the Artix-7 FPGA, (3) Nexys A7 board.

4.4. Electromagnetic Results

The success of an attack depended on the intensity of the pulse and the position along the XY plane of the FPGA of the ChipSHOUTER probe tip. There were some areas where only one pulse was needed to detect an anomalous voltage change in the FPGA by the XADC, while in other areas the injection of an EMP would reset and clear the whole FPGA.

The aim of the scheme was to determine whether pulse injection affected the internal voltage values of the FPGA. As can be seen in Figure 12, the injection of EM pulses into a zero logic signal and a clock signal altered the values of these signals, producing a disturbance of more than two volts. This test was performed over the external signals because the internal FPGA signal could not be sampled by the oscilloscope. These disturbances could be sampled by the XADC as variations of the voltage and, therefore, could detect EM attacks. To ascertain whether the EMP produced a voltage change outside of the typical values, circuitry was placed at the ADC output to sample its value each clock cycle (the clock frequency was 100 MHz). If a voltage rise or fall was detected, then the alarm was triggered. As a result, pulse injection was detected in the tests, allowing us to determine that the EM pulses altered the internal voltage and therefore were detectable by the proposed scheme. In these cases, the output of the criptocircuit was zero. The efficiency of the proposed protection scheme was 100% (all effective attacks were detected) in those cases where the injected fault caused no clearing or resetting of the FPGA.

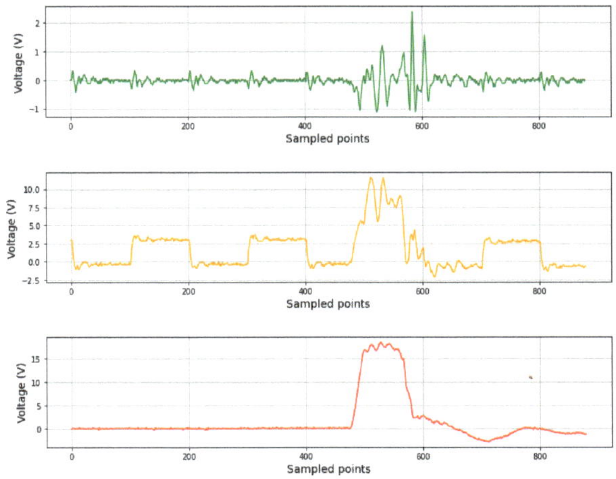

Figure 12. Result of an EM attack over the FPGA. Signal with zero logic value (green), FPGA clock signal (orange), and EM pulse injection (red).

As can be seen, the use of this methodology allows one to determine whether the system is being attacked by any of the three attack methods, representing a single detection and response scheme necessary to neutralize these attacks. As soon as the converter determines that the system is under attack, the circuit will not leak information, since it will not respond. Hence, the attacker cannot have access to the erroneous operation produced by the different fault injections into the cryptographic algorithm, and thus the attacker is unable to perform the differential analysis.

4.5. Comparison with Other Protection Schemes

With the aim of properly analyzing the methodology and protection proposed, Table 3 presents a summary of some schemes proposed in the literature for the cryptocircuit AES. As can be seen, there are different types of protections, such as information redundancy, hardware redundancy, temporal redundancy, and combined approaches. It should be noted that our protection does not belong to any of the above groups but can be compared with them in order to achieve an overview of their penalties. Table 3 presents the area overhead, frequency degradation, attack detection, and technology with which the schemes were implemented.

It is possible to establish a hierarchy of attack detection based on the fault coverage of the reported schemes, which can be compared with our scheme. On the other hand, if references [28–34] are considered, the table shows that these schemes are based on information redundancy, i.e., the protections add additional information to the data processed by the encryption circuit to detect if there has been any change. From this group, it is possible to see that the impact in terms of area is very high in the schemes of [29–31,33], which have at least 40% extra area. However, if the scheme whose area penalty is lower [28] is considered, it can be seen that its frequency degradation is very high, reaching 30%. In [35], a protection based on hardware redundancy was presented. This scheme adds part of the operations redundantly and compares the results to look for any error injection. Despite obtaining the same result for frequency degradation as the scheme presented in this paper, its area degradation is very high compared to the scheme proposed in this work, reaching an additional 86%. On the other hand, the scheme based on temporal redundancy in [36] built its protection on rerunning certain operations in search of discrepancies. As can be seen, its area degradation is slightly higher than our proposal, but, on the other hand, its frequency penalty is much higher, reaching 17%. Finally, the schemes in [37,38] employ a combination of protections. It can be observed that both have a high cost in area and their results in terms of frequency degradation are higher than the presented scheme, reaching up to 22%.

On the other hand, if attack detection is taken into account, only in the cases of [35–37] and the proposed scheme was 100% reached. Among these, the scheme of [36] was the closest in terms of area consumption, with an additional 6%. However, as mentioned above, its frequency penalty is much higher.

Finally, it should be noted that the proposed solution could be applied in parallel with any of the other protection schemes, as it operates completely independently of the circuits and any other protection countermeasures. Furthermore, the use of this system makes it possible to determine not only whether the system is under attack but also whether an extreme situation is occurring that could cause the circuit to malfunction.

Table 3. Comparison with different detection schemes.

Scheme	Type of Protection	Area Overhead	Frequency Degradation	Attack Detection	Technology
Unprotected	None	1	1		Artix-7
[28]	Information	1.08	0.7	75.6%	Virtex 1000
[29]		1.44	NIA	99.12%	NIA
[30]		1.73	0.64	88%	NIA
[31]		1.40	NIA	97%	NIA
[32]		1.32	0.97	97%	Virtex II
[33]		1.77	0.86	90%	Virtex E
[34]		1.25	0.88	98%	Virtex 5
[35]	Hardware	1.87	1	100%	Virtex 5
[36]	Temporal	1.07	0.83	100%	Virtex 4
[37]	Combination	1.38	0.78	100%	Virtex 6
[38]		1.58	0.83	93.75%	Spartan 3
Our Scheme	-	1.01	1	100%	Artix-7

NIA = No Information Available. Artix, Virtex and Spartan versions belong to Xilinx Inc. from San Jose, CA, USA.

5. Conclusions

Due to the constant development and sophistication of attacks on hardware implementations of cryptographic elements, it is necessary to develop systems to protect user information from malicious third parties. To minimize these attacks, countermeasures or protections are developed to strengthen the hardware for implementing these circuits. In this paper, we presented a design methodology that uses XADCs in the FPGA as a countermeasure to detect and thwart non-invasive active attacks, in particular those based on the supply voltage, temperature, and electromagnetic pulses. The proposed solution offers the possibility of using FPGA resources for protection with minimal resource cost in the case of FPGA technologies because the XADCs are implemented in the FPGA itself. Several temperature, voltage, and EM tests were performed, and the results show that once the ranges of temperature and voltage were defined, the scheme was able to detect any variation when the circuit was outside of its operating range. On the other hand, the results showed no frequency degradation because the scheme was implemented in parallel with the cryptocircuit, and in the case of the area overhead, the penalty was due to the attack response, which was 1%. In addition, this scheme is useful for different types of analysis because it is very versatile and can be implemented together with any cryptosystem. It should be noted that such schemes and methodologies are also applicable to Application-Specific Integrated Circuit (ASIC) technologies if an ADC is included in the circuit. This is an important point for future work, as the analog design of an ADC in ASICs will allow us to determine its impact on the design, as well as to study the variability of results with respect to those that can be obtained in FPGAs.

Author Contributions: Conceptualization, F.E.P.-O. and E.T.-S.; methodology, F.E.P.-O. and E.T.-S.; software, F.E.P.-O.; validation, F.E.P.-O., A.C.-G. and E.T.-S.; formal analysis, F.E.P.-O., A.C.-G. and

E.T.-S.; investigation, F.E.P.-O., A.C.-G. and E.T.-S.; resources, E.T.-S.; data curation, F.E.P.-O.; writing—original draft preparation, F.E.P.-O., A.C.-G. and E.T.-S.; writing—review and editing, F.E.P.-O. and E.T.-S.; visualization, F.E.P.-O., A.C.-G. and E.T.-S.; supervision, E.T.-S.; project administration, E.T.-S. All authors have read and agreed to the published version of the manuscript.

Funding: Programa Operativo FEDER 2014-2020 and Consejería de Economía, Conocimiento, Empresas y Universidad de la Junta de Andalucía under project US-1380823; and project grant PID2020-116664RB-I00 funded by MCIN/AEI/10.13039/501100011033.

Informed Consent Statement: Not applicable.

Data Availability Statement: The original contributions presented in the study are included in the article, further inquiries can be directed to the corresponding author.

Acknowledgments: The authors want to express gratitude for the SPIRS (Secure Platform for ICT Systems Rooted at the Silicon Manufacturing Process) Project, Grant Agreement No. 952622, under the European Union's Horizon 2020 research and innovation programme; Programa Operativo FEDER 2014-2020 and Consejería de Economía, Conocimiento, Empresas y Universidad de la Junta de Andalucía under project US-1380823; and project grant PID2020-116664RB-I00 funded by MCIN/AEI/10.13039/501100011033.

Conflicts of Interest: The authors declare no conflicts of interest. The funders had no role in the design of the study; in the collection, analyses, or interpretation of data; in the writing of the manuscript; or in the decision to publish the results.

Abbreviations

The following abbreviations are used in this manuscript:

ADC	Analog-to-digital converter
AES	Advanced Encryption Standard
AMS	Analog mixed signal
ASIC	Application-Specific Integrated Circuit
DES	Data Encryption Standard
DFA	Differential Fault Analysis
EM	Electromagnetic
EMFI	Electromagnetic Fault Injection
EMP	Electromagnetic pulse
FF	Flip-flop
FIA	Fault injection attack
FPGA	Field-Programmable Gate Array
IFA	Ineffective Fault Attack
IoT	Internet of Things
NIST	National Institute of Standards and Technology
PC	Personal Computer
RAM	Random-Access Memory
SCA	Side-Channel Analysis
XADC	Xilinx Analog-to-Digital Converter

References

1. Dutta, I.K.; Ghosh, B.; Bayoumi, M. Lightweight Cryptography for Internet of Insecure Things: A Survey. In Proceedings of the IEEE 9th Annual Computing and Communication Workshop and Conference (CCWC'19), Las Vegas, NV, USA, 7–9 January 2019; pp. 475–481.
2. Mahmoud, R.; Yousuf, T.; Aloul, F.; Zualkernan, I. Internet of things (IoT) security: Current status, challenges and prospective measures. In Proceedings of the International Conference for Internet Technology and Secured Transactions (ICITST'15), London, UK, 14–16 December 2015; pp. 336–341.
3. Xu, T.; Wendt, J.B.; Potkonjak, M. Security of IoT systems: Design challenges and opportunities. In Proceedings of the International Conference on Computer-Aided Design (ICCAD'14), San Jose, CA, USA, 3–6 November 2014; pp. 417–423,
4. Gunathilake, N.A.; Al-Dubai, A.; Buchana, W.J. Recent Advances and Trends in Lightweight Cryptography for IoT Security. In Proceedings of the 16th International Conference on Network and Service Management (CNSM'20), Izmir, Turkey, 2–6 November 2020; pp. 1–5.

5. Kazemi, Z.; Fazeli, M.; Hely, D.; Beroulle, V. Hardware Security Vulnerability Assessment to Identify the Potential Risks in a Critical Embedded Application. In Proceedings of the IEEE 26th International Symposium on On-Line Testing and Robust System Design (IOLTS'20), Napoli, Italy, 13–15 July 2020; pp. 1–6.
6. Bar-El, H.; Choukri, H.; Naccache, D.; Tunstall, M.; Whelan, C. The Sorcerer's Apprentice Guide to Fault Attacks. *Proc. IEEE* **2006**, *94*, 370–382. [CrossRef]
7. Mangard, S.; Oswald, E.; Popp, T. *Power Analysis Attacks: Revealing the Secrets of Smart Cards*; Springer Science & Business Media: Berlin, Germany, 2008; Volume 31.
8. Tena-Sánchez, E.; Potestad-Ordónez, F.E.; Jiménez-Fernández, C.J.; Acosta, A.J.; Chaves, R. Gate-level hardware countermeasure comparison against power analysis attacks. *Appl. Sci.* **2022**, *12*, 2390. [CrossRef]
9. Kudithi, T. An efficient hardware implementation of the elliptic curve cryptographic processor over prime field. *Int. J. Circuit Theory Appl.* **2020**, *48*, 1256–1273. [CrossRef]
10. Marchesan, G.C.; Weirich, N.R.; Culau, E.C.; Weber, I.I.; Moraes, F.G.; Carara, E.; de Oliveira, L.L. Exploring RSA Performance up to 4096-bit for Fast Security Processing on a Flexible Instruction Set Architecture Processor. In Proceedings of the IEEE International Conference on Electronics, Circuits and Systems (ICECS'18), Bordeaux, France, 9–12 December 2018; pp. 757–760.
11. Dong, X.; Zhang, L.; Gao, X. An Efficient FPGA Implementation of ECC Modular Inversion over F256. In Proceedings of the International Conference on Cryptography, Security and Privacy (ICCSP'18), Guiyang, China, 16–18 March 2018; pp. 29–33.
12. Tanougast, C. Hardware Implementation of Chaos Based Cipher: Design of Embedded Systems for Security Applications. In Proceedings of the Chaos-Based Cryptography, Studies in Computational Intelligence, Bordeaux, France, 9–12 December 2018; Kocarev, L., Lian, S., Eds.; Springer: Berlin/Heidelberg, Germany, 2011; Volume 354.
13. Biehl, I.; Meyer, B.; Müller, V. Differential fault attacks on elliptic curve cryptosystems. In Proceedings of the Annual International Cryptology Conference (CRYPTO'00), Santa Barbara, CA, USA, 20–24 August 2000; pp. 131–146.
14. Nitaj, A.; Fouotsa, E. A new attack on RSA and Demytko's elliptic curve cryptosystem. *J. Discret. Math. Sci. Cryptogr.* **2019**, *22*, 391–409. [CrossRef]
15. Dhanda, S.S.; Singh, B.; Jindal, P. Demystifying elliptic curve cryptography: Curve selection, implementation and countermeasures to attacks. *J. Interdiscip. Math.* **2020**, *23*, 463–470. [CrossRef]
16. Biham, E.; Shamir, A. Differential fault analysis of secret key cryptosystems. *Lect. Notes Comput. Sci. Adv. Cryptol.* **1997**, *1294*, 513–525.
17. Giraud, C. Dfa on aes. In Proceedings of the International Conference on Advanced Encryption Standard, Bonn, Germany, 10–12 May 2004; pp. 27–41.
18. Dussart, P. Differential Fault Analysis on A.E.S. In Proceedings of the Applied Cryptography and Network Security (ACNS'03), Kunming, China, 16–19 October 2003; pp. 293–306.
19. Pogue, T.E.; Nicolici, N. Incremental Fault Analysis: Relaxing the Fault Model of Differential Fault Attacks. *IEEE Trans. Very Large Scale Integr. (Vlsi) Syst.* **2020**, *28*, 750–763. [CrossRef]
20. Piret, G.; Quisquater, F. A Differential Fault Attack Technique against SPN Structures, with Application to the AES. In Proceedings of the International Workshop on Cryptographic Hardware and Embedded Systems (CHES'03), Cologne, Germany, 8–10 September 2003; pp. 77–88.
21. Pub, NIST FIPS. 197: Advanced Encryption Standard (AES). *Fed. Inf. Process. Stand. Publ.* **2001**, *197*, 0311.
22. Barenghi, A.; Bertoni, G.M.; Breveglieri, L.; Pellicioli, M.; Pelosi, G. Low voltage fault attacks to AES. In Proceedings of the IEEE International Symposium on Hardware-Oriented Security and Trust (HOST'10), Anaheim, CA, USA, 13–14 June 2010; pp. 7–12.
23. Potestad-Ordóñez, F.E.; Valencia-Barrero, M.; Baena-Oliva, C.; Parra-Fernández, P.; Jiménez-Fernández, C.J. Breaking Trivium stream cipher implemented in ASIC using experimental attacks and DFA. *Sensors* **2020**, *20*, 6909. [CrossRef] [PubMed]
24. Peterson, I. *Chinks in Digital Armor: Exploiting Faults to Break Smart-Card Cryptosystems*; Science News, Wiley Online Library: Hoboken, NJ, USA, 1997; Volume 151, pp. 78–79.
25. Skorobogatov, S. *Low Temperature Data Remanence in Static RAM*; University of Cambridge, Computer Laboratory: Cambridge, UK, 2002.
26. Quisquater, J.J.; Samyde, D. Eddy current for magnetic analysis with active sensor. In Proceedings of the eSMART, San Jose, CA, USA, 21–22 November 2002; pp. 185–194.
27. Baumann, R. Radiation-induced soft errors in advanced semiconductor technologies. *IEEE Trans. Device Mater. Reliab.* **2005**, *5*, 305–316. [CrossRef]
28. Wu, K.; Goessel, M. Low Cost Concurrent Error Detection for the Advanced Encryption Standard. In Proceedings of the International Test Conference (ITC'04), Charlotte, NC, USA, 26–28 October 2004; pp. 1242–1248.
29. Breveglieri, L.; Koren, I.; Maistri, P. Incorporating Error Detection and Online Reconfiguration into a a Regular Architecture for the Advanced Encryption Standard. In Proceedings of the IEEE International Symposium on Defect and Fault Tolerance in VLSI Systems (DFT'05), Monterey, CA, USA, 3–5 October 2005; pp. 72–80.
30. Yen, C.H.; Wu, B.F. Simple Error Detection Methods for Hardware Implementation of Advanced Encryption Standard. *IEEE Trans. Comput.* **2006**, *55*, 720–731.
31. Kermani, M.M.; Reyhani-Masoleh, A. Parity-based fault detection architecture of S-box for advanced encryption standard. In Proceedings of the IEEE International Symposium on Defect and Fault Tolerance in VLSI Systems (DFT'06), Arlington, VG, USA, 4–6 October 2006; pp. 572–580.

32. Mozaffari-Kermani, M.; Reyhani-Masoleh, A. A lightweight highperformance fault detection scheme for the advanced encryption standard using composite fields. *IEEE Trans. Very Large Scale Integr. (Vlsi) Syst.* **2011**, *19*, 85–91. [CrossRef]
33. Karpovsky, M.; Kulikowski, K.J.; Taubin, A.; Member, S. Robust Protection Against Fault-Injection Attacks on Smart Cards Implementing the Advanced Encryption Standard. In Proceedings of the International Conference on Dependable Systems and Networks (DSN'04), Florence, Italy, 28 June–1 July 2004; pp. 93–101.
34. Mestiri, H.; Benhadjyoussef, N.; Machhout, M.; Tourki, R. High performance and reliable fault detection scheme for the advanced encryption standard. *Int. Rev. Comput. Softw. (IRECOS'13)* **2013**, *8*, 730–746.
35. Joye, M.; Manet, P.; Rigaud, J.B. Strengthening hardware AES implementations against fault attacks. *IET Inf. Secur.* **2007**, *1*, 106–110. [CrossRef]
36. Rajendran, J.; Borad, H.; Mantravadi, S.; Karri, R. Slide-based concurrent error detection technique for symmetric block ciphers. In Proceedings of the IEEE International Symposium on Hardware Oriented Security and Trust (HOST'10), Anaheim, CA, USA, 13–14 June 2010; pp. 70–75.
37. Karri, R.; Wu, K.; Mishra, P.; Kim, Y. Concurrent error detection schemes for fault-based side-channel cryptanalysis of symmetric block ciphers. *IEEE Trans. Comput. Aided Des. Integr. Circuits Syst.* **2002**, *21*, 1509–1517. [CrossRef]
38. Chu, J.; Benaissa, M. Error detecting AES using polynomial residue number systems. *Microprocess. Microsyst.* **2013**, *37*, 228–234.

materials

MDPI

Article

Chloride Permeability of Alkali-Activated Slag Concretes after Exposure to High Temperatures

Baomeng Zhou [1,2,3], Qianmin Ma [1,2,3,*], Rongxin Guo [1,2,3] and Ping Li [1,2,3]

1 Faculty of Civil Engineering and Mechanics, Kunming University of Science and Technology, Kunming 650500, China
2 Yunnan Key Laboratory of Disaster Reduction in Civil Engineering, Kunming 650500, China
3 International Joint Laboratory for Green Construction and Intelligent Maintenance of Yunnan Province, Kunming 650500, China
* Correspondence: maqianmin666@163.com; Tel.: +86-130-9535-8933

Abstract: The number of fires in buildings and on bridges has increased worldwide in recent years. As a structural material, the strength of alkali-activated slag (AAS) concrete after exposure to high temperatures has been given much attention. However, research of its durability is still lacking, which limits the application of this type of concrete on a larger scale. In this context, as one of the most important aspects of durability, the chloride permeability of AAS concretes after exposure to high temperatures was examined in this study. The influence of the alkali concentration (Na_2O%) and the modulus (Ms) of the activator, as well as the influence of heating regimes, including the heating rate, duration of exposure to the target temperature, and cooling method, was also discussed. The results show that the chloride permeability of the AAS concretes increased with temperature elevation. Due to the interference of pore solution conductivity, the influence of the Na_2O% and the Ms of the activator on the chloride permeability of the AAS concretes was not made clear by using the ASTM C 1202 charge passed method; however, after exposure to high temperatures, AAS with a lower Na_2O% and lower Ms has lower porosity and may have lower chloride permeability, which needs further investigation. Faster heating for a longer duration at the target temperature and water cooling reduced the resistance of the AAS concretes to chloride permeability as a result of their increased porosity.

Keywords: alkali-activated slag concrete; high temperature; chloride permeability; pore structure

Citation: Zhou, B.; Ma, Q.; Guo, R.; Li, P. Chloride Permeability of Alkali-Activated Slag Concretes after Exposure to High Temperatures. *Materials* **2024**, *17*, 1028. https://doi.org/10.3390/ma17051028

Academic Editors: Yadir Torres Hernández, Ana María Beltrán Custodio and Manuel Félix Ángel

Received: 24 January 2024
Revised: 18 February 2024
Accepted: 20 February 2024
Published: 23 February 2024

1. Introduction

Alkali-activated slag (AAS) is a type of cement-free cementitious material where ground granulated blast furnace slag (GGBS) is activated using an alkali activator (such as sodium silicate solution, namely, water glass: WG) [1,2]. Due to its effective reduction in the consumption of non-renewable resources and the emission of greenhouse gases, as well as its high strength and excellent durability, there is significant potential to use AAS as an environmentally friendly cementitious material in the manufacture of structural concrete in construction.

The number of fires in buildings and bridges has increased worldwide in recent years. The temperature of a building during a fire can reach hundreds or even thousands of degrees; the structural safety and service life of the building is reduced as a result of physio-chemical phenomena such as structural expansion, shrinkage, thermal cracks, and hydrate decomposition that occur during exposure to high temperatures. This means that the design life of the general building structure, usually 50 years, cannot be reached. If a construction is to continue to be in use after a fire, especially concrete structures in chloride environments, such as residential buildings in coastal areas, and sea-crossing bridges, its residual strength and durability should be fully understood. In terms of the residual strength of concrete after exposure to high temperatures, it was found that with the elevation of temperature, the strength of Portland cement (PC) concrete decreases [3], as does AAS concrete [4–11]. Another aspect is chloride permeability as an important evaluation of durability, and studies of this also concluded that the chloride permeability of PC concrete increases generally with the elevation of temperature as a result of the deterioration of the microstructure [12–16]. With respect to AAS, Gutierrez et al. used the ASTM C 1202 test to evaluate the charge passed of AAS mortar with an increase in temperature from room temperature to 1000 °C [17]. It was found that even up to 700 °C, the charge passed of the AAS mortar increased no more than four times compared with that at room temperature (from approximately 1600 C to 5500 C), and the values were always lower than that of PC following exposure to elevated temperature. Above 1000 °C, the charge passed of the AAS mortar increased dramatically, up to approximately 21,000 C, which was nearly twice as high as that of the PC mortar. Fu et al. used the RCM method to test the chloride penetration coefficient of AAS mortar with the elevation of temperature [18]. It was found that after 200 °C, the coefficient of the AAS mortar was not only lower than that of PC mortar but also lower than its value at room temperature. However, as a result of an increase in porosity, especially the fraction of the pores with sizes of 0.1 μm to 1 μm, the coefficient value of the AAS mortar dramatically increased afterward. After 600 °C, the value was one order of magnitude higher than the value at room temperature.

From this brief review, it can be seen that the chloride permeability of AAS after exposure to high temperatures has scarcely been studied. The existing studies have only focused on mortar and not concrete specimens, which has limited the guidance on structural construction. Furthermore, as one of the crucial factors influencing the chloride permeability of AAS concrete at room temperature [19], the effect of alkali concentration (Na_2O%) and the modulus (Ms) of WG on the chloride permeability of AAS concrete after exposure to high temperatures should also be clearly identified.

On the other hand, heating regimes, such as the heating rate, duration of exposure at the target temperature, and cooling method, have a considerable influence on the strength of AAS after exposure to high temperatures. Ma et al. [20] found that extending the duration of AAS concrete at a target temperature decreased its strength, and after 400 °C, the increase in the heating rate also increased the loss of strength. The results of Akçaözoğlulu et al. show that water cooling exacerbated the temperature difference between the inside and outside of the specimen, and serious deterioration of the mechanical properties of AAS mortars was observed [21]. Ma et al. [20] showed that water cooling caused a larger loss in the strength of AAS concrete compared to natural cooling. However, the influence of heating regimes on the chloride permeability of AAS after exposure to high temperatures has been scarcely reported.

Therefore, considering the lack of research in this respect, the current study examines the effect of alkali concentration and modulus of WG, as well as different heating regimes, on the chloride permeability of AAS concrete, which can provide a research theory for the high-temperature resistance of this type of concrete and can also provide a basis for the repair and reinforcement of buildings after fire, widening the scope of AAS concrete applications in structural engineering.

2. Experimental Program

In this study, AAS concretes with alkali concentrations (Na_2O% mass of slag) of 4, 6, and 8 and moduli (SiO_2/Na_2O molar ratio) of WG of 1.0, 1.5, and 2.0 were manufactured to study the influence of high temperatures on their chloride permeability. Furthermore, the influence of heating regimes, including heating rate (5 °C/min, 10 °C/min), duration of exposure at a targeted temperature (2 h, 1 h), and cooling method (water cooling, natural cooling), on the chloride permeability of the concretes was also studied. In order to better understand the influence of high temperature on the chloride permeability of the AAS concretes, the evolution of the pore structure of AAS pastes with temperature elevation was also studied. PC counterparts with the same binder content, water/binder ratio (W/B), and sand ratio were manufactured as a reference.

2.1. Materials

GGBS from Qujing, Yunnan, China, with a grade of S75 was used to manufacture the AAS concretes and paste specimens. Chemical composition was determined using an X-ray fluorescence spectrometer; its chemical composition is given in Table 1. Ordinary Portland cement with a strength grade of 42.5, which was manufactured by Huaxin Cement Co., Ltd. in Kunming, China, was used to manufacture the PC counterparts. Its chemical composition is also given in Table 1. From Table 1, it can be seen that the chemical composition of GGBS and PC is mainly CaO, SiO_2, and Al_2O_3, accounting for more than 80% of the total chemical composition, but the CaO content of PC is much higher than that of GGBS.

Table 1. Chemical compositions of GGBS and PC (oxide%).

	CaO	SiO_2	Al_2O_3	MgO	TiO_2	SO_3	MnO	Na_2O	K_2O	Fe_2O_3
GGBS	38.26	31.40	15.71	6.90	2.94	1.43	1.23	0.96	0.53	0.38
PC	60.45	20.44	5.59	3.41	1.18	4.30	0.06	0.32	0.77	3.21

WG from Kunmin, Yunnan, China, with Na_2O% and SiO_2% of 7.51% and 22.22%, respectively, was used as an alkali activator. NaOH produced by Fuchen Chemical Reagent Co., Ltd. (Tianjin, China), with a purity of not less than 96%, was added to the WG to change the content of Na_2O% and the Ms of the WG to the required values.

Limestone with a continuous gradation of 5–25 mm was used as a coarse aggregate (CA). Manufactured sand with a fineness modulus of 2.82 was used as a fine aggregate (FA) that was sieved before mixing to remove stone powder.

Tap water was used for mixing. A polycarboxylic polymer-based superplasticizer (from Fuchen, Tianjin, China) was used to mix the PC mixtures, whose dosage was selected as 0.08% after a trial.

2.2. Mix Proportions

AAS concretes with Na_2O% of 4, 6, and 8 and Ms of water glass of 1.0, 1.5, and 2.0 were manufactured. The binder content, which was the sum of the GGBS content and the solid content of the adjusted WG, was kept constant at 380 kg/m^3 for all mixes. A W/B of 0.45 was used for all of the mixes, and the water content in the WG was taken into account. A sand ratio of 40% was used. The mix proportions of the concretes are given in Table 2. For the purpose of comparison, PC concrete was also prepared as a reference, whose binder content, W/B, and sand ratio were kept the same as those used for the AAS concretes. Paste specimens were also prepared for pore structure analysis, where a W/B of 0.3 was used.

Table 2. Mix proportion of the concretes.

Mix No. (Na_2O%-Ms)	kg/m^3							
	GGBS	PC	WG	NaOH	CA	FA	Water	Superplasticizer
4%-1.5	344.2	-	90.0	9.1	1138.8	759.2	107.8	-
6%-1.0	336.4	-	87.9	17.5	1142.7	761.8	109.2	-
6%-1.5	328.7	-	128.9	13.0	1164.5	776.3	80.5	-
6%-2.0	321.5	-	168.0	8.6	1185.4	790.2	53.0	-
8%-1.5	314.6	-	164.4	1.5	1188.0	792.0	55.5	-
PC	-	380	-	-	1079.4	719.6	171.0	2.0×10^{-3}

2.3. Specimen Preparation

Concrete mixing was conducted following the "Standard for test method of performance on ordinary fresh concrete" GB/T 50080-2016 [22]. One day before mixing, half of the additional water was used to dissolve NaOH; the solution was then mixed with WG to adjust its modulus. In order to ensure the quality of the mix proportion, all mixes achieved similar compressive strength grades at 90 days, of 60–70 MPa, and the slump was measured; the results are provided in Table 3 and Figure 1.

Table 3. Slump and 90 d compressive strength of concretes.

Mix No.	Slump (mm)	Compressive Strength (MPa)
4%-1.5	75.4 (±6.5)	61.4 (±1.9)
6%-1.0	100.2 (±4.1)	63.6 (±7.5)
6%-1.5	119.7 (±3.2)	71.9 (±7.7)
6%-2.0	114.8 (±3.9)	70.4 (±4.7)
8%-1.5	130.3 (±5.6)	74.5 (±6.7)
PC	100.2 (±7.1)	63.1 (±3.2)

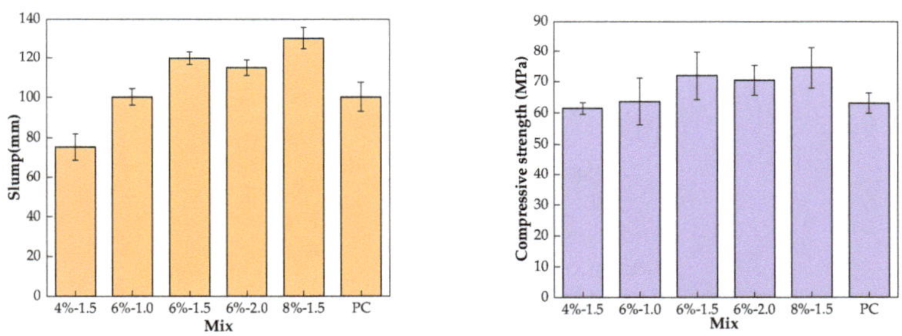

(a) (b)

Figure 1. Slump (a) and compressive strength (b).

The concrete mixture was cast into molds in two layers. After each layer casting, the mixture, with the mold, was placed on a vibration table to be vibrated until no bubbles were generated at the surface of the mixture. After demolding 24 h later, the specimens were placed in a standard curing room with a temperature of 20 ± 2 °C and a relative humidity of 95% for 90 days. The sizes of the specimens were ϕ 100 mm × 200 mm and 100 mm × 100 mm × 100 mm for the chloride permeability and compressive strength tests, respectively, and three parallel specimens were prepared for each test at each temperature. After curing, the tops and bottoms of the specimens for the chloride test were cut off and the middle parts of the specimens, with sizes of ϕ 100 mm × 50 mm, were used for the test. For the purpose of quality control, concrete specimens with sizes of 100 mm × 100 mm × 100 mm were prepared as well to test their compressive strength and slump at room temperature; the results are provided in Table 3 and Figure 1. The paste specimens were mixed in accordance with "Test methods for water requirement of normal consistency, setting time and soundness of the Portland cement" GB/T 1346-2011 [23]. After mixing, the mixture was cast into a mold measuring 25 mm × 25 mm × 25 mm, followed by vibration to make the mixture dense. The same curing regime used for the concrete was applied to the paste.

2.4. Heating Regime

The concrete and paste specimens were heated in a muffle furnace to temperatures of 200 °C, 400 °C, 600 °C, 800 °C, and 1200 °C, respectively. The heating regime benchmark was a heating rate of 5 °C/min. When the target temperature was achieved, the temperature was maintained for 2 h; then, the power was tuned off and the specimens were cooled in the furnace naturally to room temperature. In order to study the influence of the heating regime on the chloride permeability of the concretes after exposure to high temperatures, an additional regime involving a heating rate of 10 °C/min, a duration at the target temperature of 1 h, and water cooling was implemented on the AAS mix 6%-1.5 for comparison; see Figure 2. Water cooling refers to the process of taking out the specimens immediately after heating and then placing them in a sink of water until room temperature was reached.

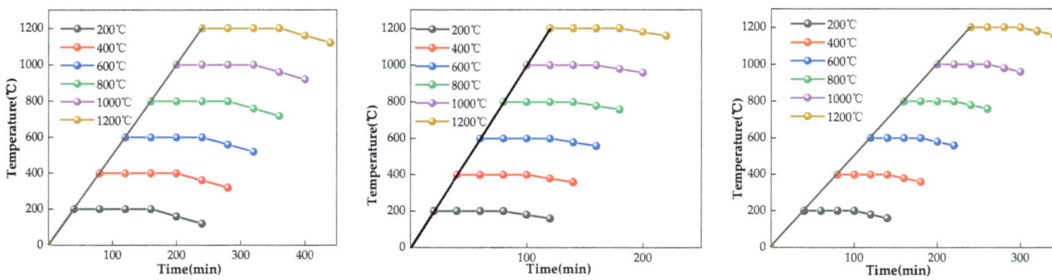

(**a**) 5 °C/min-2 h-natural (**b**) 10 °C/min-2 h-natural (**c**) 5 °C/min-1 h-natural

Figure 2. Heating regime.

2.5. Tests

In accordance with the "Standard for test methods of long-term performance and durability of ordinary concrete" GB/T 50082-2009 [24], the ASTM C 1202 method was used to evaluate the resistance of the concretes to chloride at room temperature and after temperature elevation. Before testing, the specimens were water-saturated in a vacuum. Paraffin was applied onto the surfaces of the samples except the two test surfaces. One of the test surfaces faced an upstream cell (cathode) containing 0.55 M NaCl solution, and the other surface faced a downstream cell (anode) containing 0.30 M NaOH solution. Copper electrodes were inserted into both cells. A voltage of 60 V was applied across the specimen by connecting the anode and the cathode electrodes to the negative and positive terminal leads of the power supply, respectively. The current flow was monitored during the test. The integral of the current during the first six hours of the test, namely, charge passed, was used to evaluate the chloride permeability of the specimen. A higher charge passed value usually suggests a higher chloride permeability. The result was reported as an average of three specimens.

Unfortunately, the specimens cracked significantly after 600 °C exposure, resulting in the charge passed value exceeding the upper limitation of the instrument. Therefore, the ASTM C 1202 test was not carried out on the specimens exposed to conditions of 600 °C and higher. Mercury intrusion porosimetry (MIP) was used to characterize the pore structure of the paste specimens at room temperature and after temperature elevation, where an Micromeritics Auto Pore IV 9500 model (from Micromeritics, Norcross, GA, American) was applied. The principle of the test is to press mercury into the pores of the specimen and then obtain the parameters of the pore structure by calculating the amount of mercury injected and the pressure used. The paste specimens were crushed, and a fragment from the inner part with a size of approximately 5 mm was taken and its hydration terminated by soaking in absolute ethanol for 3 days. Following this, the fragment was dried in a vacuum drying dish until the test. The PC specimen was severely damaged after 1200 °C, and as it was so loose, the MIP test could not be performed.

3. Results and Discussion

3.1. Influence of Na_2O% and Ms

The charge passed of the concrete specimens at room temperature and those after temperature elevation are given in Table 4. From the table, it can be seen that the charge passed of both the AAS and PC concretes significantly increase with the increase in the exposure temperature, indicating that the chloride permeability of these concretes increased due to heating. The porosity and pore size distribution of the corresponding paste specimens were obtained from the MIP test, and the results are expressed in Table 5 and Figure 3, respectively. From Table 5, it can be seen that with the temperature elevation, the total porosity of all mix proportions of concrete increases, which could be the reason for the increased chloride permeability. This conclusion was also put forward in the study by Fu et al. [18]. Further, from Figure 3, it can be seen that for any given specimen type, the peaks of the curves generally shift leftward with temperature, indicating that the pores have been coarsened with the increase in temperature. Through the analysis of SEM images of AAS paste after exposure to high temperatures, Ma et al. [20] also found that the pores were coarsened with the increase in the temperature, and this could also be responsible for increased chloride permeability.

Table 4. Charge passed of the concrete specimens.

Mix No.	Temperature	Charge Passed (C)
4%-1.5	25 °C	759 (±47.5)
	200 °C	2485 (±101.8)
	400 °C	5247 (±166.0)
6%-1.0	25 °C	804 (±39.0)
	200 °C	2731 (±105.8)
	400 °C	5940 (±94.9)
6%-1.5	25 °C	908 (±7.0)
	200 °C	3161 (±238.9)
	400 °C	6280 (±218.9)
6%-2.0	25 °C	1068 (±74.0)
	200 °C	3263 (±74.8)
	400 °C	6212 (±59.6)
8%-1.5	25 °C	1224 (±24.9)
	200 °C	3907 (±135.9)
	400 °C	6233 (±18.5)
PC	25 °C	827 (±38.6)
	200 °C	2431 (±140.4)
	400 °C	4801 (±258.2)

Table 5. Porosity of the paste specimens (%).

Mix No.	25 °C	400 °C	800 °C	1200 °C
4%-1.5	9.14	9.25	13.43	15.27
6%-1.0	8.57	11.66	13.01	18.31
6%-1.5	7.17	20.15	22.03	23.54
6%-2.0	7.56	18.49	31.90	41.86
8%-1.5	6.05	19.01	41.74	51.34
PC	15.54	23.18	44.58	-

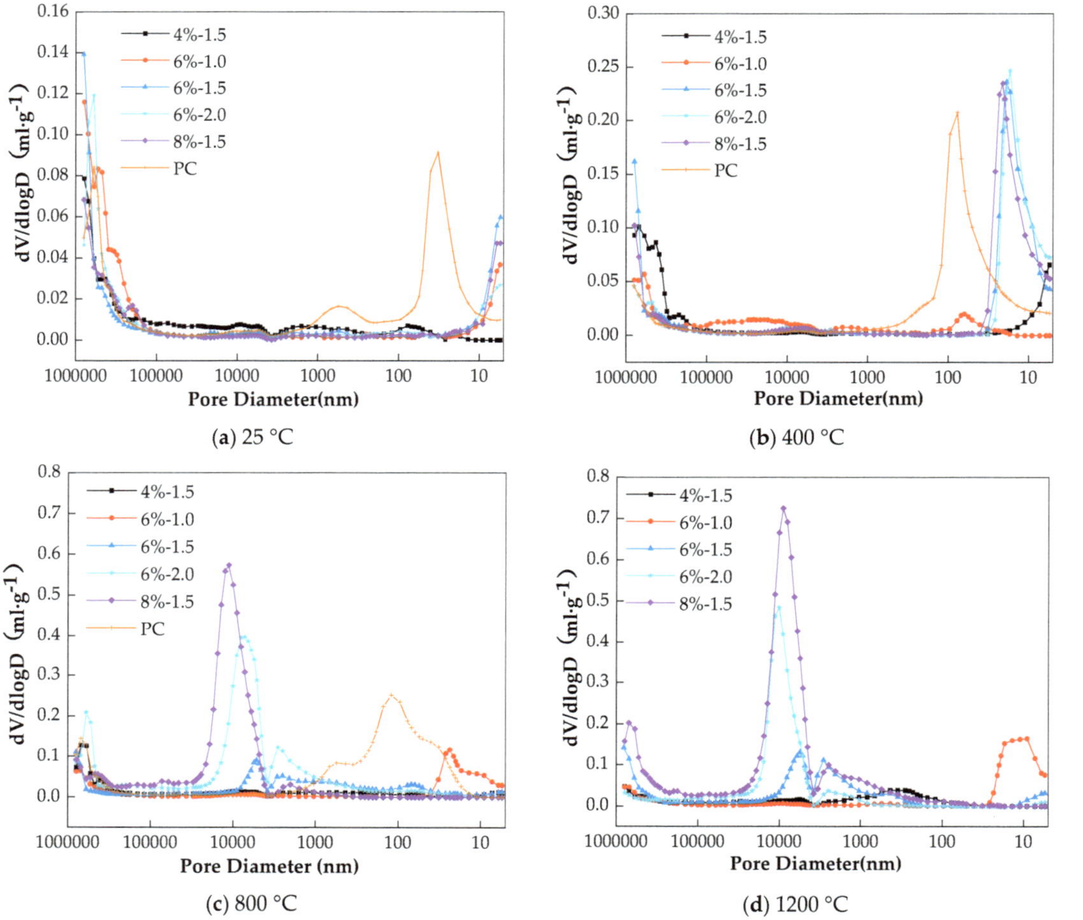

Figure 3. Pore size distribution of AAS pastes exposed to different elevated temperatures.

Compared to the PC samples, fewer pores were detected in the AAS, as shown in Table 5, and the pores in the AAS were finer at room temperature and even after 400 °C (the peaks of the AAS are on the right side of those of the PC, see Figure 3). As a result, the resistance of AAS concretes to chloride permeability should be much better than that of PC concrete both at room temperature and exposure above 400 °C. Fu et al.'s study also proposed this conclusion [18]. However, from Table 4, it can be seen that the charge passed of all AAS mixes is higher than that of PC after 400 °C. Even with room temperature storage, which is the most favorable for giving a low-permeability pore structure, the charge passed of the AAS mixes was comparable with that of PC, with some AAS mixes even exhibiting a significantly higher charge passed than that of PC. It is known that charge passed is a reflection of both the pore structure and pore solution conductivity of specimens [9]. A high charge passed value could be a result of a poorly developed pore structure and/or the high conductivity of the pore solution. It has been confirmed that AAS has higher pore solution conductivity than that of PC [19]. Therefore, the higher charge passed of the AAS concretes observed here could be caused by their higher pore solution conductivity rather than higher chloride permeability.

Similarly, as shown in Figure 4, with the increase in Na_2O%, the AAS concretes at room temperature had a lower porosity as a result of sufficient reaction [25–30], therefore, indicates that these samples should have lower chloride permeability. However, the AAS concretes with more Na_2O% yielded higher charge passed, indicating higher chloride permeability, as shown in Figure 5. In this context, it could be deduced that the pore solution conductivity dominated the charge passed rather than the pore structure, since the pore solution conductivity of AAS increased significantly with Na_2O%. After exposure to high temperatures, the pore structure of AAS became significantly damaged, as illustrated in Figure 4. Further, probably due to the AAS with higher Na_2O% being prone to shrinkage [31], the AAS concretes being exposed to high temperatures could have intensified the shrinkage. Therefore, different from room temperature, the porosity of the AAS generally increased with the increase in Na_2O% after exposure to high temperatures, which, therefore, dominated the charge passed of the AAS concrete, which seems to indicate that AAS concretes with the lowest Na_2O have the best relative performance when exposed to high temperatures. However, the experimental results of Rashad et al. [32] showed that Na_2O% had no obvious effect on the performance of AAS paste after high temperature.

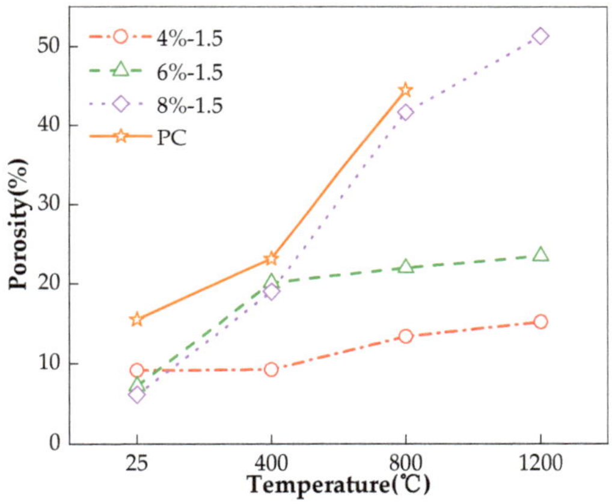

Figure 4. Effect of Na_2O% on porosity of pastes.

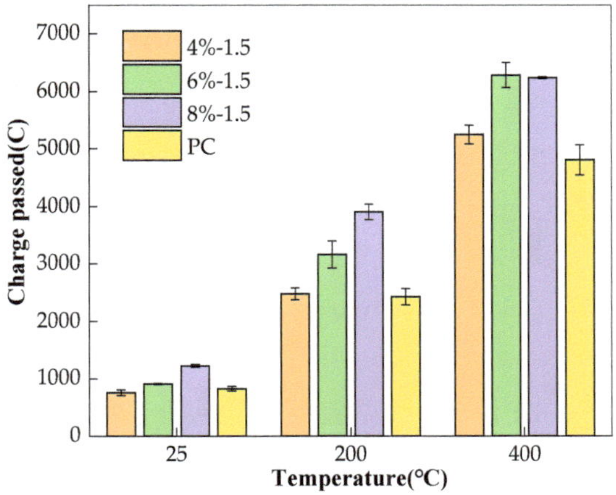

Figure 5. Effect of Na_2O% on charge passed of concretes.

This is also true for the influence of Ms on the charge passed and porosity of the AAS mixes, as given in Figures 6 and 7, respectively. At room temperature, higher Ms shortens the hydration induction period of the slag and can provide more SiO_4^{4-} ions to accelerate the production of hydration products [27], and therefore, there is lower porosity;

see Figure 6. The sample should have a lower chloride permeability. However, since the conductivity of the pore solution dominated the charge passed, the higher Ms yielded a higher charge passed, as shown in Figure 7. After exposure to high temperatures, the pore structure of the AAS concrete deteriorated; meanwhile, the higher Ms, fuller hydration reaction, and greater shrinkage of the AAS concretes led to the pore structure becoming significantly damaged, and there was a higher charge passed, as shown in Figures 6 and 7. The experimental results of Guerrieri et al. [33] also showed that Ms had no obvious effect on the performance of the AAS paste after high temperature.

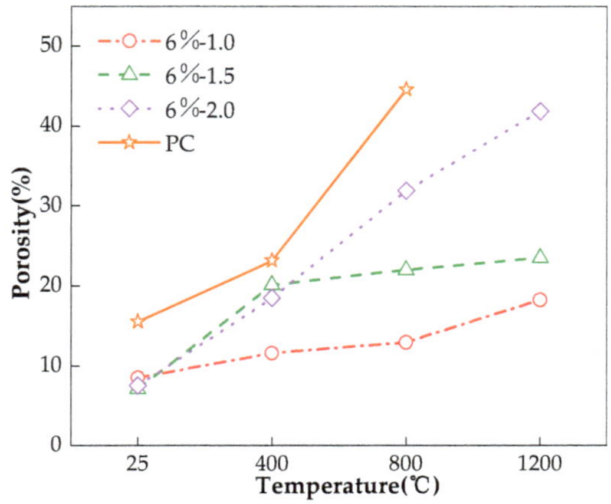

Figure 6. Effect of Ms on porosity of pastes.

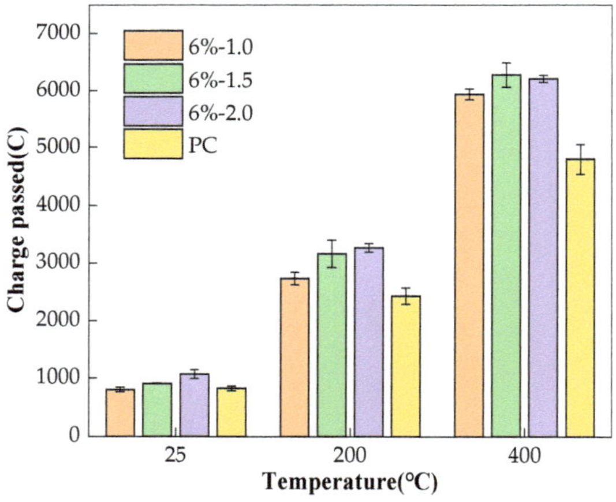

Figure 7. Effect of Ms on charge passed of concretes.

Overall, at room temperature and after high temperatures, due to the influence of the pore solution conductivity, the ASTM C 1202 method to test the chloride permeability of the AAS concretes remained unclear. However, this information may be obtained from the porosity results: AAS concrete has lower chloride permeability at lower Na_2O% and Ms when exposed to high temperatures; therefore, the influences of the activator (Na_2O% and Ms) need further investigation using a proper chloride test approach instead.

3.2. Influence of Heating Regime

The influence of the heating rate (10 °C/min, 5 °C/min), duration of exposure at the target temperature (2 h, 1 h), and cooling method (water cooling, natural cooling) on the charge passed of the AAS concretes is illustrated in Figure 8. It can be seen that the concrete that was heated faster always gave a higher charge passed, indicating a lower resistance to chloride permeability. The reason for this could be that the faster heating enlarged the difference in temperature between the surface and the internal part of the specimen. Thermal stress then formed, resulting in the deterioration of the pore structure, as shown in Figure 9, and accelerating the charge passing through; these findings are consistent with those of Fu [18]. Further, shortening the duration of the specimens' exposure to the target temperature reduced the charge passed, indicating reduced the chloride permeability. A shorter duration at the target temperature could reduce thermal damage to the structure of the specimen, leading to lower porosity [34], as shown in Figure 9. This could be the reason for the lower chloride permeability. Compared to natural cooling, water cooling sharply reduced the temperature of the surface of the specimens in a shorter time, which could have increased the temperature difference between the surface and the inner part of the specimens. Then, thermal stress could have been generated to induce cracking to allow more charge passing (see Figures 8 and 9), as illustrated in Figure 8 [20,35,36].

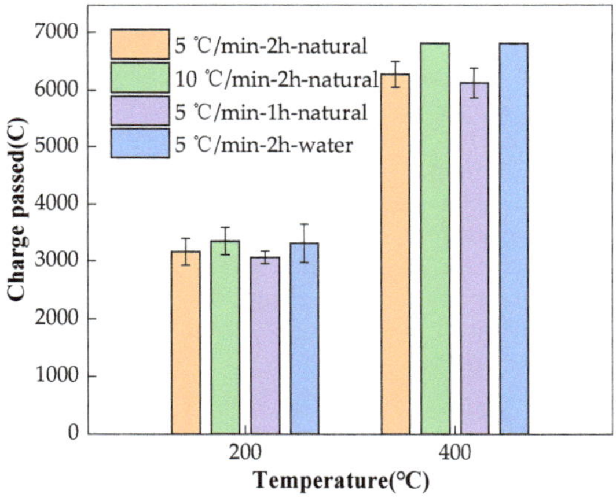

Figure 8. Influence of heating regime on charge passed of the AAS concretes.

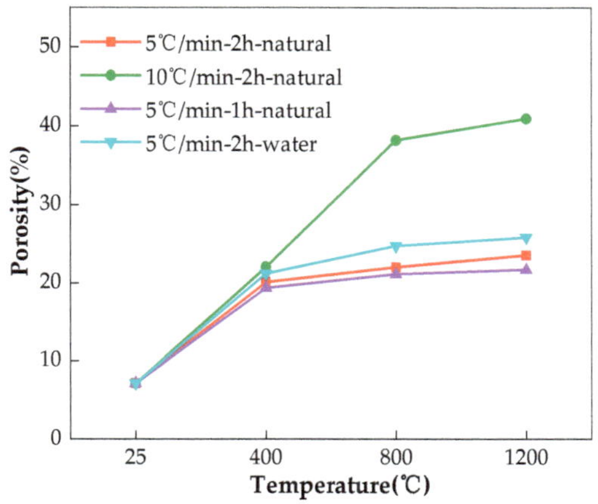

Figure 9. Influence of heating regime on porosity of the AAS pastes.

Figure 9 further shows that the heating rate was the most responsible for the thermal deterioration of the pore structure, followed by the cooling method. The duration at the target temperature had the smallest influence. These findings were especially true when the temperature increased to 800 °C and above. Though the charge passed test was only conducted up to 400 °C, similar results still began to emerge, and it is not difficult to presume that if the test could be carried out after a higher temperature exposure, the findings would be much clearer.

4. Conclusions

This study focused on the chloride permeability of AAS concretes exposed to high temperatures, particularly with regard to the effect of the Na_2O%, Ms, and heating regime on the chloride permeability. Based on the experimental program used in this study, the following findings could be concluded:

(1) The resistance of AAS concretes to chloride ingress was reduced with the increase in temperature as a result of both increased total porosity and pore coarsening.
(2) The porosity results imply that, after exposure to high temperatures, AAS concrete with a lower Na_2O% and lower Ms may exhibit lower chloride permeability; however, the charge passed of the AAS concretes examined in this study could have been caused by pore solution conductivity rather than chloride permeability, and it is therefore not suitable to apply the ASTM C 1202 charge passed method to compare the chloride permeability of concretes with different cementitious systems, such as PC and AAS, or different AAS concrete mixes. These findings should be confirmed using a proper approach instead, such as the non-steady-state chloride migration test (NT BUILD 492).
(3) Faster heating and water cooling following the high-temperature exposure in this study could have increased the temperature gradient between the surface and the internal part of the AAS concrete, aggravating the coarsening of the pore structure and then reducing the resistance of the concrete to chloride ingress. Shortening the duration of the AAS concretes' exposure to the target temperature may have reduced the thermal damage to the pore structure, thus yielding better resistance to chloride ingress.

5. Future Prospects

(1) Future studies may use a proper approach to confirm the results of this study, such as the non-steady-state chloride migration test (NT BUILD 492).
(2) After a large number of studies have been conducted, artificial neural network models could be utilized in the prediction and evaluation of the impermeability performance of AAS concretes after exposure to high temperatures.

Author Contributions: Conceptualization, Q.M. and B.Z.; investigation, Q.M., B.Z., P.L. and R.G.; resources, Q.M.; data curation, Q.M.; writing—original draft preparation, Q.M. and B.Z.; visualization, Q.M. and B.Z.; supervision, Q.M. and B.Z.; project administration, Q.M.; funding acquisition, Q.M. All authors have read and agreed to the published version of the manuscript.

Funding: This research was funded by the National Natural Science Foundation of China, grant number 52068038; the Yunnan Provincial Department of Science and Technology, grant number 202101AT070089; the Analysis and Test Foundation of Kunming University of Science and Technology, grant number 2022M20212210042; and Top-notch Innovative Talent Programme of Kunming University of Science and Technology, grant number CA23107M124A. All support is greatly appreciated.

Institutional Review Board Statement: Not applicable.

Informed Consent Statement: Not applicable.

Data Availability Statement: Data are contained within the article.

Acknowledgments: The authors express thanks to all members of the laboratory team for their technical support.

Conflicts of Interest: The authors declare that they have no known competing financial interests or personal relationships that could have appeared to influence the work reported in this paper.

Abbreviations

PC	Portland cement
AAS	Alkali-activated slag
GGBS	Ground granulated blast furnace slag
Na_2O%	Alkali concentration of activator
Ms	Modulus of activator
CA	Coarse aggregate
FA	Fine aggregate
WG	Water glass
W/B	Water/binder ratio
C	Charge passed

References

1. Rivera, O.G.; Long, W.R.; Weiss, C.A., Jr.; Moser, R.D.; Williams, B.A.; Torres-Cancel, K.; Gore, E.R.; Allison, P.G. Effect of elevated temperature on alkali-activated geopolymeric binders compared to portland cement-based binders. *Cem. Concr. Res.* **2016**, *90*, 43–51. [CrossRef]
2. Zhang, M.; Ma, Q.; Chen, Y.; Liu, Z.; Zhou, H. Thermal-Related Stress–Strain Behavior of Alkali Activated Slag Concretes under Compression. *Materials* **2023**, *16*, 3470. [CrossRef] [PubMed]
3. Ma, Q.; Guo, R.; Zhao, Z.; Lin, Z.; He, K. Mechanical properties of concrete at high temperature—A review. *Constr. Build. Mater.* **2015**, *93*, 371–383. [CrossRef]
4. Yang, Y.; Huang, L.; Xu, L.; Yu, M.; Ye, H.; Chi, Y. Temperature-dependent compressive stress-strain behaviors of alkali-activated slag-based ultra-high strength concrete. *Constr. Build. Mater.* **2022**, *357*, 129250. [CrossRef]
5. Yang, S.; Wang, J.; Dong, K.; Zhang, X.; Sun, Z. A predictive solution for fracture modeling of alkali-activated slag and fly ash blended sea sand concrete after exposure to elevated temperature. *Constr. Build. Mater.* **2022**, *329*, 127111. [CrossRef]
6. Cai, R.; Ye, H. Clinkerless ultra-high strength concrete based on alkali-activated slag at high temperatures. *Cem. Concr. Res.* **2021**, *145*, 106465. [CrossRef]
7. Dener, M.; Karatas, M.; Mohabbi, M. High temperature resistance of self compacting alkali activated slag/portland cement composite using lightweight aggregate. *Constr. Build. Mater.* **2021**, *290*, 123250. [CrossRef]
8. Behfarnia, K.; Shahbaz, M. The effect of elevated temperature on the residual tensile strength and physical properties of the alkali-activated slag concrete. *J. Build. Eng.* **2018**, *20*, 442–454. [CrossRef]
9. Manjunath, R.; Narasimhan, M.C.; Umesha, K.M. Studies on high performance alkali activated slag concrete mixes subjected to aggressive environments and sustained elevated temperatures. *Constr. Build. Mater.* **2019**, *229*, 116887. [CrossRef]
10. Ray, T.; Mohanta, N.R.; Kumar, M.H.; Macharyulu, I.S.; Samantaray, S. Study of effect of temperature on behavior of alkali activated slag concrete. *Mater. Today Proc.* **2021**, *43*, 1352–1357. [CrossRef]
11. Tran, T.T.; Kwon, H.M. Influence of activator Na_2O concentration on residual strengths of alkali-activated slag mortar upon exposure to elevated temperatures. *Materials* **2018**, *11*, 1296. [CrossRef] [PubMed]
12. Choinska, M.; Khelidj, A.; Chatzigeorgiou, G.; Pijaudier-Cabot, G. Effects and interactions of temperature and stress-level related damage on permeability of concrete. *Cem. Concr. Res.* **2007**, *37*, 79–88. [CrossRef]
13. Kodur, V.K.R.; Agrawal, A. Effect of temperature induced bond degradation on fire response of reinforced concrete beams. *Eng. Struct.* **2017**, *142*, 98–109. [CrossRef]
14. Memon, S.A.; Shah, S.F.A.; Khushnood, R.A.; Balochet, W.L. Durability of sustainable concrete subjected to elevated temperature–A review. *Constr. Build. Mater.* **2019**, *199*, 435–455. [CrossRef]
15. Touil, B.; Ghomari, F.; Bezzar, A.; Khelidj, A.; Bonnet, S. Effect of temperature on chloride diffusion in saturated concrete. *Mater. J.* **2017**, *114*. [CrossRef]
16. Mathews, M.E.; Kiran, T.; Naidu, V.C.H.; Jeyakumaret, G.; Anand, N. Effect of high-temperature on the mechanical and durability behaviour of concrete. *Mater. Today Proc.* **2021**, *42*, 718–725. [CrossRef]
17. De Gutierrez, R.M.; Maldonado, J.; Gutiérrez, C. Performance of alkaline activated slag at high temperatures. *Mater. Constr.* **2004**, *54*, 87–92.
18. Fu, H.; Mo, R.; Wang, P.; Wang, Y.; Cao, Y.; Guang, W.; Ding, Y. Influence of Elevated Temperatures and Cooling Method on the Microstructure Development and Phase Evolution of Alkali-Activated Slag. *Materials* **2022**, *15*, 2022. [CrossRef]
19. Ma, Q.; Nanukuttan, S.V.; Basheer, P.A.M.; Bai, Y.; Yang, C. Chloride transport and the resulting corrosion of steel bars in alkali activated slag concretes. *Mater. Struct.* **2016**, *49*, 3663–3677. [CrossRef]

20. Ma, Q.; Liu, Q.; Li, L.; Qing, X.; Zhang, M.; Shi, T. Influence of temperature elevation on residual compressive strength of alkali activated slag concretes. *J. Civ. Environ. Eng.* **2023**, 1–8. (In Chinese)
21. Akçaözoğlu, S.; Çiflikli, M.; Bozkaya, Ö.; Atiş, C.D.; Uluet, C. Examination of mechanical properties and microstructure of alkali activated slag and slag-metakaolin blends exposed to high temperatures. *Struct. Concr.* **2022**, *23*, 1273–1289. [CrossRef]
22. *GB/T 50080-2016*; Standard for Test Method of Performance on Ordinary Fresh Concrete. China Standards Press: Beijing, China, 2016.
23. *GB/T 1346-2011*; Test Methods for Water Requirement of Normal Consistency, Setting Time and Soundness of the Portland Cement. China Standards Press: Beijing, China, 2011.
24. *GB/T 50082-2009*; Standard for Test Methods of Long-Term Performance and Durability of Ordinary Concrete. China Standards Press: Beijing, China, 2009.
25. Bahrami, H.; Cheng, Y.P.; Bai, Y. Effect of modulus and dosage of waterglass on early age shrinkage of sodium silicate activated slag paste. In Proceedings of the International Workshop on Innovation in Low-Carbon Cement & Concrete Technology, London, UK, 21–24 September 2016.
26. Vu, T.H.; Dang, L.C.; Kang, G.; Sirivivatnanon, V. Chloride induced corrosion of steel reinforcement in alkali activated slag concretes: A critical review. *Case Stud. Constr. Mater.* **2022**, *16*, e01112.
27. Fu, Q.; Bu, M.; Zhang, Z.; Xu, W.; Yuan, Q.; Niu, D. Hydration characteristics and microstructure of alkali-activated slag concrete: A review. *Engineering* **2023**, *20*, 162–179. [CrossRef]
28. Amer, I.; Kohail, M.; El-Feky, M.S.; Rashad, A.; Khalaf, A.M. A review on alkali-activated slag concrete. *Ain. Shams. Eng. J.* **2021**, *12*, 1475–1499. [CrossRef]
29. Ji, X.; Wang, X.; Zhao, X.; Wang, Z.; Zhang, H.; Liu, J. Properties, Microstructure Development and Life Cycle Assessment of Alkali-Activated Materials Containing Steel Slag under Different Alkali Equivalents. *Materials* **2023**, *17*, 48. [CrossRef]
30. Lima, V.M.E.; Basto, P.A.; Henrique, M.A.; Almeida, Y.M.; de Melo Neto, A.A. Optimizing the concentration of Na_2O in alkaline activators to improve mechanical properties and reduce costs and CO_2 emissions in alkali-activated mixtures. *Constr. Build. Mater.* **2022**, *344*, 128185. [CrossRef]
31. Sadeghian, G.; Behfarnia, K.; Teymouri, M. Drying shrinkage of one-part alkali-activated slag concrete. *J. Build. Eng.* **2022**, *51*, 104263. [CrossRef]
32. Rashad, A.M.; Zeedan, S.R.; Hassan, A.A. Influence of the activator concentration of sodium silicate on the thermal properties of alkali-activated slag pastes. *Constr. Build. Mater.* **2016**, *102*, 811–820. [CrossRef]
33. Guerrieri, M.; Sanjayan, J.G. Behavior of combined fly ash/slag-based geopolymers when exposed to high temperatures. *Fire Mater.* **2010**, *34*, 163–175. [CrossRef]
34. Choe, G.; Kim, G.; Yoon, M.; Hwang, E.; Nam, J.; Guncunski, N. Effect of moisture migration and water vapor pressure build-up with the heating rate on concrete spalling type. *Cem. Concr. Res.* **2019**, *116*, 1–10. [CrossRef]
35. Shoaib, M.M.; Ahmed, S.A.; Balaha, M.M. Effect of fire and cooling mode on the properties of slag mortars. *Cem. Concr. Res.* **2001**, *31*, 1533–1538. [CrossRef]
36. Bi, J.; Liu, P.; Gan, F. Effects of the cooling treatment on the dynamic behavior of ordinary concrete exposed to high temperatures. *Constr. Build. Mater.* **2020**, *248*, 118688. [CrossRef]

sensors

MDPI

Article

Identification of Olives Using In-Field Hyperspectral Imaging with Lightweight Models

Samuel Domínguez-Cid *, Diego Francisco Larios , Julio Barbancho , Francisco Javier Molina, Javier Antonio Guerra and Carlos León

Department of Electronic Technology, Escuela Politecnica Superior, Universidad de Sevilla, 41011 Seville, Spain
* Correspondence: sdcid@us.es

Abstract: During the growing season, olives progress through nine different phenological stages, starting with bud development and ending with senescence. During their lifespan, olives undergo changes in their external color and chemical properties. To tackle these properties, we used hyperspectral imaging during the growing season of the olives. The objective of this study was to develop a lightweight model capable of identifying olives in the hyperspectral images using their spectral information. To achieve this goal, we utilized the hyperspectral imaging of olives while they were still on the tree and conducted this process throughout the entire growing season directly in the field without artificial light sources. The images were taken on-site every week from 9:00 to 11:00 a.m. UTC to avoid light saturation and glitters. The data were analyzed using training and testing classifiers, including Decision Tree, Logistic Regression, Random Forest, and Support Vector Machine on labeled datasets. The Logistic Regression model showed the best balance between classification success rate, size, and inference time, achieving a 98% F1-score with less than 1 KB in parameters. A reduction in size was achieved by analyzing the wavelengths that were critical in the decision making, reducing the dimensionality of the hypercube. So, with this novel model, olives in a hyperspectral image can be identified during the season, providing data to enhance a farmer's decision-making process through further automatic applications.

Keywords: hyperspectral imaging; olives; precision agriculture; machine learning; pattern recognition

Citation: Domínguez-Cid, S.; Larios, D.F.; Barbancho, J.; Molina, F.J.; Guerra, J.A.; León, C. Identification of Olives Using In-Field Hyperspectral Imaging with Lightweight Models. *Sensors* **2024**, *24*, 1370. https://doi.org/10.3390/s24051370

Academic Editor: Yiannis Ampatzidis

Received: 22 January 2024
Revised: 15 February 2024
Accepted: 19 February 2024
Published: 20 February 2024

1. Introduction

Over the past three decades, global olive production has significantly increased from 2.403 million to 5.671 million tons between 1990/1991 and 2020/2021 [1]. This production is divided into table olives and olive oil products, with the majority of these products coming from Mediterranean basin countries such as Spain, Turkey, Greece, or Italy. Olive cultivation is a significant industry in these countries, responsible for producing 32.6%, 15.1%, 14%, and 9.6%, respectively, of the world's olives [1]. Monitoring the status of olives to ensure optimal nutrition, water, and fertilizer is crucial for producing a high-quality final product. Additionally, experts have traditionally classified olives using the well-known Maturity Index (MI) through physical inspection [2–4]. This physical approach requires the destruction of the olives, picking the olives from the tree and cutting them to see their inside. An automatic technique should be developed to eliminate the need for constant monitoring by an expert in the field. Precision agriculture, a technology-driven approach to farming, has been increasingly utilized in recent years to achieve this goal.

Precision agriculture involves managing farm inputs to maximize output. In this regard, the objective of the management policy could be to increase the yield and quality or even to reduce the inputs to maximize profits [5]. Typical farm inputs include fertilizers, irrigation water, herbicides, or plant additives. The plants' requirements are monitored to determine the precise quantity and type of product to apply [6]. To utilize this approach, it is necessary to invest in sensor networks. Various types of sensors, including water flow sensors, humidity sensors, and soil sensors, must be deployed to measure the inputs [7]. Thus, these sensors must send the information using wireless communication to a platform which will manage the information. This information can be process on the edge (edge computing) or on the cloud (cloud computing). Nevertheless, a wireless sensor network or the Internet of Things (IoT) paradigm is required to acquire the information of the physical systems [8]. After aggregating the information and using support decision algorithms, the next step is to make the necessary adjustments to the system. In this proposal, the Cyber–Physical System (CPS) paradigm plays a major role [9]. A CPS digitalizes the characteristics of a physical entity through the sensor monitoring of its variables. The information is then sent to the digital world and processed using specific models and algorithms. The decision criteria are applied to the control units to modify the entity's characteristic variables and change its operational state. Therefore, remote sensing is a feasible technique for monitoring fields.

Three sources of information are available in the remote sensing paradigm: satellites, aerial devices such as unmanned aerial vehicles (UAVs), and in-field devices [10]. Remote sensing can be used to manage crops and develop the precision agriculture paradigm. The use of satellites is particularly extensive and has been employed since the launch of Landsat 1 in 1972. The most recent Landsat satellite, Landsat 9, was launched in 2021 [11]. Landsat is a satellite program of NASA, but there are also missions in other countries, such as the Sentinel program in Europe [12]. These satellite missions collect data that are available for the community, such as VIIRS [13], MODIS [14], or HSL [15], which contain a collaboration between Landsat and Sentinel. The monitoring of natural environments and some crops is available through these satellite missions. The first satellite mission had RGB imaging capabilities, but nowadays a wide bandwidth of the spectrum is available for using with these tools. Using these data, several indicators such as NVDI, WSI, and red-edge slope are defined to evaluate and correlate the crops' properties. This approach is used to obtain general field information, but image resolution is not high enough to identify individual olives. In our case, other approaches with higher resolution must be considered.

The use of UAVs or drones is a common approach in precision agriculture [16]. These platforms improve data resolution and quality [17], and reduce the waiting period for information compared to satellite data [18]. In addition, flights can be organized and programmed to acquire all necessary field details. However, this approach has several issues; for instance, it consumes a lot of power, which calls for the development of low-consumption portable systems. Although the autonomy of UAVs has increased in recent years, it may still be necessary to stop and change the battery of the system when monitoring a crop and acquiring information [19]. Additionally, weather conditions can affect the flight plan, and wind can impact data acquisition. The resolution of the UAV is limited, making it insufficient for certain applications as it is in our case. Monitoring fruit, in particular, can be challenging due to the high resolution required and the obstruction caused by canopies.

A methodology exists for analyzing fields directly on the surface. This is called in-field monitoring, which involves using cameras or hand-held sensors to register the crops [20,21]. The purpose of this approach is to gather the required data by taking the equipment directly to the field of interest. This ensures that resolution is not a problem when selecting which area of the crops to monitor. This technique is particularly useful for applications where it is not possible to detect and obtain information about fruits by flying UAVs over the canopy. However, it is important to note that this technique can be costly due to the equipment required and the need to deploy cameras or sensors in the field.

After monitoring the field and crop, the next step is to identify areas of interest or the focus of the study. Classical image processing techniques can be used to identify regions of interest in images. For example, patterns can be used to identify olive trees in satellite images [10,22,23]. Using several indicators from satellite images, it is possible to provide decision support to farmers. In addition to classical methods, patterns can be searched for in the images. This is particularly useful in specific applications such as fruit tree plantations, where the trees follow a specific pattern. However, in cases where there is no discernible pattern, such as in our case, these methods are not appropriate. In addition to these solutions, it is possible to use artificial intelligence to identify objects in images [24,25]. This approach involves processing the images to find a pattern that has been previously identified in training. For RGB images, there are solutions based on Artificial Neural Networks (ANNs) that are trained with the image and perform several operations to identify patterns of interest. Convolutional Neural Networks (CNNs) are an evolution of Artificial Neural Networks (ANNs) that are specifically designed for image processing [26]. They use a kernel to reduce the dimension of the image and identify patterns. Some state-of-the-art models based on CNNs include Yolo [27,28], ResNet [29], and VGG [30], among others. However, these models are often criticized for their large size and slow processing time [31]. For this purpose, we searched for a model that takes into account both size and execution time. Moreover, images that have a similar background color to the detected object may not provide sufficient information. This is especially problematic when detecting green olives against a green background caused by leaves. To address this chromatic issue, hyperspectral technology is used to gather more spectral information. Multiple wavelengths in the visible and NIR spectra are used to evaluate the quality of the olive, instead of relying solely on the three bands that contain red, blue, and green values. This approach provides additional information [32].

In this type of model, a pretrained model is utilized and then the internal weights of the network are adjusted for the specific application through transfer learning [33]. Additionally, there are transformer-based approaches for remote sensing applications in image classification [34,35], as well as other approaches for regression [36,37], rather than natural language processing. These models are modifications of the attention-based model presented by Vaswani [38]. These models exhibit high performance, but they are computationally intensive. Our objective is to develop a lightweight model capable of identifying olives in images. Since the monitoring will be conducted directly in the field, the deployed processing units must be able to load and execute inferences from the model on the edge. For this purpose, spectral information is utilized along with machine learning models such as Decision Trees, Logistic Regressors, and ANN with a MultiLayer Perceptron. The MultiLayer Perceptron is a classical model with a few hidden layers and neurons [39].

The main contributions of the paper are as follows:

(1) A lightweight classifier is proposed for the real-time identification of olives in images taken in olive groves, which is a problem that has not been extensively addressed in the literature.
(2) Spectral band analysis is used for wise dimensional reduction. A spectral band analysis was performed to identify the critical bands for the identification problem through wise dimensional reduction using wavelengths in the visible and NIR spectra.
(3) Comparison with other state-of-the-art techniques. We compared the state-of-the-art techniques for real-time in-field object identification using a lightweight classifier.

2. Materials and Methods

Hyperspectral images of olives on the tree were captured periodically throughout the olive season using a non-invasive, in-field approach. with the Specim IQ model from Specim [40]. This device permits the acquisition of images with 512 $\times$ 512 pixels and 204 bands, covering a range of 397 to 1004 nm of the spectrum, with a resolution of 7 nm for each band. Apart from the hyperspectral camera, a 5-megapixel RGB camera is available in the device to acquire the scene information. The images of the olives were acquired at the end of the fruit set phase, and the growth process was recorded until the olives reached the maturity stage, characterized by their outer color turning to purple. The study is focused on green olives, so the monitoring stage concludes with the color change that takes place during the ripening phase. The monitoring period was conducted from May to September, with images taken at a distance of 1 to 2 m from the tree and avoiding direct sunlight. Specifically, they were collected between 9 a.m. and 11 a.m. UTC during the years 2021 and 2022. The olive grove is located in Andalusia, southern Spain, near the city of Seville. Its GPS coordinates are 37.393999, −6.122260.

A white reference tile made of PolyTetraFluoroEthilene (PTFE) was included in all images to standardize illumination conditions. The tile measures 20 cm × 20 cm. PTFE exhibits a uniform response across the electromagnetic spectrum in the 400 to 1000 nm range [41], making it detectable by the camera.

2.1. Dataset and Data Management

The dataset consists of hyperspectral images, denoted by $\mathcal{H} \in R^{512\times512\times204}$, which are referred to as hypercubes in the literature. The hypercube is formed by the two coordinates of an image and a third dimension that has the information of the bands. While an RGB image has 3 bands, hyperspectral images have more bands with reflectance values of the wavelength of the spectral. In this paper, we denote the horizontal coordinate of the image as x, and the vertical coordinate as y, while the wavelength of the image is denoted as z. So, regarding these criteria, we have 3 vectors: $x \in R^{512}$, $y \in R^{512}$, and $z \in R^{204}$. As described before, the images are taken from May to September, creating a database of 400 hyperspectral images during the season of 2022. We evaluated the images at pixel level, to identify if the pixel belongs to an olive or not, only by its spectral signature. In this regard, the models that we discuss in this paper are not aware of the context. Each image contains 204 values per pixel and has a resolution of 512 × 512 pixels, resulting in a total of 53.477×10^6 values per hyperspectral image that need to be processed.

The olives in the images were manually classified, and a mask file was generated to indicate the classes. The two classes identified in this novel approach, which was conducted under non-controlled lighting conditions, are 'olive' and 'non-olive', denoted by 1 and 0, respectively. The resulting mask image, $\mathcal{M} \in R^{512\times512}$, is used as the target class, so, for a specific pixel of the image, we have:

$$i, j, k \in \mathbb{N} \text{ and } i, j \in [0,\ 511] \text{ and } k \in [0,\ 203]$$

$$\mathcal{H}_{i,j} = \{z_0, z_1, z_2,\ \ldots,\ z_{202}, z_{203}\},\ \text{where } z_k \in \mathbb{R}$$

$$\mathcal{M}_{i,j} \in \{0,\ 1\}$$

The output is an artifact we denoted as $\mathcal{O} \in R^{512\times512}$, which contains elements that belong to two classes resulting from binary classification, $\mathcal{O}_{i,j} \in \{0,\ 1\}$. Three datasets were created for the algorithm using images from the 2022 season. The training dataset contained approximately 70% of the images, while the validation dataset contained 20% of the total images. The validation dataset images were also taken during the 2022 season, the same season as the training. Finally, the last 10% of the images were from a different season than the training set, which we refer to as the cross-validation dataset. This dataset contained hyperspectral images from the 2021 season, selected randomly to ensure a representative sample. We also made sure to include samples from different days for both training and validation. All images acquired during the 2021 season were used for testing, despite their lower quantity compared to the 2022 season.

2.2. Evaluation Metrics

As we sought to develop a classification model, the analysis of model performance would be performed using the metrics extracted from the confusion table. Precision, accuracy, and F1-score [42] were selected as the evaluation metrics to characterize the classification performance. Additionally, we considered size and inference time to determine which model performed better with fewer resources or less time. As is commonly understood, the confusion matrix identifies four categories: True Positive (TP), False Positive (FP), True Negative (TN), and False Negative (FN) [43]. To measure resource usage, we compared model inference time in milliseconds and model size in bytes.

3. Framework Presentation

The framework we present consists of three sections: data acquisition and preparation, training loop, and validation and testing. These sections are illustrated in Figure 1, which shows the workflow of information. The arrows indicate the flow of information between points. The dataset is generated offline, and this process is used to train and validate algorithms and models.

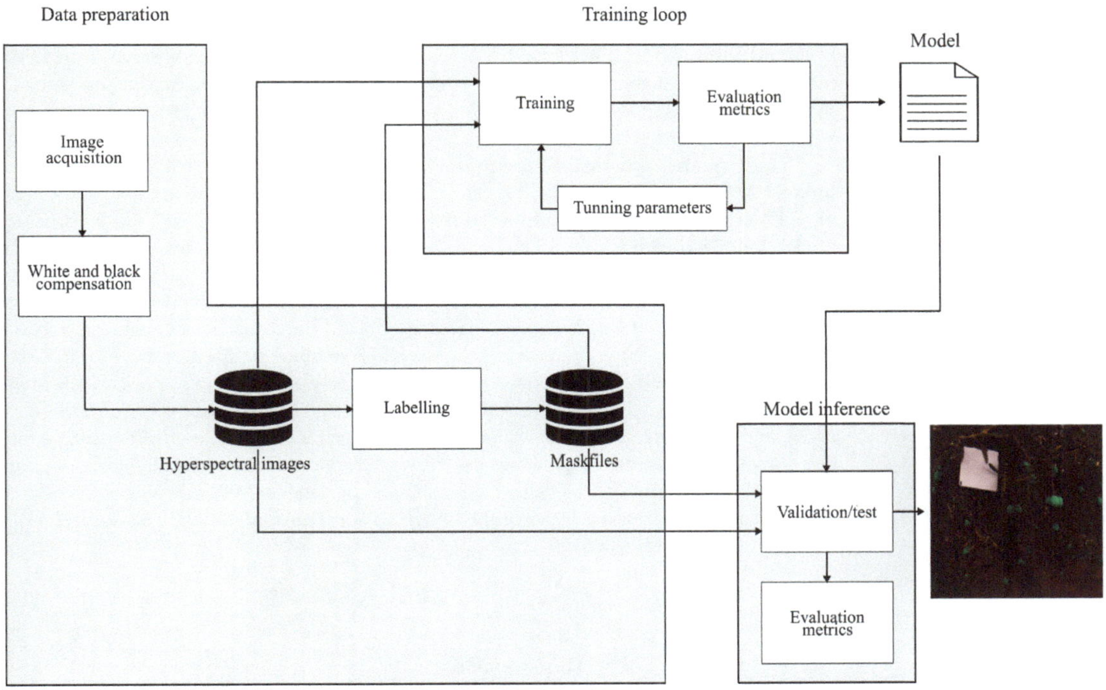

Figure 1. Framework of the proposed model for the identification of the Gordal green olive variety in hyperspectral images.

3.1. Data Acquistion and Preparation

The image was preprocessed to compensate the light conditions and then processed with the algorithm. This process was performed by using the white and black reflectance. These data were available in any image since there was a white PTFE plate present in all images. The compensation expression is the following:

$$\mathcal{H}_{norm_{i,j,k}} = \frac{\mathcal{H}_{i,j,k} - w_k}{w_k - b_k} \; \forall\, i,j,k \in \mathbb{N}_0 \wedge k < 204 \wedge i,j < 512$$

where $w, b \in R^{204}$, $\mathcal{H}$ is the raw hyperspectral image, $\mathcal{H}_{norm}$ is the compensate hyperspectral image, w is the mean reflectance values in the white reference plate, and b represents the mean black values of the camera with the objective closed, acquired before the image acquisition. $\mathcal{H}$ and $\mathcal{H}_{norm}$ are hypercubes, as described in Section 2.1.

After compensation, the image was stored in a database for the labeling process. The olives in the hyperspectral images were manually identified during this process using image processing software developed by our research team. The software is similar to state-of-the-art solutions for image labeling, such as LabelMe (Release3.0, v5.4.1) for Yolo Images [44]. The software is designed for hyperspectral images following the ENVI standard, while

LabelMe is intended for RGB images. It generates a mask file associated with each processed image, which is stored in a database.

3.2. Proposed Models and Training

In this section, several models were used to train the images and their associated mask files. The models were evaluated using normalization techniques to tune the training hyperparameters and improve performance based on evaluation metrics. As the model needs to process a large amount of information, we looked for lightweight machine learning-based classifiers. In this regard, models such as Decision Tree (DT), Logistic Regression (LR), and Support Vector Machine (SVM) as a classifier, Random Forest (RF), and clustering models such as k-means or Self-Organizing Maps (SOM) are suitable for our problem. The SVM algorithm used was a C-SVM using a squared L2 penalty and a squared exponential kernel in the isotropic variant using a length score of 0.004902.

These models were trained with the hyperspectral images, extracting the region of interest labelled and reducing the information per image. In this regard, the dataset $\mathcal{S} = \{X, y\} / X \in R^{N \times 204} \wedge y \in R^N$ where X are the pixels identified as olives or non-olives, and $\boldsymbol{y}$ is the binary target value of the mask, with 1 being an olive and 0 a non-olive. N is the number of pixels that composed the dataset.

As shown in Figure 1, a normalization process was carried out for the model execution. Normalization can be performed in various ways, and the decision of classification can be improved by using normalization techniques [45]. We followed the analysis of the normalization techniques used with hyperspectral images and applied them to our classifiers with the normalized data. The normalization techniques used in this study were Z-norm, Bandnorm, and Max-min. Z-norm involved calculating the mean and standard deviation of the reflectance values across all bands in the image. Bandnorm involved calculating the mean and standard deviation of the values for each band per image. Max-min involved using the maximum and minimum values of the image without normalization. The normalization expressions are as follows:

$$X_{zscore_{i,j,k}} = \frac{X_{i,j,k} - \mu}{\sigma} \; \forall \, i,j,k \in \mathbb{N}_0 \wedge i,j < 512 \wedge k < 204$$

$$X_{maxmin_{i,j,k}} = \frac{X_{i,j,k} - minval}{maxval - minval} \; \forall \, i,j,k \in \mathbb{N}_0 \wedge i,j < 512 \wedge k < 204$$

$$X_{bandnorm_{i,j,k}} = \frac{X_{i,j,k} - \mu_k}{\sigma_k} \; \forall \, i,j,k \in \mathbb{N}_0 \wedge i,j < 512 \wedge k < 204$$

where X is the hypercube of the hyperspectral image, $X \in R^{512 \times 512 \times 204}$; $\mu, \sigma \in R$ are the mean value and the standard deviation of the whole hypercube. The variables minval and maxval are the minimum and maximum values of the hypercube, $minval \in R : \nexists \, x \; in \; X \mid x < minv$ $maxval \in R : \nexists \, x \; in \; X \mid x > maxval$. Then, $\mu, \sigma \in R^{204}$ are the mean and standard deviation of the image for each of the bands of the image.

Several iterations were performed using different models, tuning their hyperparameters to achieve the best results. Each of these well-known models has its own advantages in terms of size, inference time, and classification performance. The criteria we established for selecting the model was the one that achieved high classification performance with the lowest cost. We defined cost as the time and size required to execute the model. The cost associated with the model increases as its size or execution time increases. We defined three classes in the image: olives, non-olive, and white reference. The white reference class is present in all images and has a known signature pattern, which reduces variability with the other classes. The decision is made for each pixel of the image. Therefore, each of the 512×512 pixels is evaluated with the model and assigned to one of three classes: olive, non-olive, or white.

3.3. Test and Validation

After training and saving the model, it was evaluated using validation images that were not seen during training. The evaluation metrics were calculated for each model and compared to determine their performance. The validation dataset consisted of hyperspectral images from several days of the same season as the model's training. Additionally, to assess the models' performance across multiple seasons, they were tested using hyperspectral images from a different season. This is referred to as cross-validation data. The results, including evaluation metrics for both the validation and test datasets, are presented in the results section.

4. Results and Discussion

4.1. Model Results

In this section, we present the results of the models comparing the classifiers results with the data normalized using the techniques described in Section 3.2. Table 1 contains information about the evaluation metrics of precision accuracy and F1-score. The F1-score is the best metric for comparing the models and selecting the best performance. Accuracy is a simple metric that can cause problems with classification in situations of imbalanced data. Precision only uses the ratio of true cases and is very sensitive to data distribution. Instead, the F1 metric should be used as it provides information on the functionality of the classifier and is less sensitive to imbalanced data than accuracy and precision [42]. For this study, three classes were defined for the classification: white reference, olive, and non-olive. It is worth noting that the number of non-olive pixels is significantly higher than that of olive and white reference pixels in every image. Additionally, we compare the data from one season to another season dataset. Table 1 includes information on the cross-validation dataset, labeled as the 'test' column.

Table 1. Evaluation metrics of the validation and test metrics of the models using normalization techniques with the validation and test dataset.

Normalization	Classifier	Precision (%)		Accuracy (%)		F1-Score (%)	
		Validation	Test	Validation	Test	Validation	Test
Bandnorm	LR	98.598	98.988	98.791	99.232	98.641	99.139
	DT	95.777	95.381	96.209	95.142	95.731	94.431
	SVM	98.361	99.008	98.521	99.203	98.335	99.105
	RF	93.481	85.647	94.500	87.569	93.872	86.606
Maxmin	LR	99.225	99.056	99.188	98.835	99.084	98.680
	DT	97.065	96.460	96.613	95.175	96.120	94.367
	SVM	98.967	98.181	99.004	98.218	98.877	97.987
	RF	94.252	96.306	93.772	95.090	92.793	94.273
Z-score	LR	98.865	87.855	99.006	89.534	98.882	88.867
	DT	97.362	93.765	96.855	93.786	96.396	92.882
	SVM	98.720	95.197	98.861	96.068	98.719	95.621
	RF	97.159	92.495	96.246	93.025	95.660	92.081
No normalization	LR	98.949	98.298	99.112	98.501	99.002	98.314
	DT	97.605	96.182	97.777	95.792	97.493	95.173
	SVM	98.348	98.741	98.535	98.831	98.352	98.684
	RF	93.524	85.916	94.518	87.759	93.887	86.889

As it can be seen, the results of the models are very promising, with results in the classification higher than the 95% in F1-score. This metrics were calculated from a multiclass confusion matrix, averaging the precision in the classification of the three classes. The white reference class validated the image, indicating that it was well taken. The non-olive and white reference class pixels were merged into a single class called non-olive.

Two examples of the confusion matrix of Table 1 were obtained and can be seen in Figure 2. Figure 2a stands for the results of the LR classifier using the Bandnorm technique in the image, while Figure 2b displays the results of the LR classifier without using normalization. Figure 2c,d show the confusion matrix of the SVM with and without normalization, respectively. Furthermore, Figure 3 presents the results obtained from the test dataset.

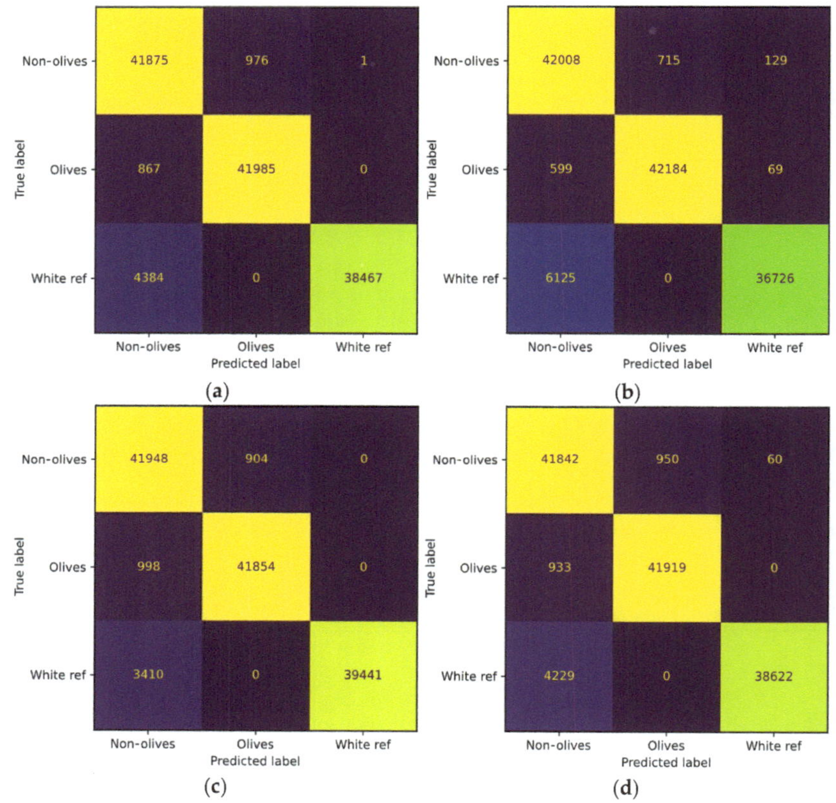

Figure 2. Confusion matrix with the validation dataset. (**a**) LR classifier with the Bandnorm technique. (**b**) LR model without normalization. (**c**) SVM model with the Bandnorm technique. (**d**) SVM model without normalization.

The data suggest that adjusting image values after white compensation did not significantly enhance classification. The Z-score evaluates the variation of the signatures around a mean and the variation of the values of the images. Then, the bandnorm technique normalizes over each of the bands, so the information of different values of illumination produces major changes which causes an increase in the detection failures. Therefore, illumination conditions have a great impact on this kind of normalization technique. The normalization of the bands affects the values of each band, which can have a negative impact on images with varying illumination conditions. Depending on the dataset, the normalization technique should be evaluated using hyperspectral images [45].

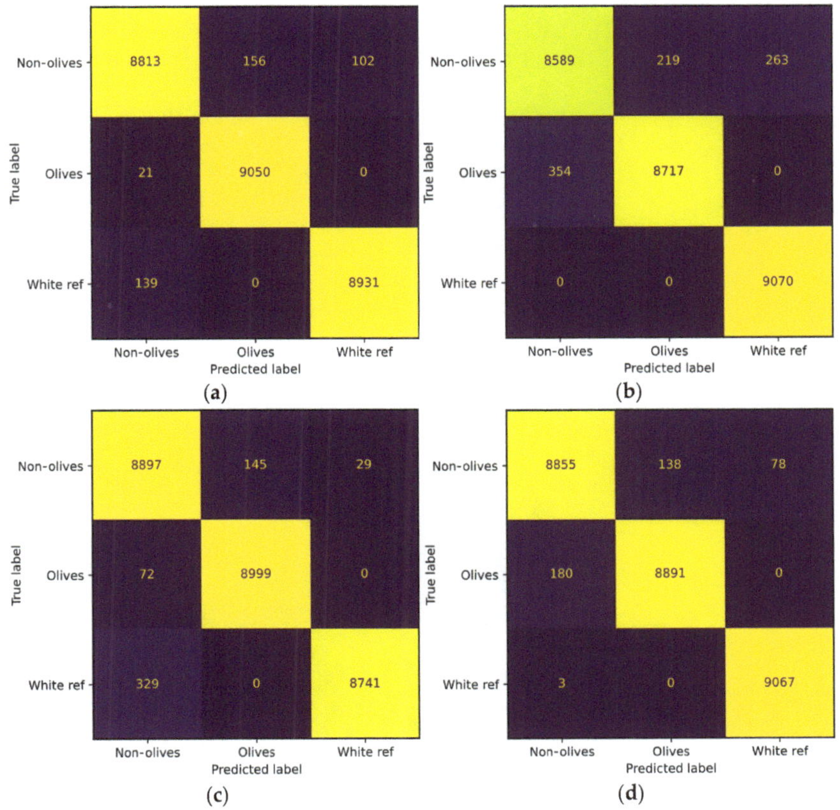

Figure 3. Confusion matrix with the test dataset. (**a**) LR classifier with the Bandnorm technique. (**b**) LR model without normalization. (**c**) SVM model with the Bandnorm technique. (**d**) SVM model without normalization.

After analyzing the classification performance, we conducted an analysis of the cost of the classifier. To do this, we evaluated each classifier during the inference of the validation dataset on the same machine and with the same input. We compared the performance of the classifiers and evaluated their relative time with respect to the lowest value. Table 2 displays the results, indicating that the LR classifier is the lightest and fastest. The system used to perform the inference was a personal computer with an Intel Core i5-8265U CPU with four cores and eight threads working at 1.60 GHz with 8 GB of Ram and an integrated graphical unit.

Table 2. Comparison of the evaluation metrics of the models without the normalization technique.

Classifier	F1-Score (%)	Inference Time (ms)	Size (KB)
DT	95.17	164.63	245.97
SVM	98.35	720,248.31	20,841.96
LR	99.00	231.91	5.51
RF	93.89	4855.78	739.96

The results indicate that the SVM classifier is the slowest and heaviest model, while the RF classifier is lighter and faster than the SVM but not as fast as the DT and LR classifiers. In terms of classification performance, the RF model is the worst, while the DT and LR models are both better suited for olive identification due to their low resource usage and high classification performance. The DT classifier is the fastest, although its F1-score is lower than that of the SVM and LR models. The LR model, despite being 44 times smaller than the DT model, has a 31% lower inference time. However, it shows a better F1-score with almost a 4% improvement compared to the DT model. Therefore, for this purpose, the LR model is the best option to use. It is a lightweight model, only 5.5 KB, with an inference time of 231 ms for each hyperspectral image, and a 99% classification F1-score. As an example of this model's inference, Figure 4a,b display the RGB image converted from a hyperspectral image. The model classifies into three distinct categories: olives, non-olives, and white reference. The non-olives and white reference categories are combined and colored red for visualization purposes, while the olive category is colored green. Figure 4b shows the result of overlaying this mask image onto Figure 4a. The model was tested using data from different seasons. Open data on table olives from the 2021 season are available on the internet [46]. The model was also evaluated to determine if olives from different seasons can be identified. The results of the model can be seen in Figure 5a,b. Furthermore, Figure 6 shows several images of the validation and test datasets overlapped with the classifier's segmentation.

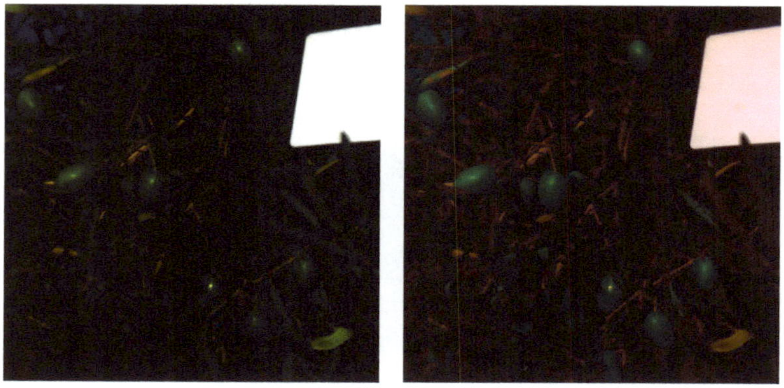

(a) (b)

Figure 4. Images of the hyperspectral image converted to RGB (**a**) and the overlapped image with the model results (**b**) of the validation dataset.

(a) (b)

Figure 5. Images of the hyperspectral image converted to RGB (**a**) and the overlapped image with the model results (**b**) of the test dataset, cross-season.

Figure 6. LR classifier output segmentation overlapped with the images of the validation and test dataset.

4.2. Critical Band Evaluation

Although the model is small, working with hyperspectral images requires high computing capabilities and hyperspectral sensors. To reduce computing requirements and the number of bands used, an analysis of the most significant bands was carried out, and the contribution of each band was measured for the LR classifier without normalization. This process involves calculating the difference between a baseline metric and a permutation of features. The metrics were calculated by permuting a column of features from the dataset 1000 times, with each feature being changed. Figure 7 depicts the contribution of the T1 significant bands for the LR model. For visualization purposes, only one third of all bands, ranked in order of importance to the classification, were plotted. These values were randomly selected to determine if there are significant differences between bands.

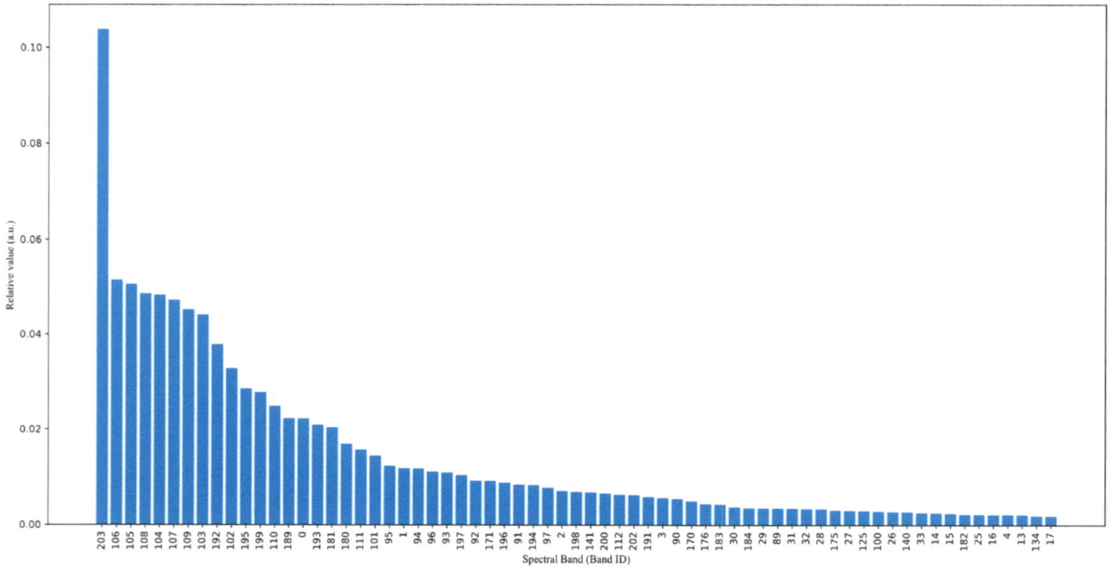

Figure 7. Importance of the 33.33% most significant bands using the LR model without normalization.

Figure 7 shows that only a few bands contribute significantly to the normalization process. Therefore, we can infer the information using only the upper bands without significantly decreasing the prediction performance. This reduces the model size and the required data for information inference. To test the hypothesis, the models underwent training with a reduced feature set. The original 204 bands were reduced to one third, one quarter, one tenth, and the fifth percentile of the most significant bands. The evaluation metrics and cost were calculated for the models using 33%, 25%, 10%, and 5% of the most significant bands, denoted as P33, P25, P10, and P5 in this paper. The model used for this experiment was the LR model without normalizing the data, as depicted in Table 3. Table 3 displays the results, indicating that the F1-score increases with the number of bands. The SVM model had the least reduction in F1-score, but also the longest inference time, as previously discussed. On the other hand, the LR model remains the smallest and has the shortest inference time when only the top 5% most significant bands are used as input. These bands are 204, 106, 105, 108, 104, 107, 109, 103, 192, and 102, which, respectively, stand for the following wavelengths in nm: 1003.58, 705.57, 702.58, 711.56, 699.60, 708.57, 714.55, 696.61, 966.55, and 693.62.

Table 3. Comparison of the evaluation metrics with the different models reducing the feature set of the data without normalizing.

Percentile	Classifier	F1-Score (%)	Inference Time (ms)	Size (KB)
P33	DT	97.094	87.063	270
	LR	98.839	109.156	2.29
	SVM	98.462	470,580.587	6500
	RF	94.287	4854.568	756
P25	DT	97.423	89.960	281
	LR	98.641	94.093	1.92
	SVM	98.330	446,215.157	5110
	RF	93.960	4711.796	768
P10	DT	96.193	81.218	367
	LR	98.079	83.616	1.19
	SVM	98.106	515,363.557	2840
	RF	92.933	5137.557	770
P5	DT	94.375	85.020	462
	LR	98.034	59.558	0.982
	SVM	98.188	762,950.853	2460
	RF	91.202	5259.811	766

In this regard, it is possible to reduce the model size to 1 KB and achieve a 98% F1-score for olive detection using 10 bands. The same images used in Figure 6 were also tested with the P5 LR model to observe the impact of a 1% reduction in F-score, as shown in Figure 8. Furthermore, precision agriculture approaches utilize visible images to detect objects such as tomatoes, apples, and strawberries. A visible image is a three-dimensional artifact that represents the color space. By using visible and NIR wavelengths, we can obtain more information about the objects present in the images. According to the analysis carried out in Table 3, these methods should be efficient in terms of time and size. However, as shown in Table 4, a comparison with state-of-the-art techniques was conducted. This table compares the approach of other researchers to acquiring in-field images using deep learning techniques. As far as we know, there are no studies on the identification of olives directly on the field, so other crops were used to compare our model. The aim of our research is different as we focus on real-time applications and use machine learning models to reduce size and computing time. In our case, identification is performed at the pixel level, while other researchers create sub-images using CNNs to classify the presence of olives.

Table 4 demonstrates that our model has been reduced in size, processing time, and F1-score. Although these results are promising, this novel approach should be tested under different conditions and seasons, with varying light conditions. To improve the robustness of this novel model, it is necessary to increase the variability of images under different light conditions, as a reduction in light conditions can increase the probability of misclassification. As exposure time decreases, the measurement range also decreases. Objects in the background may not provide accurate information due to the shorter exposure time.

Table 4. Comparison of the evaluation metrics of the models with state-of-the-art models.

Object Detection	F1-Score (%)	Parameters
InceptionV3 *	83.89	23.90 M
Inception-ResNetV2 *	84.01	55.90 M
Our model	98.03	30.00

* Reference of these results: [24].

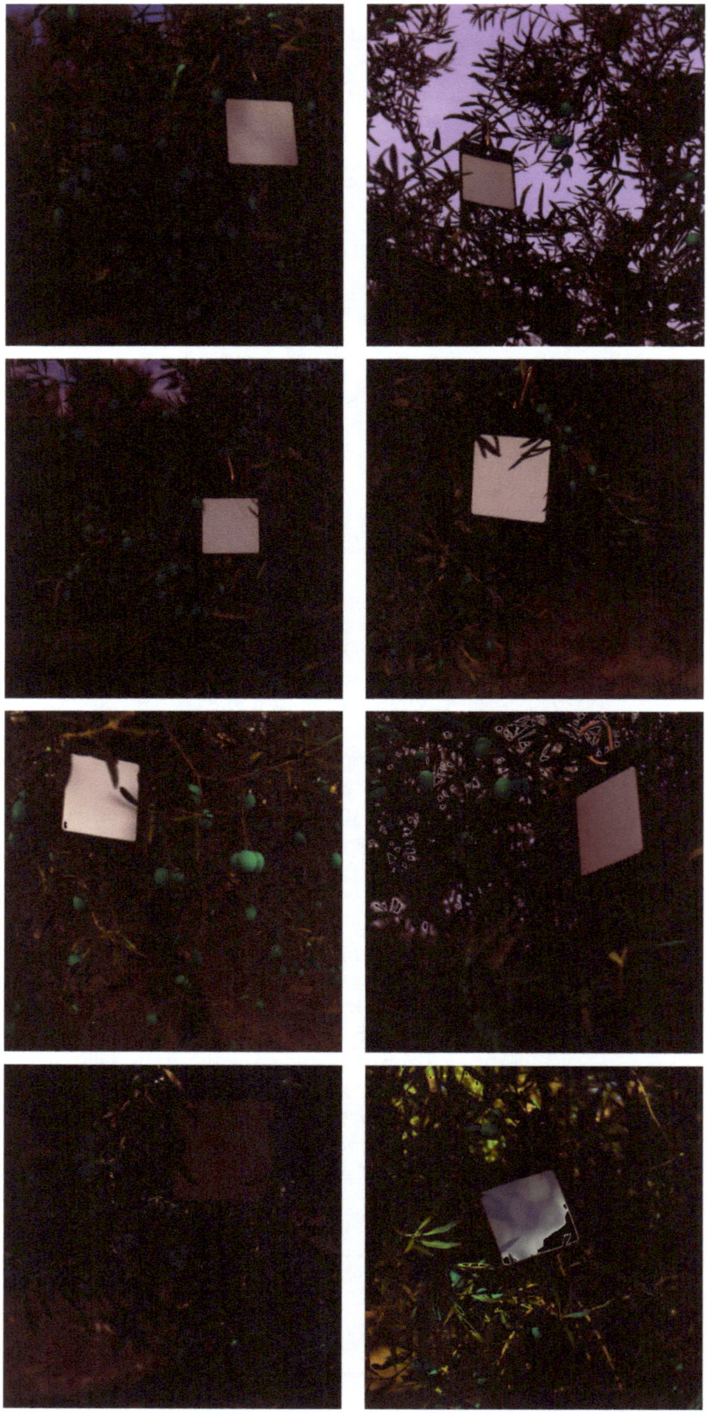

Figure 8. LR classifier output segmentation overlapped with the images of the validation and test dataset using 5% of the spectral bands.

5. Conclusions

In this work, a framework for identifying olives in hyperspectral images was presented. The hyperspectral images were acquired directly in the field without controlled lighting conditions. The reflectance values were compensated using a white PTFE reference, and the image was normalized before the inference of the ML model. The model generated a binary classification output at the pixel level, which was used to segment the image into olive and non-olive categories. A comparison of the DT, LR, SVM, and RF models was conducted to evaluate their performance using normalization techniques. Precision, accuracy, and F1-score metrics were compared. The models achieved a success rate of around 98% in classifying the presence of olives. However, it was observed that the way the image was taken was crucial, as several images identified leaves as olives. The analysis focused on the most important bands in the hyperspectral images. It was found that by using only the top 5% of significant bands, the LR size was reduced by 80% and the inference time was reduced by 74%. However, the model showed a 1% reduction in F1, so it is important to evaluate this F1 reduction for real-time applications.

Future Work

Subsequent studies should focus on comparing olive identification using only RGB information. Additionally, the dataset should be expanded to include different light conditions and multiple years, creating a comprehensive database of olives during their respective seasons. State-of-the-art complex models can be analyzed to determine the most effective method for detecting olives directly in the field. The dataset can be expanded to include more seasons and different varieties, such as Manzanilla, Hojiblanca, or Arbequina. Other potential applications include evaluating olive maturity, determining the number and size of olives in the image, or implementing global parameters to assist farmers in decision-making.

Author Contributions: Conceptualization, J.B. and F.J.M.; methodology, S.D.-C.; software, J.A.G.; validation, D.F.L., J.A.G. and S.D.-C.; formal analysis, J.B.; investigation, F.J.M.; resources, D.F.L.; data curation, S.D.-C.; writing—original draft preparation, S.D.-C.; writing—review and editing, D.F.L.; visualization, J.B.; supervision, C.L.; project administration, C.L.; funding acquisition, C.L. All authors have read and agreed to the published version of the manuscript.

Funding: This research was funded by the projects "HYperspectral Precision Agriculture based on Tracked Information for Andalusian crops—HYPATIA", grant number PYC20 RE 090 US; "Ia y Teledetección Avanzada portatil para incrementar la Calidad en la producción de Aceite de oliva—ITACA", grant number 802C2000097; and "Sistema Móvil Autónomo Hiperespectral para la Mejora del Olivar—SCOUT", grant number 2021/C005/00147860. Samuel Dominguez-Cid is funded and supported by the VI PPIV-US (VI Plan Propio de Investigación de la Universidad de Sevilla).

Institutional Review Board Statement: Not applicable.

Informed Consent Statement: Not applicable.

Data Availability Statement: Data are contained within the article.

Acknowledgments: This research was supported by Hermanos Donaire Ibáñez Agrícola, SC, and the Regulatory Council of the PGI Manzanilla and Gordal Olives from Seville.

Conflicts of Interest: The authors declare no conflicts of interest.

References

1. International Olive Council. IOC—STATISTICS. 2023. Available online: https://www.internationaloliveoil.org/wp-content/uploads/2022/12/IOC-Olive-Oil-Dashboard-2.html#production-2data (accessed on 22 January 2024).
2. Dag, A.; Kerem, Z.; Yogev, N.; Zipori, I.; Lavee, S.; Ben-David, E. Influence of time of harvest and maturity index on olive oil yield and quality. *Sci. Hortic.* **2011**, *127*, 358–366. [CrossRef]
3. Trapani, S.; Migliorini, M.; Cherubini, C.; Cecchi, L.; Canuti, V.; Fia, G.; Zanoni, B. Direct quantitative indices for ripening of olive oil fruits to predict harvest time. *Eur. J. Lipid Sci. Technol.* **2015**, *118*, 1202–1212. [CrossRef]

4. Perna, C.; Sarri, D.; Pagliai, A.; Priori, S.; Vieri, M. Assessment of Soil and Vegetation Index Variability in a Traditional Olive Grove: A Case Study. In Proceedings of the 2022 Conference of the Italian Association for Agricultural Engineering (AIIA): Biosystems Engineering Towards the Green Deal, Palermo, Italy, 19–22 September 2022; pp. 835–842.
5. Mulla, D.J. Twenty five years of remote sensing in precision agriculture: Key advances and remaining knowledge gaps. *Biosyst. Eng.* **2013**, *114*, 358–371. [CrossRef]
6. Balafoutis, A.; Beck, B.; Fountas, S.; Vangeyte, J.; Wal, T.; Soto, I.; Gómez-Barbero, M.; Barnes, A.; Eory, V. Precision Agriculture Technologies Positively Contributing to GHG Emissions Mitigation, Farm Productivity and Economics. *Sustainability* **2017**, *9*, 1339. [CrossRef]
7. Saiz-Rubio, V.; Rovira-Más, F. From Smart Farming towards Agriculture 5.0: A Review on Crop Data Management. *Agronomy* **2020**, *10*, 207. [CrossRef]
8. Tzounis, A.; Katsoulas, N.; Bartzanas, T.; Kittas, C. Internet of Things in agriculture, recent advances and future challenges. *Biosyst. Eng.* **2017**, *164*, 31–48. [CrossRef]
9. Griffor, E.R.; Greer, C.; Wollman, D.A.; Burns, M.J. *Framework for Cyber-Physical Systems: Volume 1, Overview*; National Institute of Standards and Technology: Gaithersburg, MD, USA, 2017.
10. Khan, A.; Khan, U.; Waleed, M.; Khan, A.; Kamal, T.; Marwat, S.N.K.; Maqsood, M.; Aadil, F. Remote Sensing: An Automated Methodology for Olive Tree Detection and Counting in Satellite Images. *IEEE Access* **2018**, *6*, 77816–77828. [CrossRef]
11. United States Geological Survey. Landsat Data Access, Landsat Data Access. U.S. Geological Survey. 2023. Available online: https://www.usgs.gov/landsat-missions/landsat-data-access (accessed on 23 October 2023).
12. European Space Agency. Sentinel-3 Olci—Technical Guide—Sentinel Online, Sentinel Online. 2023. Available online: https://sentinels.copernicus.eu/web/sentinel/technical-guides/sentinel-3-olci (accessed on 23 October 2023).
13. Vermote, E.; Franch, B.; Claverie, M. VIIRS/NPP Surface Reflectance 8-Day L3 Global 500m SIN Grid V002. 2023. Available online: https://data.nasa.gov/dataset/VIIRS-NPP-Surface-Reflectance-8-Day-L3-Global-500m/emiq-s47e/about_data (accessed on 22 January 2024).
14. Wang, D. MODIS/Terra+Aqua Surface Radiation Daily/3-Hour L3 Global 1km SIN Grid V061. 2021. Available online: https://lpdaac.usgs.gov/products/mcd18a1v061/ (accessed on 22 January 2024).
15. Masek, J.; Ju, J.; Roger, J.-C.; Skakun, S.; Vermote, E.; Claverie, M.; Dungan, J.; Yin, Z.; Freitag, B.; Justice, C. HLS Operational Land Imager Surface Reflectance and TOA Brightness Daily Global 30m v2.0. 2021. Available online: https://cmr.earthdata.nasa.gov/search/concepts/C2021957657-LPCLOUD/35 (accessed on 22 January 2024).
16. Maes, W.H.; Steppe, K. Perspectives for Remote Sensing with Unmanned Aerial Vehicles in Precision Agriculture. *Trends Plant Sci.* **2019**, *24*, 152–164. [CrossRef] [PubMed]
17. Adão, T.; Hruška, J.; Pádua, L.; Bessa, J.; Peres, E.; Morais, R.; Sousa, J. Hyperspectral Imaging: A Review on UAV-Based Sensors, Data Processing and Applications for Agriculture and Forestry. *Remote Sens.* **2017**, *9*, 1110. [CrossRef]
18. Rasmussen, J.; Azim, S.; Boldsen, S.K.; Nitschke, T.; Jensen, S.M.; Nielsen, J.; Christensen, S. The challenge of reproducing remote sensing data from satellites and unmanned aerial vehicles (UAVs) in the context of management zones and precision agriculture. *Precis. Agric.* **2020**, *22*, 834–851. [CrossRef]
19. Boukoberine, M.N.; Zhou, Z.; Benbouzid, M. A critical review on unmanned aerial vehicles power supply and energy management: Solutions, strategies, and prospects. *Appl. Energy* **2019**, *255*, 113823. [CrossRef]
20. Gao, Z.; Shao, Y.; Xuan, G.; Wang, Y.; Liu, Y.; Han, X. Real-time hyperspectral imaging for the in-field estimation of strawberry ripeness with deep learning. *Artif. Intell. Agric.* **2020**, *4*, 31–38. [CrossRef]
21. Wang, F.; Zhao, C.; Yang, H.; Jiang, H.; Li, L.; Yang, G. Non-destructive and in-site estimation of apple quality and maturity by hyperspectral imaging. *Comput. Electron. Agric.* **2022**, *195*, 106843. [CrossRef]
22. Haboudane, D. Hyperspectral vegetation indices and novel algorithms for predicting green LAI of crop canopies: Modeling and validation in the context of precision agriculture. *Remote Sens. Environ.* **2004**, *90*, 337–352. [CrossRef]
23. Ruiz, L.A.; Almonacid-Caballer, J.; Crespo-Peremarch, P.; Recio, J.A.; Pardo-Pascual, J.E.; Sánchez-Garc, E. Automated classification of crop types and condition in a mediterranean area using a fine-tuned convolutional neural network. *Int. Arch. Photogramm. Remote Sens. Spat. Inf. Sci.* **2020**, *XLIII-B3-2020*, 1061–1068. [CrossRef]
24. Aquino, A.; Ponce, J.M.; Andújar, J.M. Identification of olive fruit, in intensive olive orchards, by means of its morphological structure using convolutional neural networks. *Comput. Electron. Agric.* **2020**, *176*, 105616. [CrossRef]
25. Mart, S.S.; Gila, D.M.; Beyaz, A.; Ortega, J.G.; Garc, J.G. A computer vision approach based on endocarp features for the identification of olive cultivars. *Comput. Electron. Agric.* **2018**, *154*, 341–346.
26. Chauhan, N.K.; Singh, K. A Review on Conventional Machine Learning vs Deep Learning. In Proceedings of the 2018 International Conference on Computing, Power and Communication Technologies (GUCON), Greater Noida, India, 28–29 September 2018.
27. Jiang, P.; Ergu, D.; Liu, F.; Cai, Y.; Ma, B. A Review of Yolo Algorithm Developments. *Procedia Comput. Sci.* **2022**, *199*, 1066–1073. [CrossRef]
28. Yan, B.; Fan, P.; Lei, X.; Liu, Z.; Yang, F. A Real-Time Apple Targets Detection Method for Picking Robot Based on Improved YOLOv5. *Remote Sens.* **2021**, *13*, 1619. [CrossRef]
29. Dhillon, A.; Verma, G.K. Convolutional neural network: A review of models, methodologies and applications to object detection. *Prog. Artif. Intell.* **2019**, *9*, 85–112. [CrossRef]

30. Applalanaidu, M.V.; Kumaravelan, G. A Review of Machine Learning Approaches in Plant Leaf Disease Detection and Classification. In Proceedings of the 2021 Third International Conference on Intelligent Communication Technologies and Virtual Mobile Networks (ICICV), Tirunelveli, India, 4–6 February 2021.
31. Figorilli, S.; Violino, S.; Moscovini, L.; Ortenzi, L.; Salvucci, G.; Vasta, S.; Tocci, F.; Costa, C.; Toscano, P.; Pallottino, F. Olive Fruit Selection through AI Algorithms and RGB Imaging. *Foods* **2022**, *11*, 3391. [CrossRef]
32. Guzmán, E.; Baeten, V.; Pierna, J.A.F.; García-Mesa, J.A. Infrared machine vision system for the automatic detection of olive fruit quality. *Talanta* **2013**, *116*, 894–898. [CrossRef]
33. Zhuang, F.; Qi, Z.; Duan, K.; Xi, D.; Zhu, Y.; Zhu, H.; Xiong, H.; He, Q. A Comprehensive Survey on Transfer Learning. *Proc. IEEE* **2021**, *109*, 43–76. [CrossRef]
34. Qing, Y.; Liu, W.; Feng, L.; Gao, W. Improved Transformer Net for Hyperspectral Image Classification. *Remote Sens.* **2021**, *13*, 2216. [CrossRef]
35. Bazi, Y.; Bashmal, L.; Rahhal, M.M.A.; Dayil, R.A.; Ajlan, N.A. Vision Transformers for Remote Sensing Image Classification. *Remote Sens.* **2021**, *13*, 516. [CrossRef]
36. Domínguez-Cid, S.; Larios, D.F.; Barbancho, J.; Salvador, A.G.; Quintana-Ortí, E.S.; León, C. TEFNEN: Transformer for Energy Forecasting in Natural Environment. In Proceedings of the 2023 3rd International Conference on Electrical, Computer, Communications and Mechatronics Engineering (ICECCME), La Laguna, Spain, 19–21 July 2023.
37. Li, Y.; Wang, H.; Li, J.; Liu, C.; Tan, J. ACT: Adversarial Convolutional Transformer for Time Series Forecasting. In Proceedings of the 2022 International Joint Conference on Neural Networks (IJCNN), Padua, Italy, 18–23 July 2022.
38. Vaswani, A.; Shazeer, N.; Parmar, N.; Uszkoreit, J.; Jones, L.; Gomez, A.N.; Kaiser, L.; Polosukhin, I. Attention Is All You Need. In Proceedings of the Thirty-First Annual Conference on Neural Information Processing Systems (NIPS), Long Beach, CA, USA, 4–9 December 2017.
39. Bishop, C.M. *Pattern Recognition and Machine Learning (Information Science and Statistics)*; Springer: Berlin/Heidelberg, Germany, 2006.
40. S.I.L. SPECIM. SPECIM IQ. Available online: https://www.specim.com/iq/ (accessed on 22 January 2024).
41. Tsai, B.K.; Allen, D.W.; Hanssen, L.M.; Wilthan, B.; Zeng, J. A comparison of optical properties between solid PTFE (Teflon) and (low density) sintered PTFE. *SPIE Proc.* **2008**, *7065*, 70650Y.
42. He, H.; Garcia, E.A. Learning from Imbalanced Data. *IEEE Trans. Knowl. Data Eng.* **2009**, *21*, 1263–1284.
43. Cichosz, P. Assessing the quality of classification models: Performance measures and evaluation procedures. *Open Engineering* **2011**, *1*, 132–158. [CrossRef]
44. MIT Computer Science & Artificial Intelligence Laboratory. LabelMe. Available online: http://labelme.csail.mit.edu/Release3.0/ (accessed on 22 January 2024).
45. Cao, F.; Yang, Z.; Ren, J.; Jiang, M.; Ling, W.-K. Does Normalization Methods Play a Role for Hyperspectral Image Classification? *arXiv* **2017**, arXiv:1710.02939.
46. Dom, S.; Barbancho, J.; Larios, D.F.; Molina, F.J.; Gómez, A.; León, C. In-field hyperspectral imaging dataset of Manzanilla and Gordal olive varieties throughout the season. *Data Brief* **2023**, *46*, 108812.

foods

MDPI

Article

Development and Characterization of Edible Films Based on Cassava Starch Modified by Corona Treatment

Carlos Mauricio Otálora González [1,2], Manuel Felix [3], Carlos Bengoechea [3,*], Silvia Flores [1,2] and Lía Noemí Gerschenson [1,2]

[1] Departamento de Industrias, Facultad de Ciencias Exactas y Naturales, Universidad de Buenos Aires, Intendente Guiraldes 2620, Buenos Aires 1428, Argentina; camaota@yahoo.es (C.M.O.G.); skflores14@gmail.com (S.F.); lia@di.fcen.uba.ar (L.N.G.)

[2] Instituto de Tecnología de Alimentos y Procesos Químicos (ITAPROQ), CONICET—Universidad de Buenos Aires, Buenos Aires 1428, Argentina

[3] Departamento de Ingeniería Química, Escuela Politécnica Superior, Universidad de Sevilla, 41011 Sevilla, Spain; mfelix@us.es

* Correspondence: cbengoechea@us.es

Abstract: Corona treatment (CT), a surface treatment widely used in the plastic industry, can be used to alter the properties of cassava starch. In the present work, CT was performed on dry granular starch (DS), water-suspended humid granular starch (HS), and gelatinized starch (GS). Different properties and structural characteristics of treated starches were studied. A lowering in pH was generally observed after CT and the rheological properties depended on the starch presentation. A reinforcement of DS and HS samples after CT was deduced from higher viscosity values in flow assays and viscoelastic moduli, but weak gels were obtained when CT was applied to GS. Changes in the A-type polymorphic structure, as well as a drop in relative crystallinity, were produced by CT for DS and HS. Additionally, changes in O-H and C-O-C FTIR bands were observed. Therefore, CT can be applied for starch modification, producing predominantly cross-linking in the DS and depolymerization in the HS. Casting films made from the modified DS showed higher tensile strength and lower hydrophilicity, solubility, water absorption capacity, and water vapor permeability. Thus, the DS cross-linking induced by CT improved mechanical characteristics and hydrophobicity in edible films, which can be better used as packaging materials.

Keywords: starch; corona treatment; modification; edible films

Citation: Otálora González, C.M.; Felix, M.; Bengoechea, C.; Flores, S.; Gerschenson, L.N. Development and Characterization of Edible Films Based on Cassava Starch Modified by Corona Treatment. *Foods* **2024**, *13*, 468. https://doi.org/10.3390/foods13030468

Academic Editor: Seyed Mohammad Hashem Hosseini

Received: 10 January 2024
Revised: 26 January 2024
Accepted: 29 January 2024
Published: 2 February 2024

1. Introduction

Starch, constituted by linear polysaccharide molecules (20 to 25% (*w*/*w*) amylose) and branched molecules (75 to 80% (*w*/*w*) amylopectin), is an important renewable resource for the industry. Amylose and amylopectin naturally and hierarchically assemble into granular forms ranging in size from 1 to 100 μm [1]. Some physicochemical and functional properties of native starches, like their limited solubility, resistance to gel formation, or retrogradation, restrict their direct application for certain applications [2]. Thus, starch can be modified chemically, physically, or enzymatically to broaden its applicability [3]. Chemical methods generally provide high efficiency in starch modification, but they have disadvantages related to cost, chemical waste management, and time consumption [4]. The growing interest in environmental protection requires more eco-friendly processes, such as physical starch modification, which is a fast and non-polluting production technique [5]. Plasma treatment is distinguished as an innovative method for starch modification. Corona electrode discharge treatment is an emerging, non-thermal, high-efficiency, lower-energy, and zero-waste technology that can be used to change the functionality of starch [5,6]. This treatment is often used as a pre-treatment for several industrial applications, such as polymer matrices, plastics, or textiles, resulting in modifications in the surface energy and polarity of the treated material [7].

Corona treatment (CT) is performed on the polymer surface through the action of electrodes that generate an electrical discharge (10–50 kV; 1–4 kHz). This discharge causes a partial ionization of the surrounding atmosphere and produces different energetic species (i.e., electrons, ions, photons, free radicals), collectively known as plasma [7,8]. Plasma reactive species interact with starch molecules, forming new covalent bonds and functional groups, like ether, carbonyl, or carboxyl, as well as forming new compounds [9]. Plasma treatments used for starch modification are generally operated under low-pressure conditions through the application of different plasma source gases (e.g., argon, oxygen, hydrogen, methane, ethylene, ammonia) [6,10,11]. Recently, atmospheric pressure cold plasma (ACP) has been employed for decontamination purposes in the food industry [12,13]. Thus, ACP offers many innovative applications, such as increasing shelf-life, improving seed germination rate, promoting microbial inactivation, altering enzyme activity, adjusting hydrophilic or hydrophobic properties, etching or deposition of thin films, etc. [14]. Depending on the plasma type and conditions used, these treatments can modify starch to obtain different functional properties through specific surface reactions. The modifications can be, for example, surface energy increase, hydrophilicity increase or decrease, oxidation, bond breaking, molecular de-polymerization, cross-linking, and the introduction of new functional groups [6,7,10,15].

Lately, the interest in biodegradable materials has increased due to the non-renewable and mostly non-biodegradable character of petroleum-derived materials. Even if edible films generally present many advantages (e.g., degradability, affordability, recyclability readiness) [12], their technological application has found restrictions because of their inherent poor mechanical and barrier properties, high water sensitivity, low surface functionality, or poor printability and adhesiveness [16]. Hence, different treatments are necessary to improve these features [7,17].

At present, no systematic research report concerning the modification of cassava starch using a corona electrode at atmospheric pressure could be found. Therefore, the goals of the present study were to modify cassava starch using CT, determine the physicochemical, thermal, rheological, and structural characteristics, and study the chemical structures of the modified starch. And, finally, simple edible films based on plasma-modified starch were developed and characterized. Three different forms of starch were considered: two for intact granules (dry and in aqueous suspension) and one as a disrupted granule (gelatinized). Thus, the effects of the polysaccharide presentation on the reactions occurring during CT were investigated.

2. Materials and Methods

2.1. Materials

Native cassava starch (13.1% moisture) was food-grade (Bernesa S.A., Lomas de Zamora Buenos Aires, Argentina). The starch used was characterized by a degree of purity of 92–98% w/w. The amylose content was 19 % w/w, the amylopectin content was 81 % w/w, and the average molecular weight was 68×10^6 Da for amylopectin and 0.8×10^6 Da for amylose [18]. Analytical-grade glycerol (Biopack, Zárate, Buenos Aires, Argentina) and potassium sorbate (Sigma, St. Louis, MI, USA) were used. Deionized water was employed.

2.2. Sample Preparation for Plasma Treatment

Three forms of starch presentation were used: dry (DS), humid (HS) (water-suspended), and gelatinized (GS) starch. For DS, the granule without further treatment was used. The HS was obtained by mixing cassava starch in deionized water at a 1:4 ratio (w/w) for 4 h at 25 °C using a magnetic stirrer (300 rpm). Regarding GS, a suspension of 5% (w/w) was made by dispersing 5 g of cassava starch in 95 g of distilled water under mechanical stirring (300 rpm) and the system was heated at a rate of 2 °C/min up to 82 °C. The GS was cooled to 25 °C for 10 min before applying the plasma treatment.

Corona Treatment

A lab-scale corona unit (Model BD-20, Electro-Technic Products INC, Chicago, IL, USA) was employed. This unit has a voltage output that could be adjusted from 10,000 to 48,000 volts at frequencies from 4 to 5 MHz. Previous tests were performed to establish the proper treatment time. The DS samples (1 g) were supported on a round glass (5 cm diameter) base fixed to an aluminium square mobile platform (12 cm × 12 cm) and were treated with the 3-inch wire field-effect electrode for 2 min at 5 mm from the sample surface. The samples resting on the movable base were displaced at 1.0 cm/s in different directions (i.e., from right to left and from back to front) to ensure that the whole surface received the treatment conveniently. On the other hand, both the HS and GS were treated with a 2-inch-long electrode, located 5 mm from the sample surface, for 8 min. During the treatment, samples were continuously stirred using a magnetic stirrer (300 rpm). After treatment, the HS and GS samples were dried, the former by centrifugation followed by storage in an oven at 40 °C for 24 h, and the latter through freeze-drying. Both dried samples were eventually milled. Untreated (DS-U, HS-U, and GS-U) and treated (DS-T, HS-T, and GS-T) samples were kept at 4 °C.

2.3. *Fourier Transform Infrared Spectroscopy (FTIR)*

A Jasco FTIR 4200 spectrometer (Jasco Analytical Instrument, Tokyo, Japan) was used to obtain FTIR spectra in a wavenumber range of 400–4000 cm^{-1} at a resolution of 4 cm^{-1} in the transmission mode. The tests were carried out on samples after being pressed into KBr pellets.

2.4. *Determination of pH, Water Retention Capacity (WRC), and Solubility in Water (S)*

The pH was measured for 0.5% w/w aqueous starch suspension. WRC and S were determined according to Alzate et al. [18]. Briefly, 0.1000 g samples were re-suspended in distilled water (9 g) into pre-weighed centrifuge tubes which were placed in a hot water bath (85 °C, 30 min) and then centrifuged (Centrifuge 5804 R, Eppendorf, Hamburgo, Germany) at 5000 rpm for 15 min at 5 °C. A volume of supernatant (5 mL) was dispensed into weighed Petri dishes and dried in a vacuum oven (100 °C, 24 h). The total solubilized mass of starch was expressed as g of solubilized starch per 100 g of original sample (%). The WRC (water per g of sample) was calculated as the ratio of the weight gain of the swelled pellet and the initial weight of the sample discounted by the corresponding soluble fraction.

2.5. *Pasting Properties*

The pasting behavior of untreated and treated samples was measured with a Modular Compact Rheometer (Model MCR 102e, Anton Paar, Seiersberg, Graz-Umgebung, Austria), which was controlled by the RheoCompass software (21 CFR Part 11, 1.21.805 version, Anton Paar). The starch suspension (5%, w/w) was prepared by suspending 2.5 g cassava starch in 50 g distilled water and stirring for 5 min before subjecting it to the pasting assay. The racket initially rotated at 960 rpm for 10 s, and then was kept at 160 rpm up to the end of the pasting test. All samples were first kept at 50 °C for 1 min, and then heated up to 95 °C at 6 °C/min and held isothermally for 5 min. Eventually, all samples were cooled to 50 °C at 6 °C/min and held isothermally for 2 min. Parameters recorded were peak viscosity (PV, Pa·s), breakdown viscosity (BV, Pa·s), final viscosity (FV, Pa·s), setback viscosity (SV, Pa·s), and pasting temperature (PT, °C). All measurements were performed in duplicate.

2.6. *Differential Scanning Calorimetry*

A TA Instruments Q20 DSC (TA Instrument, New Castle, DE, USA) was employed for the thermal analysis of the starch samples. Different starch samples (3 mg) were mixed with 9 μL of distilled water, sealed in an aluminium capsule, and equilibrated at room temperature for 24 h before analysis. Thermograms were acquired by heating the sample and a reference aluminium capsule (empty capsule) from 10 to 150 °C at a heating rate of 10 °C/min in a nitrogen atmosphere (flow rate, 100.0 mL/min). Onset, peak, and conclusion temperatures of gelatinization (T_o, T_p, and T_c) along with the gelatinization enthalpy (ΔH, J/g) were quantified using Universal Analysis from Advantage Q Series software 5.1 version (TA Instruments, New Castle, DE, USA).

2.7. Thermogravimetric Analysis

A SDT Q600 device (TA Instruments, Wakefield, MA, USA) was employed for the thermogravimetric analyses carried out from 30 to 600 °C at 10 °C/min. The analysis was performed for about 5–10 mg of every sample; samples were placed in a non-reactive Pt pan under N_2 purge flow at 100 mL/min. Percent weight loss was calculated by the percent weight difference between 270 °C and 350 °C.

2.8. X-ray Diffraction Analysis (XRD)

The X-ray diffraction tests were performed on the samples in powder form in a Bruker D8 Advance X-ray diffractometer (Bruker, Manning Park Billerica, MA, USA) with monochromatic CuKα radiation at 40 kV and 30 mA. A scanning range (2θ) of 5 to 40° with a step size of 0.05° and a counting time of 15 s per step were applied. The crystallinity degree (CD) was estimated from the area of crystalline peaks divided by the global area, and multiplied by 100 [18,19].

2.9. Rheological Analysis

The samples were heated to 83 °C at 5 °C/min, and then the cooled rheological properties of the pastes were assessed after starch suspensions (5%, (*w*/*w*)) stirred at 160 rpm were first cooled at 3 °C/min to room temperature. The flow behavior and dynamic viscoelastic properties were evaluated by means of an AR2000 rheometer (TA Instruments, New Castle, DE, USA) employing a 40 mm diameter parallel plate geometry with a gap of 0.5 mm. The temperature was kept at 20.0 °C using a Peltier system.

Flow behavior was determined in starch pastes through rotational experiments obtaining the viscosity (η) of the samples after achieving the steady state as a function of shear rate ($\dot{\gamma}$) (0.01–1000 s^{-1}). Experimental data were fitted to the Carreau–Yasuda model:

$$\eta(\dot{\gamma}) = \eta_\infty + \frac{(\eta_0 - \eta_\infty)}{\left[1 + (\lambda \cdot \dot{\gamma})^a\right]^{\frac{1-n}{a}}}$$

where $\dot{\gamma}$ is the shear rate (s^{-1}); η_0(Pa·s) is the zero-shear-rate viscosity; η_∞ (Pa·s) is the high-shear-rate viscosity; λ (s) is the characteristic time constant for the fluid that determines the shear rate at which the transition from the zero-shear-rate plateau to the power-law region takes place; n is the flow index that describes the slope of the power-law region; and *a* is a dimensionless constant affected by the shape of the zero-shear rate-to-power-law transition.

Strain sweep tests (1 Hz at 25 °C) were first performed to determine the linear viscoelastic region (LVR) of samples. Then, frequency sweep tests were performed from 0.1 to 10 Hz at 25 °C, and the mechanical spectra were obtained. Thus, elastic (G′) and viscous (G″) moduli, as well as the loss tangent (tan δ = G″/G′), were determined.

2.10. Film Fabrication

Simple films were fabricated by casting from an aqueous gelatinized solution of the modified starches DS-T and GS-T (FDS-T, FGS-T, respectively) or from an untreated starch (FDS-U) (5% *w*/*w*) as a control system. Films were produced following the procedure informed by Otálora et al. [7]. The suspensions also containing glycerol (2% *w*/*w*), potassium sorbate (0.2% *w/w*), and distilled water up to 100 g were mixed (400 rpm) under heating (rate: 5 °C/min) up to 83 °C. Each film-forming solution (24 g) was cast into silicone molds (diameter = 7 cm) (Silikomart, Venice, Italy) and dried in a controlled temperature chamber at 45 °C (FAC, Buenos Aires, Argentina) for 15 h. Once the films were constituted, they were stabilized over a saturated NaBr solution (water activity, $a_w \cong 0.575$) for 5 days at 25.0 ± 0.5 °C, before characterization.

2.10.1. Characterization of the Films

Water Uptake Capacity and Soluble Matter Loss

Water uptake capacity (WUC) tests were performed following a modified protocol based on the ASTM D570 [20]. Soluble material loss (SML) was estimated according to Otálora et al. [19].

Contact Angle and Water Vapor Permeability (WVP)

The contact angle was evaluated with an interfacial tensiometer (Sinterface, Berlin, Germany) using the sessile drop method according to Otálora et al. [7]. The WVP was determined twice gravimetrically at 25 °C, using the procedure established in ASTM E96-00 [21], adapted by Gennadios et al. [22] for hydrophilic films.

Mechanical Properties

Tensile properties were determined at room temperature in a universal Instron testing machine 3345 (Instron Corp, Norwood, MA, USA), equipped with a 100 N load cell and a strain rate of 0.8 mm/s, The method described by the D882-02 standard [23] was followed with some modifications. Stress–strain curves were first obtained, and then characteristic tensile parameters were calculated: maximum stress (σ_{max}) and strain at break (ε_{max}). Ten replicates for every film formulation were analyzed.

2.10.2. Scanning Electron Microscopy (SEM)

The microstructure of the films was observed employing a Scanning Electron Microscope ZE122 SEM Supra 40 (Carl Zeiss, Jena, Germany). The cross-sections and surface of the films were gold-coated before any observation. Images analysis was conducted with SmartSEM® V05.06 software (Carl Zeiss, Jena, Germany).

2.11. Experimental and Statistical Analysis

The Prism 5 software (GraphPad, Boston, MA, USA) was used for non-linear fits and statistical evaluations. The statistical significance was evaluated using ANOVA with a level α of 0.05. The Tukey test was used as an "a posteriori test" [24]. All measurements of starch samples and edible films were performed at least three times, with results reported as averages and standard deviations.

3. Results and Discussion

The CT at atmospheric pressure is an unconventional procedure for the chemical modification of cassava starch. During plasma treatment, different changes can occur: starch cross-linking, starch de-polymerization, and modification of the surface (damages) of starch granules, all of these due to the action of the energy evolved from the process as well as due to the species present in the plasma generated [5,6,15,25]. According to the prevalence of these reactions, different effects can be generated affecting the characteristics and properties of the starches involved. The form of presentation of starch samples to CT as well as the experimental arrangement used for the CT can also exert an important influence on the trends observed.

3.1. Starch Characterization

3.1.1. FTIR

Figure 1 shows the FTIR spectra of untreated and plasma-treated starch samples. The starch samples had three main IR peaks at 3100–3600 cm^{-1} (O-H stretching), 3000–2800 cm^{-1} (C-H stretching), and 1640 cm^{-1} (O-H bending of bound water) [11,19]. Asymmetric C-O-C, C-O, and C-C skeleton stretching vibrations lead to the occurrence of absorption peaks at 1163 cm^{-1}, 1089 cm^{-1}, and 1017 cm^{-1}, respectively. On the other hand, the peak at 926 cm^{-1} was due to the skeletal mode vibrations of the C-O-C α-1,4-glycosidic bonds. Nemţanu and Braşoveanu [25] and Gao et al. [11] reported similar peaks. No new absorption peaks were distinguished for starches after plasma treatment.

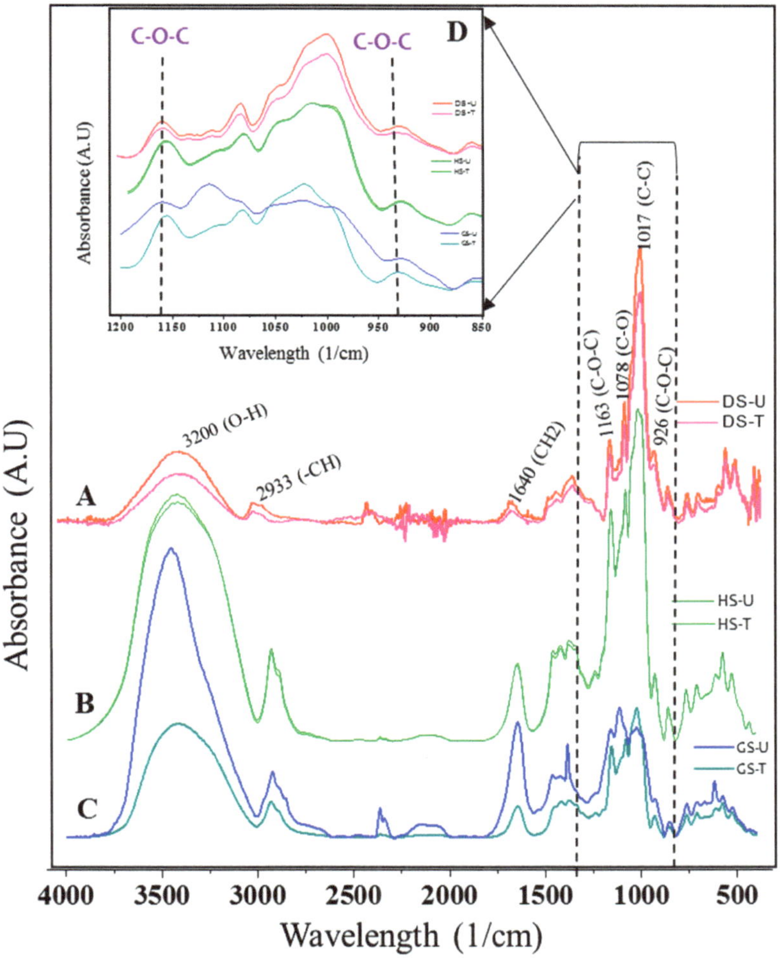

Figure 1. FTIR spectra of untreated and treated samples of starch. (**A**) DS-U versus DS-T, (**B**) HS-U versus HS-T, (**C**) GS-U versus GS-T, and (**D**) changes in C-O-C band intensities between DS-T and GS-T with respect to DS-U and GS-U.

A decrease in the intensity of OH groups is observed in the FTIR spectra, as seen from the peak at wavenumber 3200 cm^{-1} and 1640 cm^{-1} after plasma treatment for the DS-T and GS-T samples (Figure 1A,C). The decrease in the bands indicates the loss of OH groups in the modified starch, which would indicate that a plasma-induced cross-linking mechanism occurred. This mechanism forced an O-H bond and a C-OH bond to break before cross-linking, generating a new C-O-C glycosidic bond between two starch chains and extracting a water molecule. This would explain the decrease in OH groups. Khorram et al. [9] reported a decrease in the OH group in modified starch using argon and oxygen glow discharge plasma, revealing that cross-linking occurred.

In DS-T and GS-T samples, changes in the intensities of the bands at 1163 cm^{-1} and 926 cm^{-1} were observed (Figure 1D), probably indicating the modifications of C-O-C bonds, which could mean that starch cross-linking by plasma treatment is the prevailing reaction for the occurrence of these bonds. The cross-linking of the starch molecules because of the increase in the glycosidic bond was confirmed by an increase in the absorption peak at 926 cm^{-1} in cassava starches. Free radicals and energetic electrons formed during plasma generation may induce that cross-linking or intercrossing among the starch molecules [26]. Zou et al. [27] previously reported the cross-linking of starch in argon glow discharge plasma with a mechanism like that reported before.

The FTIR spectra of the HS-T samples were like the control sample (HS-U) (Figure 1B). According to Guo et al. [28], water can be activated by means of plasma application and can be used to enhance agriculture production. Therefore, the energy generated by corona electrode might have been at least partially used by the water present in this system to be activated, diminishing the availability of energy and active species to perform changes in the studied starch.

3.1.2. pH, Water Retention Capacity (WRC), and Solubility in Water (S)

After the plasma treatment, a lowering in the pH of the starch solution (Table 1) was detected, indicating the formation of chemical groups with acidic character, such as carboxyl, carbonyl, or peroxide groups [4], due to the oxidation of starch. However, in the FTIR spectrum (Figure 1), no observable change was detected for the peak at around 1700 cm^{-1}, which corresponds to the carboxylic acid functional group. Thirumdas et al. [13] have previously reported that the decrease in pH could not be confirmed from the changes in FTIR spectra of rice starch.

Table 1. pH, water retention capacity (WRC), and solubility (S) of treated and untreated starches.

Sample	pH	WRC (g/g)	S (%)
DS-U	6.49 ± 0.04^{a}	18.8 ± 0.8^{a}	26 ± 1^{a}
DS-T	4.87 ± 0.06^{b}	9.10 ± 0.4^{b}	6.2 ± 0.4^{b}
HS-U	5.67 ± 0.04^{a}	18.2 ± 0.5^{a}	14 ± 1^{a}
HS-T	3.45 ± 0.07^{b}	22.0 ± 1.0^{b}	20 ± 2^{b}
GS-U	5.68 ± 0.03^{a}	11.5 ± 0.6^{a}	42 ± 2^{a}
GS-T	3.89 ± 0.02^{b}	9.50 ± 0.6^{b}	37 ± 1^{b}

Data are expressed as mean ± SD. Means between untreated and treated starches (DS-U vs. DS-T, HS-U vs. HS-T, and GS-U vs. GS-T) with different letters in the same column are significantly different ($p < 0.05$).

It was observed that there was a significant decrease in WRC and S in the DS-T and GS-T modified starch when compared with untreated starch (DS-U, GS-U) (Table 1). A decrease in the water solubility of corn starch was observed using corona discharge plasma by Nemtanu and Minea [29], and this decrease could be attributed to cross-linking between starch molecules due to the effect of the active species of the plasma treatment that reduced the number of hydroxyl groups, producing starch with a lower capacity to interact with water [4]. The reduction in –OH groups could be evidenced by the significantly lower-intensity bands of O-H stretching for DS-T and GS-T samples in FTIR spectra (Figure 1A,C).

In contrast, HS-T showed an increase in S and WRC after treatment (Table 1). This could be attributed to the de-polymerization of starch, which could lead to the formation of smaller fragments that have higher water absorption power, resulting in higher solubility [10]. In addition, the CT energy could cause surface damage to the starch granules, which might have enhanced water penetration, resulting in higher WRC and S. Thirumdas et al. [14] demonstrated that cold air plasma increased the water absorption rate and solubility of rice starch.

The different trends observed in this study for cold plasma application could be attributed to the different presentation of samples used as well as to different treatment conditions, such as the type of electrode used for the different samples. It is important to remark that plasma in contact with water also promotes the formation of water with chemically reactive oxygen and nitrogen species that can modify the starch properties [29,30].

3.1.3. Pasting Properties

Pasting of starch involves different phenomena, such as granular swelling, transformation from ordered to disordered state, exudation of certain molecular components, and finally, the disruption of the granules [31,32]. The processing quality of starch depends greatly on its pasting properties. For example, understanding pasting behavior can be used to optimize the concentration of the ingredients, as well as the processing conditions (i.e., temperature, pressure, shear limits) when manufacturing the desired product [31]. Pasting properties are often determined from pasting curves obtained through a rapid visco-analyzer (RVA), which is a temperature-controlled viscometer that monitors the resistance of a starch granule sample to a specific shear. The pasting curves of the untreated (DS-U, HS-U, GS-U) and treated (DS-T, HS-T, GS-T) cassava starch samples are shown in Figure 2 and their pasting parameters are summarized in Table 2.

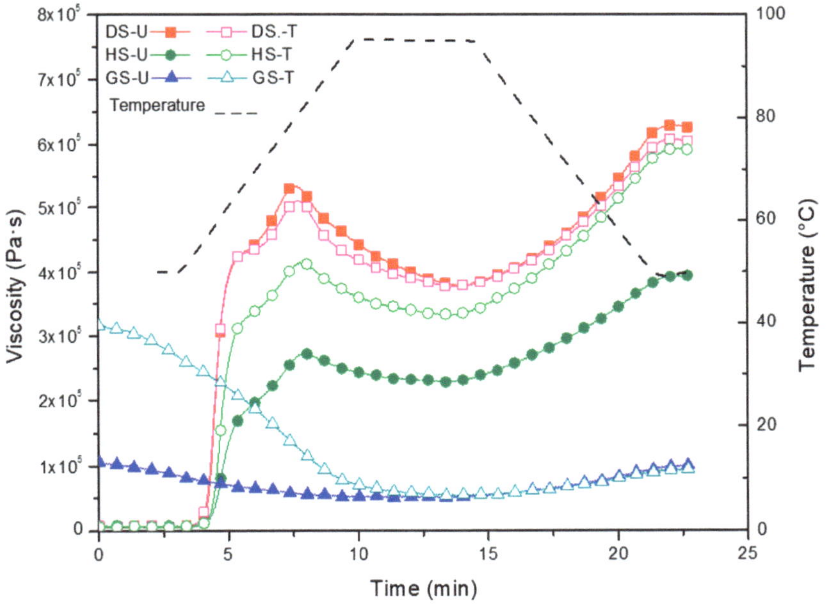

Figure 2. Pasting curves of untreated and plasma-modified starches.

Table 2. Pasting properties and gelatinization properties of the untreated and corona electrical discharge-treated cassava starches.

Samples	Pasting Properties					Gelatinization Properties			
	PT (°C)	PV × 10³ (Pa·s)	BV × 10³ (Pa·s)	FV × 10³ (Pa·s)	SV × 10³ (Pa·s)	T_o (°C)	T_p (°C)	T_c (°C)	ΔH (J/g)
DS-U	67.9 ± 0.5 [a]	534 ± 1 [a]	156.4 ± 0.8 [a]	625 ± 1 [a]	91.1 ± 0.8 [a]	61.3 ± 0.5 [a]	69.4 ± 0.7 [a]	84.7 ± 0.9 [a]	81 ± 1 [a]
DS-T	66.5 ± 0.2 [a]	510 ± 2 [b]	135 ± 2 [b]	604 ± 2 [b]	93.7 ± 0.1 [b]	82.0 ± 1.0 [b]	85.0 ± 2.0 [b]	96.0 ± 1.0 [b]	34 ± 2 [b]
HS-U	68.5 ± 0.4 [a]	273.9 ± 0.3 [a]	45.8 ± 0.4 [a]	391 ± 1 [a]	117.8 ± 0.7 [a]	58.9 ± 0.4 [a]	63.9 ± 0.7 [a]	73.3 ± 0.5 [a]	80.3 ± 0.6 [a]
HS-T	67.9 ± 0.8 [a]	415.7 ± 0.5 [b]	84.4 ± 0.8 [b]	592 ± 1 [b]	176.7 ± 0.5 [b]	62.6 ± 0.7 [b]	68.8 ± 0.8 [b]	76.8 ± 0.4 [b]	0.63 ± 0.03 [b]
GS-U	ND	ND	51.9 ± 0.5 [a]	99.1 ± 0.7 [a]	3.61 ± 0.04 [a]	ND	ND	ND	ND
GS-T	ND	ND	257.4 ± 0.8 [b]	94.1 ± 0.6 [b]	216.7 ± 0.7 [b]	ND	ND	ND	ND

ND, not detectable; PT, pasting temperature PV, peak viscosity; BV, breakdown viscosity; FV, final viscosity; SV, setback viscosity; T_o, onset temperature; T_p peak temperature; T_c, conclusion temperature; ΔH, enthalpy of gelatinization. Data are expressed as mean ± SD. Means of untreated and treated starches (DS-U vs. DS-T, HS-U vs. HS-T, and GS-U vs. GS-T) with different letters in the same column are significantly different ($p < 0.05$).

The pasting temperature (PT) indicates the temperature at which the viscosity abruptly increases during the continuous heating process. The PT did not change after CT for DS-T and HS-T (Table 2).

Regarding dry starch, plasma treatment significantly reduced the peak viscosity (PV, maximum viscosity during pasting) of dry starch from 5.34×10^5 Pa·s (DS-U) to 5.10×10^5 Pa·s (DS-T), as the energy involved in the treatment probably produced the modification of the starch macromolecules [5]. This should also be associated with the decrease observed in swelling capacity (Table 1). A significant decrease in breakdown viscosity (BV) was also detected in the treated starch (DS-T). BV is used to evaluate the shear thinning response of starches [10]. As displayed in Table 2, plasma-treated DS starches showed a good resistance to shear thinning when heated because cross-linking would have prevailed after plasma treatment. Chaiwat et al. [33] reported that the formation of covalent cross-links between starch chains strengthened the swollen granules, which minimized the viscosity loss under high temperature and shear conditions. Table 2 shows that SV and FV in DS-T samples decreased significantly ($p < 0.05$) compared to untreated starch (DS-U). Setback values (SVs) are defined as the degree of re-association between starch molecules involving amylose [11]. Zhang et al. [34] observed a global reduction in viscosity after potato starches were processed with nitrogen and helium glow plasma.

In humid starch, there was an increase in viscosities after treatment (Table 2). An increase in the viscosity of rice starch after cold plasma treatment was reported by Thirumdas et al. [14], as previously mentioned. Maximum viscosity represented the equilibrium point between swelling and breaking of starch granules. The increase in PV correlates with an enhancement in solubility and WRC after plasma treatment (HS-T) (Table 1). The higher FV in HS-T during the cooling process could be because amylose began to retrograde and the bonds between the chain molecules were reformed. In addition, the high-energy plasma species result in some degree of molecular alteration of the starch that altered the final viscosity [14,35].

The treatment of gelatinized starch with plasma (GS-T) increased significantly both BV and SV (Table 2). The incorporation of water molecules is facilitated during starch modification, resulting in increased viscosities (BV and SV). Therefore, plasma treatment causes de-polymerization of starch molecules due to the breaking of hydrogen bonds [14,33].

3.1.4. Gelatinization Properties

DSC analysis was performed to determine the gelatinization properties of DS-U, DS-T, HS-U, and HS-T samples. Since GS-U and GS-T samples were already gelatinized, the DSC technique was not applied. The onset, peak, and conclusion temperature (T_o, T_p, and T_c) of DS-T and HS-T increased with plasma treatment (Table 2). Bie et al. [35] reported an increase in the gelatinization temperature of cassava starch treated with helium plasma. A decrease in gelatinization enthalpy (ΔH) was detected after treatment for DS, being 81 and 34 J/g for DS-U and DS-T, respectively. In the case of wet starch, the decrease in ΔH was more drastic, from 80.3 to 0.63 J/g for HS-U and HS-T, respectively (Table 2). Some degree of crystallinity is probably lost during treatment. The decrease in enthalpy shows that plasma-treated starches consume less energy for gelatinization. Wang et al. [36] reported that the gelatinization enthalpy (ΔH) shows the loss of the double helix structure of starch molecules. Carmona-Garcia et al. [37] produced the cross-linking of banana starch with sodium trimetaphosphate (STMP)/sodium tripolyphosphate (STPP) and demonstrated an increase in gelatinization temperatures with cross-linking accompanied by a slight decrease in ΔH.

3.1.5. Thermogravimetric Analysis

Figure 3 shows the weight loss curves when untreated starches and plasma-treated starches were heated at 10 °C/min for a period of time. The TGA profiles of the starch samples show a region of weight loss in the range of 270 to 350 °C, which refers to the thermal decomposition of starch [15]. These results showed that the weight losses of plasma-modified starches (DS-T, HS-T, and GS-T) were lower than those of unmodified starches (DS -U, HS -U, and GS -U) (Figure 2), which could be due to the prevalence in the results observed of cross-linking reactions that took place during plasma treatment.

This implies a better thermal stability, a relatively good thermal stability, and a relatively stronger structure after plasma treatment, which could be associated with a cross-linking reaction of the modified starch. Singh and Nath [38] reported that cross-linked sago starch exhibited a higher thermal stability when compared to native sago starch.

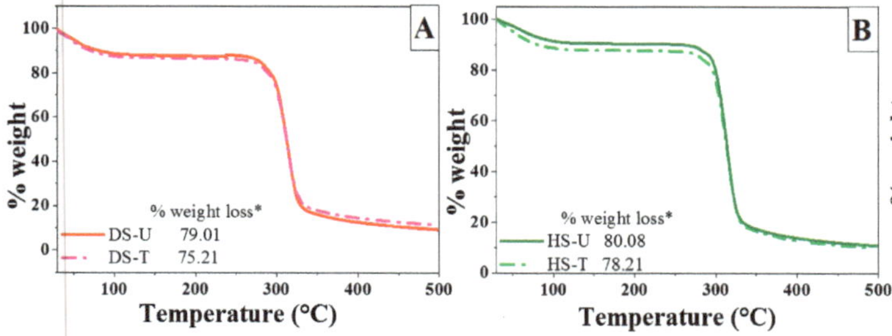

Figure 3. TGA thermograms and %weight loss of the untreated starches (DS-U, HS-U, and GS-U) and plasma-treated starches (DS-T, HS-T, and GS-T). * %weight loss = %weight at 270 °C − %weight at 350 °C. (**A**) DS-U and DS-T. (**B**) HS-U and HS-T. (**C**) GS-U and GS-T.

3.1.6. RX Diffraction

The XRD patterns of cassava starches before and after plasma treatment are shown in Figure 4. Characteristic X-ray diffraction peaks, which correspond to the A-type crystal pattern of cassava starch [18,39], at 2θ of 15°, 17°, 19°, and 23° appear in samples DS-U, DS-T, and HS-U. The crystallinity slightly decreases with plasma treatment in DS-T. It was observed that the % degree of crystallinity (DC) decreased from 29% (untreated starch) to 22% (DS-T) after treatment in agreement with DSC results. The decrease in crystallinity is due to the loss of ordered structures of starch molecules caused by reactive species in plasma [34]. When studying the influence of oxygen and helium plasma on the crystallinity of cassava starch, Bie et al. [35] observed an 8% decrease in the DC after helium gas plasma treatment, while a 12% decrease was observed after nitrogen gas plasma. In the present study, a 7% decrease in CD after atmospheric air plasma treatment was observed in the dried starch samples. It can be concluded that both the treatment time and the type of feed gas employed for plasma generation affect the rate of decrease in crystallinity.

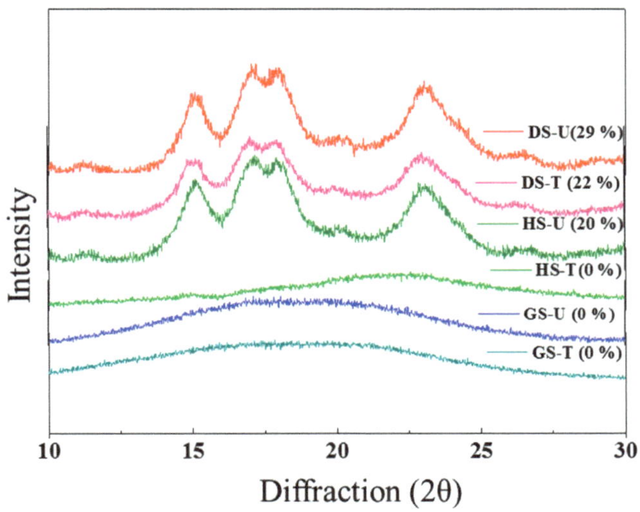

Figure 4. X-ray diffraction patterns and degree of crystallinity (DC) (in parentheses) of starch samples.

In systems where the starch granule was treated in suspension with water (HS-T), the absence of XRD patterns was observed after treatment. This change could be due to an increase in temperature during CT that had determined important progress in the gelatinization of the starch granule (Figure A1 shows the SEM images of the starch granules employed), as could also be inferred from a very low ΔH value (Table 2) for this sample. Previous work reported a reduction in rice starch crystallinity after plasma treatment [6,40]. The decrease in the degree of crystallinity of modified starch DS-T and HS-T coincides with the decrease in the gelatinization enthalpy (ΔH) (Table 2), that is, a lower energy requirement for starch gelatinization. Thus, plasma treatment affects the amorphous and crystalline regions of starch molecules.

3.1.7. Rheological Properties of Starch Paste

The study of the rheological parameters is generally of great importance for the starch industries. In this section, the rheological properties of the pastes obtained after the sample preparation described in Section 2.9 are analyzed. Thus, the viscosity as a function of the shear rate for every sample is displayed in Figure 5A–C. As expected for shear-thinning fluids, viscosity decreased as shear rate increased, which can be explained on the basis of a structural rearrangement. So, intra- and intermolecular bonding in the starch network is disrupted under shear stress, and then aggregates can align in the direction of the force through rearrangement or deformation. This would finally result in a reduction in the internal resistance and a viscosity reduction [14]. The DS-T and HS-T curves (Figure 5A,B) showed the highest viscosity throughout the shear rate range. In contrast, the GS-T samples (Figure 5C) displayed lower viscosity values than GS-U because of CT. It has been reported that the viscosity of starch was either increased or decreased by plasma treatment, depending on the types of starch and plasma as well as the treatment conditions [29,35]. In this study, the Carreau-Yasuda model was used to fit the flow curves of starch pastes with a coefficient of determination (R^2) between 0.9975 and 0.9983. The rheological parameters obtained for plasma-treated starches after fitting to the Carreau-Yasuda model are presented in Table 3. All starch samples showed pseudoplastic behavior, with values of flow index n less than 1. The Newtonian zero-shear-rate viscosity (η_o) of (DS-U) showed a value of 220.4 Pa·s and that of DS-T showed a significantly higher value of 1992.7 Pa·s, demonstrating that plasma treatment of dry starch granules probably produced a cross-linking phenomenon that not only increased η_o, but also increased the relaxation time ($\lambda\tau_c$) (Table 3). This same trend was observed for the samples that were humid-treated (HS-T). On the other hand, samples that were first gelatinized and then treated with plasma showed the opposite behavior, as a decrease in the flow curve parameters was observed after treatment.

Table 3. Rheological properties of the untreated and corona electrical discharge-treated cassava starches.

Samples	Steady Flow Test						Dynamic Viscoelastic Test		
	η_0(Pa·s)	η_∞ (Pa·s)	λ (s)	n	a	R^2	$G'_{1\,Hz}$(Pa)	$G''_{1\,Hz}$ (Pa)	tan $\delta_{1\,Hz}$
DS-U	220.4 ± 0.7 ᵃ	0.072 ± 0.001 ᵃ	39.4 ± 0.8 ᵃ	0.250 ± 0.07 ᵃ	8.08 ± 0.3 ᵃ	0.9975	3.2 ± 0.8 ᵃ	2.1 ± 0.3 ᵃ	34 ± 1 ᵃ
DS-T	1992.7 ± 0.2 ᵇ	0.123 ± 0.04 ᵇ	220.0 ± 0.2 ᵇ	0.146 ± 0.002 ᵇ	24.3 ± 0.1 ᵇ	0.9962	29 ± 0.3 ᵇ	6.9 ± 0.3 ᵇ	13.2 ± 0.2 ᵇ
HS-U	157.3 ± 0.3 ᵃ	0.048 ± 0.003 ᵃ	19.9 ± 0.5 ᵃ	0.27 ± 0.01 ᵃ	1.52 ± 0.05 ᵃ	0.9993	6.1 ± 0.4 ᵃ	2.7 ± 0.1 ᵃ	24.1 ± 0.6 ᵃ
HS-T	1212.3 ± 0.5 ᵇ	0.058 ± 0.005 ᵃ	84.4 ± 0.3 ᵇ	0.24 ± 0.01 ᵃ	14.5 ± 0.5 ᵇ	0.9997	21.7 ± 0.7 ᵇ	5.6 ± 0.5 ᵇ	14.4 ± 0.2 ᵇ
GS-U	302.4 ± 0.2 ᵃ	0.047 ± 0.006 ᵃ	63.4 ± 0.5 ᵃ	0.22 ± 0.02 ᵃ	4.44 ± 0.4 ᵃ	0.9998	6.2 ± 0.2 ᵃ	2.2 ± 0.1 ᵇ	20.3 ± 0.7 ᵃ
GS-T	22.52 ± 0.8 ᵇ	0.048 ± 0.001 ᵃ	33.9 ± 0.3 ᵇ	0.39 ± 0.06 ᵇ	68.0 ± 0.1 ᵇ	0.9983	1.5 ± 0.4 ᵇ	1.3 ± 0.2 ᵇ	41.3 ± 0.3 ᵇ

Data are expressed as mean ± SD. Means between untreated and treated starches (DS-U vs. DS-T, HS-U vs. HS-T, and GS-U vs. GS-T) with different letters in the same column are significantly different ($p < 0.05$).

Small Amplitude Oscillatory Shear (SAOS) tests provide information of the elastic modulus (G') and viscous modulus (G'') as a function of frequency (i.e., mechanical spectra), which can be related to the sample structure. All analyzed paste samples could be classified as weak gels because $G' > G''$ within the frequency range studied and the moduli showed a slight slope (Figure 5D–F). The G' and G'' of the treated starch DS-T showed higher values of moduli than the untreated dry starch (DS-U) (Figure 5D), whereas those of the HS-U paste were almost the same as those of the treated (HS-U) paste (Figure 5E). In addition, the DS-T paste showed a lower magnitude of loss tangent (tan δ) than the control (DS-U) paste (Table 3). This would indicate a stronger gel structure for the treated sample (DS-T), probably due to cross-linking reactions. This is supported by the greater dependency of G' and G'' on frequency found for the DS-U sample, even finding a tendency for a crossover at frequencies higher than 1 Hz. This is not observed for the DS-T sample, for which much higher values for both viscoelastic moduli are found, with a much more prevailing elastic character. The gelatinization of the starch followed by the CT is responsible of this strengthening, which is coherent with the flow behavior previously studied. Chaiwat et al. [33] found that G' of starch cross-linked by low-pressure argon plasma treatment was higher than that of native starch. In general, an increase in G' accompanied by a decrease in tan δ could be related to cross-linking formation of the polymer [15].

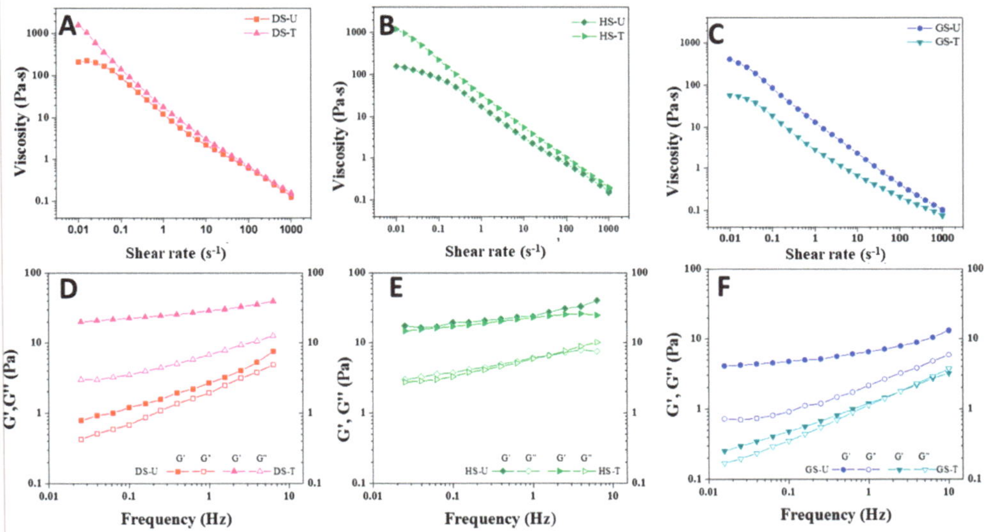

Figure 5. Flow curves (**A–C**) and mechanical spectra (**D–F**) for untreated and plasma-modified starches.

3.2. Characterization of the Films

In general, starch is used in the manufacture of edible films due to its ability to form flexible films and its acceptable barrier to gases such as oxygen and carbon dioxide. The technological application of these films has been restricted due to some inherent disadvantages, such as inadequate mechanical and water vapor barrier properties, high water sensitivity, and poor printability and adhesiveness. Therefore, it is necessary to use different treatments to improve its characteristics, such as CT. As previously stated, DS-T and GS-T modified starches showed lower WRC and S values in comparison with ones for untreated samples, probably due to the predominance of cross-linking reactions promoted by TC. The opposite was the case with the modified HS-T starch, where there was a higher WRC and S after CT. Therefore, DS-T and GS-T modified starches were selected to produce edible films based on their physicochemical characteristics.

3.2.1. Water Vapor Permeability, Water Uptake Capacity, and Contact Angle of Films

One of the main restrictions of the application of cassava starch films is their mostly hydrophilic nature, which leads to high water transfer rates due to the clustering of water molecules and their diffusion through microcavities. The WVP for the control cassava starch film (FDS-U) was found to be $4.4 \pm 0.3 \times 10^9$ g s^{-1} m^{-1}·Pa^{-1}. When films were manufactured from GS-T, a significant increase ($p > 0.05$) in WVP was observed, whereas when DS-T was used, there was a very marked decrease in WVP (Figure 6A). These results coincide with the decrease in WUC in the FDS-T (Figure 6B). This decrease may be due to the reduction of hydrophilic groups (-OH) observed in FTIR analyses (Figure 1A) and to the polymeric morphological characteristics in addition to starch cross-linking. In previous studies, a decrease in WVP of 58% was observed when the surface of cassava starch-based films was modified with CT [7]. Plasma treatment decreased the WVP of starch/PCL and starch/PLA composite films by approximately 94% according to Heidemann et al. [41].

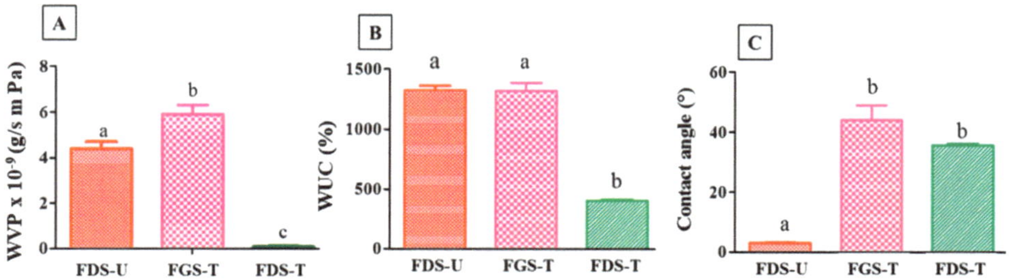

Figure 6. (**A**) Water vapor permeability (WVP), (**B**) water uptake capacity (WUC), and (**C**) contact angle of films (FDS-U) and treated films (FGS-T and FDS-T). The average and standard deviation are reported. Same letter means non-significant differences ($p < 0.05$) between samples.

The contact angle of a liquid is related to its wettability and hydrophilicity, as it measures the tendency of a drop of that fluid to spread and adhere to the surface of materials. The FDS-U showed a contact angle of 3° (Figure 6C). When FGS-T and FDS-T were produced, the contact angle increased significantly to a value of 44° (FGS-T) or to a value of 35.7° (FDS-T), which means that the hydrophilicity of the surface of the film was lower (Figure 6C), probably due to the changes suffered by the starch after CT [7]. Lyytikäinen et al. [42] reported that oxidation compounds can be formed during treatment (i.e., hydroxyl groups of starch can be oxidized to carbonyls), increasing the hydrophobic character of the polysaccharide and of the films constituted with it. Pankaj et al. [43] used plasma treatments on cassava starch films and observed an increase in hydrophobicity. Bastos et al. [44] showed that the contact angle increased after CT and more hydrophobic groups were obtained in the surface layer of the starch films.

3.2.2. Physical, Mechanical, and Morphological Properties

Table 4 shows the physical and mechanical properties of the films. A decrease in moisture content is observed in FDS-T compared to the control (FDS-U). This might be ascribed to the cross-linking of starch, thus limiting the water sorption capacity, and coincides with the decrease in WUC in the FDS-T. Solubility in water is an important property of edible films. Potential applications may require insolubility to improve product integrity and water resistance. In Table 4, SML is lower in FDS-T and FGS-T with respect to FDS-U, which indicates greater water resistance. This result coincides with the decrease in WUC (Figure 6C). The energy of the treatment probably promoted the cross-linking of the starch macromolecules [5], resulting in strengthened structure with poor solubility.

The mechanical characteristics of edible films are also key factors in protecting the quality and integrity of food products during different stages (i.e., handling, transportation, storage). Results obtained indicated that the FGS-T films had weak mechanical properties since the stress values decreased from 1.62 to 1.3 MPa. In the FDS-T films, an improvement in mechanical resistance was observed since the stress values increased from 1.62 to 2.0 MPa (Table 4). Guo et al. [45] treated potato starch with dielectric barrier discharge plasma treatment and observed that this treatment gave rise to films with better tensile stress and Young's modulus.

Table 4. Moisture, solubility in water, and mechanical parameters of control films (FDS-U) and treated films (FGS-T and FDS-T).

Physical Parameters	FDS-U	FGS-T	FDS-T
Moisture (% db)	24.7 ± 0.2^{a}	23 ± 1^{a}	20.2 ± 0.1^{b}
SML (%)	40.1 ± 0.7^{a}	35 ± 2^{b}	32 ± 1^{b}
Mechanical properties			
ε_{max} (mm/mm)	1.25 ± 0.07^{a}	1.1 ± 0.1^{a}	1.4 ± 0.1^{a}
σ_{max} (MPa)	1.62 ± 0.03^{a}	1.3 ± 0.1^{b}	2.0 ± 0.2^{c}

The mean and standard deviation are reported. Same letter in the data reported in a row means non-significant differences ($p < 0.05$).

Cracks can be observed in the surface and cross-section micrographs of control films, FDS-U (Figure 7(A1,A2)), which explained the higher WVP and WUC and the lower contact angle. Meanwhile, FGS-T (Figure 7(B1,B2)) and FDS-T (Figure 7(C1,C2)) showed a compact structure, without cracks and phase separation, which can explain the improvement in stress in the FDS-T systems. The observed morphological changes could be a consequence of starch cross-linking [5]. Other researchers reported similar trends in relation to the results of the use of plasma-treated starch on the production of films [10]. An improvement in structural properties of films based on plasma-treated starch has been attributed to the effect of the bombardment of plasma species such as electrons, ions, and radicals on the polysaccharide, favoring its cross-linking [5,43,45]. Similarly, this may explain the increase in contact angle, which means an increase in hydrophobicity (Figure 6C).

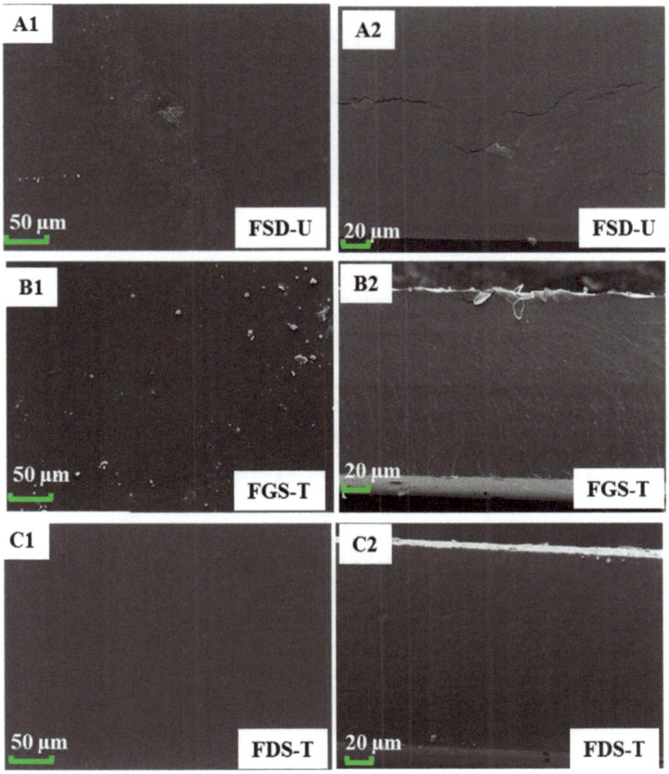

Figure 7. SEM images of the surface (**A1**,**B1**,**C1**) and cross-section (**A2**,**B2**,**C2**) of FDS-U, FGS-T, and FDS-T.

4. Conclusions

Results highlight corona treatment (CT) at atmospheric pressure as a useful alternative to chemical modifications commonly employed for cassava starch. CT produces changes in the functional, thermal, and rheological properties of starch. The study of the behavior of films produced through casting from plasma-modified cassava starch showed notable improvements in their mechanical properties when compared to those of native starch films, as they were more resistant and had lower water absorption capacity and greater hydrophobicity. Changes also occurred in the structure of the films when observed by SEM, contributing to the generation of a more compact and homogeneous structure when films were based on CT-modified starch. Thus, CT is a sustainable technology that enables the industrial processing of edible films based on modified cassava starch with adequate properties. These new materials could be a viable and environmentally friendly alternative to be used as food packaging material.

Future investigations must consider (i) surface analysis of the films developed by XPS to evaluate the surface modifications produced by atmospheric CT, (ii) evaluation of the ability of the developed films to carry pigments and antimicrobial and antioxidant compounds. All of this will help to determine the specific applications of CT treatment. The application of CT to other starches is also of interest to analyze whether the conclusions obtained for these treatments can be generalized.

Author Contributions: Conceptualization, L.N.G., C.M.O.G., S.F. and C.B.; Data curation, C.M.O.G.; Formal analysis, L.N.G., C.M.O.G., S.F., M.F. and C.B.; Funding acquisition, L.N.G., S.F. and C.B.; Investigation, C.M.O.G. and C.B.; Methodology, L.N.G., C.M.O.G., M.F., S.F. and C.B.; Project administration, L.N.G., S.F. and C.B.; Resources, L.N.G. and C.B.; Supervision, L.N.G., S.F. and C.B.; Writing—original draft, C.M.O.G.; Writing—review & editing, L.N.G., C.M.O.G., S.F. and C.B. All authors have read and agreed to the published version of the manuscript.

Funding: This work was financed by the project PID2021-124294OB-C21 funded by MCIN/AEI/10.13039/501100011033/ and by "ERDF A way of making Europe". This work was supported by grants from University of Buenos Aires (20020130100550BA; 20020170100229BA), National Research Council of Argentina, CONICET (11220120100507, 11220210100712) and National Agency of Scientific and Technical Research (PICT 2015-2109; PICT 2017-1146; PICT 2019-1842).

Institutional Review Board Statement: Not applicable.

Informed Consent Statement: Not applicable.

Data Availability Statement: The original contributions presented in the study are included in the article, further inquiries can be directed to the corresponding author.

Acknowledgments: The authors acknowledge University of Buenos Aires, National Research Council of Argentina, CONICET, National Agency of Scientific and Technical Research, MCIN/AEI/10.13039/501100011033/and ERDF A way of making Europe for financial support.

Conflicts of Interest: The authors declare no conflict of interest.

Appendix A

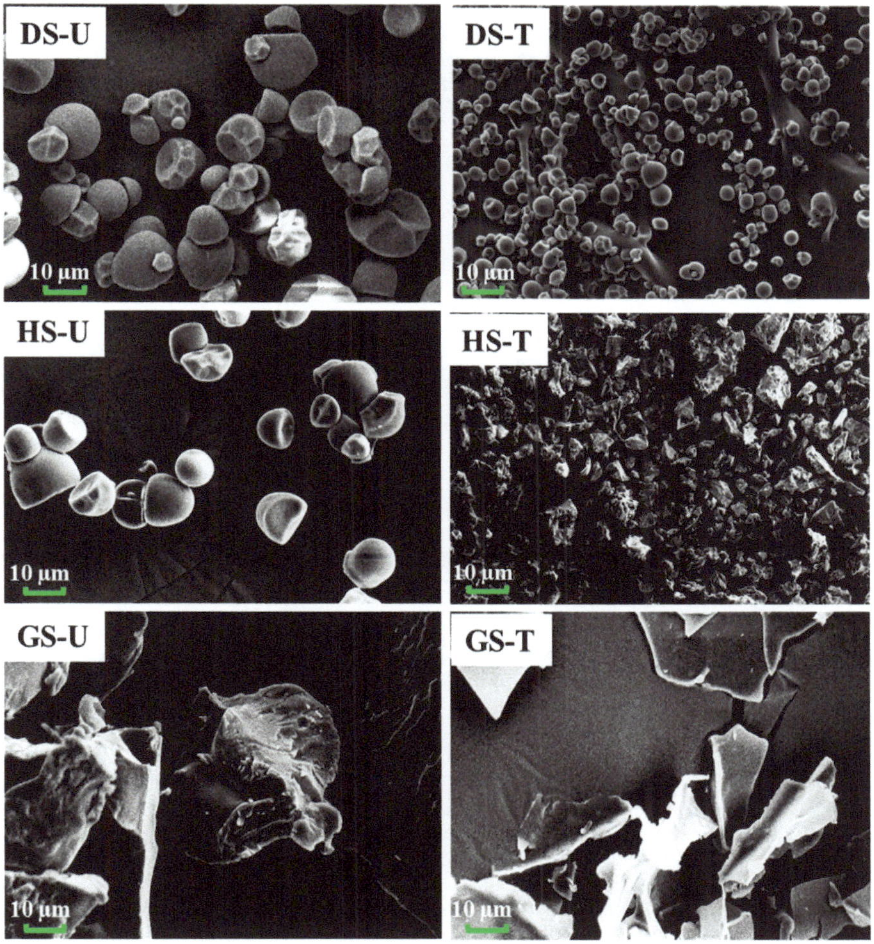

Figure A1. SEM images of the untreated and treated starch samples.

References

1. Pérez, S.; Bertoft, E. The molecular structures of starch components and their contribution to the architecture of starch granules: A comprehensive review. *Starch-Stärke* **2010**, *62*, 389–420. [CrossRef]
2. Bashir, K.; Aggarwal, M. Physicochemical, structural and functional properties of native and irradiated starch: A review. *J. Food Sci. Technol.* **2019**, *56*, 513–523. [CrossRef]
3. BeMiller, J.N.; Huber, K.C. Physical modification of food starch functionalities. *Annu. Rev. Food Sci. Technol.* **2015**, *6*, 19–69. [CrossRef]
4. Xiao, H.; Lin, Q.; Liu, G.Q.; Wu, Y.; Tian, W.; Wu, W.; Fu, X. Physicochemical properties of chemically modified starches from different botanical origin. *Sci. Res. Essays* **2011**, *6*, 4517–4525. [CrossRef]
5. Zhu, F. Plasma modification of starch. *Food Chem.* **2017**, *232*, 476–486. [CrossRef] [PubMed]
6. Thirumdas, R.; Sarangapani, C.; Annapure, U.S. Cold plasma: A novel non-thermal technology for food processing. *Food Biophys.* **2015**, *10*, 1–11. [CrossRef]
7. Otalora, C.M.; Flores, S.K.; Basanta, M.F.; Gerschenson, L.N. Effect of beetroot (*Beta vulgaris* L. var conditiva) fiber filler and corona treatment on cassava starch films properties. *Food Packag. Shelf Life* **2020**, *26*, 100605. [CrossRef]
8. Bogaerts, A.; Neyts, E.; Gijbels, R.; Van der Mullen, J. Plasmas de descarga de gases y sus aplicaciones. *Spectrochim. Acta Part B Espectroscopía Atómica* **2002**, *57*, 609–658. [CrossRef]

9. Khorram, S.; Zakerhamidi, M.S.; Karimzadeh, Z. Polarity functions' characterization and the mechanism of starch modification by DC glow discharge plasma. *Carbohydr. Polym.* **2015**, *127*, 72–78. [CrossRef] [PubMed]
10. Wu, T.Y.; Sun, N.N.; Chau, C.F. Application of corona electrical discharge plasma on modifying the physicochemical properties of banana starch indigenous to Taiwan. *J. Food Drug Anal.* **2018**, *26*, 244–251. [CrossRef]
11. Gao, S.; Liu, H.; Sun, L.; Liu, N.; Wang, J.; Huang, Y.; Wang, F.; Cao, J.; Fan, R.; Zhang, X.; et al. The effects of dielectric barrier discharge plasma on physicochemical and digestion properties of starch. *Int. J. Biol. Macromol.* **2019**, *138*, 819–830. [CrossRef]
12. Pankaj, S.K.; Bueno-Ferrer, C.; Misra, N.N.; Milosavljević, V.; O'Donnell, C.P.; Bourke, P.; Keener, K.M.; Cullen, P.J. Applications of cold plasma technology in food packaging. *Trends Food Sci. Technol.* **2014**, *35*, 5–17. [CrossRef]
13. Misra, N.N.; Kaur, S.; Tiwari, B.K.; Kaur, A.; Singh, N.; Cullen, P.J. Atmospheric pressure cold plasma (ACP) treatment of wheat flour. *Food Hydrocoll.* **2015**, *44*, 115–121. [CrossRef]
14. Thirumdas, R.; Trimukhe, A.; Deshmukh, R.R.; Annapure, U.S. Functional and rheological properties of cold plasma treated rice starch. *Carbohydr. Polym.* **2017**, *157*, 1723–1731. [CrossRef]
15. Wongsagonsup, R.; Deeyai, P.; Chaiwat, W.; Horrungsiwat, S.; Leejariensuk, K.; Suphantharika, M.; Fuongfuchat, A.; Dangtip, S. Modification of tapioca starch by non-chemical route using jet atmospheric argon plasma. *Carbohydr. Polym.* **2014**, *102*, 790–798. [CrossRef]
16. Romani, V.P.; Olsen, B.; Collares, M.P.; Oliveira, J.R.M.; Prentice, C.; Martins, V.G. Plasma technology as a tool to decrease the sensitivity to water of fish protein films for food packaging. *Food Hydrocoll.* **2019**, *94*, 210–216. [CrossRef]
17. Bahrami, R.; Zibaei, R.; Hashami, Z.; Hasanvand, S.; Garavand, F.; Rouhi, M.; Jafari, S.M.; Mohammadi, R. Modification and improvement of biodegradable packaging films by cold plasma; a critical review. *Crit. Rev. Food Sci. Nutr.* **2022**, *62*, 1936–1950. [CrossRef]
18. Alzate, P.; Gerschenson, L.; Flores, S. Ultrasound application for production of nano-structured particles from esterified starches to retain potassium sorbate. *Carbohydr. Polym.* **2020**, *247*, 116759. [CrossRef]
19. Otálora, C.M.; Alvarez Castillo, E.; Flores, S.; Gerschenson, L.N.; Bengoechea, C. Effect of plasticizer composition on the properties of injection molded cassava starch-based bioplastics. *Food Packag. Shelf Life* **2023**, *40*, 101218. [CrossRef]
20. *ASTM D570-00*; Standard Method for Water Absorption of Plastics. American Society for Testing and Materials: Philadelphia, PA, USA,, 2002. [CrossRef]
21. *ASTM E96*; Standard Test Method for Water Vapour Transmission of Materials. American Society for Testing and Materials: Philadelphia, PA, USA, 2000. [CrossRef]
22. Gennadios, A.; Weller, C.L.; Gooding, C.H. Measurement errors in water vapor permeability of highly permeable, hydrophilic edible films. *J. Food Eng.* **1994**, *21*, 395–409. [CrossRef]
23. *ASTM D882-02*; Standard Method for Tensile Properties of Thin Plastic Sheeting. American Society for Testing and Materials: Philadelphia, PA, USA, 2002. [CrossRef]
24. Sokal, R.; Rohlf, J. Biometry. In *The Principles and Practice of Statistics in Biological Research*; W. H. Freeman and Company: San Francisco, CA, USA, 2000.
25. Nemtanu, M.R.; Braşoveanu, M. Exposure of starch to combined physical treatments based on corona electrical discharges and ionizing radiation. Impact on physicochemical properties. *Radiat. Phys. Chem.* **2021**, *184*, 109480. [CrossRef]
26. Carvalho, A.P.M.G.; Barros, D.R.; da Silva, L.S.; Sanches, E.A.; Pinto, C.d.C.; de Souza, S.M.; Clerici, M.T.P.S.; Rodrigues, S.; Fernandes, F.A.N.; Campelo, P.H. Dielectric barrier atmospheric cold plasma applied to the modification of Ariá (*Goeppertia allouia*) starch: Effect of plasma generation voltage. *Int. J. Biol. Macromol.* **2021**, *182*, 1618–1627. [CrossRef]
27. Zou, J.J.; Liu, C.J.; Eliasson, B. Modification of starch by glow discharge plasma. *Carbohydr. Polym.* **2004**, *55*, 23–26. [CrossRef]
28. Guo, D.; Liu, H.; Zhou, L.; Xie, J.; He, C. Plasma-activated water production and its application in agriculture. *J. Sci. Food Agric.* **2021**, *101*, 4891–4899. [CrossRef] [PubMed]
29. Nemtanu, M.R.; Minea, R. Functional properties of corn starch treated with corona electrical discharges. In *Macromolecular Symposia*; Wiley-VCH Verlag: Weinheim, Germany, 2006; Volume 245, pp. 525–528. [CrossRef]
30. Okyere, A.Y.; Rajendran, S.; Annor, G.A. Cold plasma technologies: Their effect on starch properties and industrial scale-up for starch modification. *Curr. Res. Food Sci.* **2022**, *5*, 451–463. [CrossRef]
31. Kumar, R.; Khatkar, B.S. Thermal, pasting and morphological properties of starch granules of wheat (*Triticum aestivum* L.) varieties. *J. Food Sci. Technol.* **2017**, *54*, 2403–2410. [CrossRef] [PubMed]
32. Nawaz, M.A.; Fukai, S.; Bhandari, B. Effect of different cooking conditions on the pasting properties of flours of glutinous rice varieties from Lao people's democratic republic. *Int. J. Food Prop.* **2016**, *19*, 2026–2040. [CrossRef]
33. Chaiwat, W.; Wongsagonsup, R.; Tangpanichyanon, N.; Jariyaporn, T.; Deeyai, P.; Suphantharika, M.; Fuongfuchat, A.; Nisoa, M.; Dangtip, S. Argon plasma treatment of tapioca starch using a semi-continuous downer reactor. *Food Bioprocess Technol.* **2016**, *9*, 1125–1134. [CrossRef]
34. Zhang, B.; Chen, L.; Li, X.; Li, L.; Zhang, H. Understanding the multi-scale structure and functional properties of starch modulated by glow-plasma: A structure-functionality relationship. *Food Hydrocoll.* **2015**, *50*, 228–236. [CrossRef]
35. Bie, P.; Li, X.; Xie, F.; Chen, L.; Zhang, B.; Li, L. Supramolecular structure and thermal behavior of cassava starch treated by oxygen and helium glow-plasmas. *Innov. Food Sci. Emerg. Technol.* **2016**, *34*, 336–343. [CrossRef]
36. Wang, L.; Xie, B.; Shi, J.; Xue, S.; Deng, Q.; Wei, Y.; Tian, B. Physicochemical properties and structure of starches from Chinese rice cultivars. *Food Hydrocoll.* **2010**, *24*, 208–216. [CrossRef]

37. Carmona-Garcia, R.; Sanchez-Rivera, M.M.; Méndez-Montealvo, G.; Garza-Montoya, B.; Bello-Pérez, L.A. Effect of the cross-linked reagent type on some morphological, physicochemical and functional characteristics of banana starch (*Musa paradisiaca*). *Carbohydr. Polym.* **2009**, *76*, 117–122. [CrossRef]
38. Singh, A.V.; Nath, L.K. Synthesis and evaluation of physicochemical properties of cross-linked sago starch. *Int. J. Biol. Macromol.* **2012**, *50*, 14–18. [CrossRef]
39. Dang, K.M.; Yoksan, R. Thermoplastic starch blown films with improved mechanical and barrier properties. *Int. J. Biol. Macromol.* **2021**, *188*, 290–299. [CrossRef]
40. Sarangapani, C.; Devi, Y.; Thirundas, R.; Annapure, U.S.; Deshmukh, R.R. Effect of low-pressure plasma on physico-chemical properties of parboiled rice. *LWT-Food Sci. Technol.* **2015**, *63*, 452–460. [CrossRef]
41. Heidemann, H.M.; Dotto, M.E.; Laurindo, J.B.; Carciofi, B.A.; Costa, C. Cold plasma treatment to improve the adhesion of cassava starch films onto PCL and PLA surface. *Colloids Surfaces A Physicochem. Eng. Asp.* **2019**, *580*, 123739. [CrossRef]
42. Lyytikäinen, J.; Ovaska, S.-S.; Soboleva, E.; Rinkunas, R.; Lozovski, T.; Maldzius, R.; Sidaravicius, J.; Johansson, L.-S.; Backfolk, K. Optimizing electric corona treatment for hydroxypropylated starch-based coatings. *Carbohydr. Polym.* **2018**, *197*, 359–365. [CrossRef] [PubMed]
43. Pankaj, S.K.; Bueno-Ferrer, C.; Misra, N.N.; O'Neill, L.; Tiwari, B.K.; Bourke, P.; Cullen, P.J. Dielectric barrier discharge atmospheric air plasma treatment of high amylose corn starch films. *LWT-Food Sci. Technol.* **2015**, *63*, 1076–1082. [CrossRef]
44. Bastos, D.C.; Santos, A.E.; da Silva, M.L.; Simão, R.A. Hydrophobic corn starch thermoplastic films produced by plasma treatment. *Ultramicroscopy* **2009**, *109*, 1089–1093. [CrossRef] [PubMed]
45. Guo, Z.; Gou, Q.; Yang, L.; Yu, Q.L.; Han, L. Dielectric barrier discharge plasma: A green method to change structure of potato starch and improve physicochemical properties of potato starch films. *Food Chem.* **2022**, *370*, 130992. [CrossRef] [PubMed]

polymers

MDPI

Article

Assessing Formability and Failure of UHMWPE Sheets through SPIF: A Case Study in Medical Applications

Ana Rosa-Sainz [1,2,*], M. Beatriz Silva [3], Ana M. Beltrán [2], Gabriel Centeno [1] and Carpóforo Vallellano [1]

[1] Departamento de Ingeniería Mecánica y Fabricación, Escuela Técnica Superior de Ingeniería, Universidad de Sevilla, 41092 Sevilla, Spain; gaceba@us.es (G.C.); carpofor@us.es (C.V.)
[2] Departamento de Ingeniería y Ciencia de los Materiales y del Transporte, Escuela Politécnica Superior, Universidad de Sevilla, 41011 Sevilla, Spain; abeltran3@us.es
[3] Instituto de Engenharia Mecânica (IDMEC), Instituto Superior Técnico, University of Lisbon, Av. Rovisco Pais, 1049-001 Lisboa, Portugal; beatriz.silva@tecnico.ulisboa.pt
* Correspondence: arosa@us.es

Abstract: This work presents a comprehensive investigation of an experimental study conducted on ultra-high molecular weight polyethylene (UHMWPE) sheets using single point incremental forming (SPIF). The analysis is performed within a previously established research framework to evaluate formability and failure characteristics, including necking and fracture, in both conventional Nakajima tests and incremental sheet forming specimens. The experimental design of the SPIF tests incorporates process parameters such as spindle speed and step down to assess their impact on the formability of the material and the corresponding failure modes. The results indicate that a higher step down value has a positive effect on formability in the SPIF context. The study has identified the tool trajectory in SPIF as the primary influencing factor in the twisting failure mode. Implementing a bidirectional tool trajectory effectively reduced instances of twisting. Additionally, this work explores a medical case study that examines the manufacturing of a polyethylene liner device for a total hip replacement. This investigation critically analyses the manufacturing of plastic liner using SPIF, focusing on its formability and the elastic recovery exhibited by the material.

Keywords: SPIF; UHMWPE; formability; failure; twisting; springback

Citation: Rosa-Sainz, A.; Silva, M.B.; Beltrán, A.M.; Centeno, G.; Vallellano, C. Assessing Formability and Failure of UHMWPE Sheets through SPIF: A Case Study in Medical Applications. *Polymers* **2023**, *15*, 3560. https://doi.org/10.3390/polym15173560

Academic Editor: Andrea Sorrentino

Received: 31 July 2023
Revised: 22 August 2023
Accepted: 23 August 2023
Published: 27 August 2023

1. Introduction

In recent decades, the paradigm of Industry 4.0, characterized by an orientation towards flexible, low to medium volume production processes with a strong emphasis on product customization [1], has instigated profound transformations in manufacturing processes. One representation of this progressive trend can be recognized in the manufacturing process known as Incremental Sheet Forming (ISF), more specifically, in its simpler, dieless variant termed as Single Point Incremental Forming (SPIF). SPIF is a flexible process that relies on simple tool geometries, making it cost-effective. By eliminating the need for a die, significant reductions in tool costs are achieved. In addition, the same setup can be used to produce various geometries, allowing easy customization at minimal costs.

During the last decades, the use of SPIF in the context of polymeric materials has seen a significant increase. This trend can be attributed not only to the aforementioned benefits of the process, but also to experimental validations that demonstrate the strains accomplished by SPIF exceed the Forming Limit Curve (FLC) without any material failure [2–4]. As described in Marques et al. [2] for polymeric sheets, the occurrence of necking is often postponed or even suppressed. Consequently, traditional Forming Limits Curve (FLC) is not applicable in this context. Instead, Fracture Forming Line (FFL) is commonly used to determine the formability limits of polymeric sheets within the principal strain space, considering this unique behavior.

In this context, Martins et al. [5] and Marques et al. [2] proposed a methodology to determine the FLC and FFL in thermoplastic materials based on the Circle Grid Analysis (CGA) method. However, it is important to note that the CGA method only provides a static analysis at the end of the test, which limits its ability to accurately track the propagation of necking during typical polymer deformation. Due to the lack of systematic procedures for determining the FLC and FFL in polymeric materials, the authors' previous work established a comprehensive methodology by adapting techniques used in sheet metal forming to polymer sheets. These methodologies were employed to determine the Forming Limit Diagram (FLD) in 1 mm [6] and 2 mm [7] thickness polycarbonate (PC) sheets, as well as in 2 mm thickness polyether ether ketone (PEEK) sheets [8].

Martins et al. [5] identified the main principal modes of failure in the SPIF process of polymers for conical geometry. The first mode, known as circumferential crack, involves cracks opening along the circumferential direction. The second mode, known as twisting or wrinkling, occurs when the material undergoes twisting within the cone, leading to the formation of wrinkles on the wall. Lastly, the third mode, termed oblique cracking, is characterized by cracks opening along the bisector direction in the inclined wall of the part. Typically, the most common modes of failure observed in SPIF are circumferential cracking and twisting. Le et al. [9] observed both of these failure modes in a cone-shaped part with a circular arc generatrix using a 3 mm thickness polypropylene (PP) sheet. Similarly, Davarpanah et al. [10] conducted tests on two different SPIF geometries, namely variable wall angle and fixed angle conical geometries, using a 0.7 mm thickness polylactic acid (PLA) sheet. In both geometries, circumferential cracking and twisting failures were identified. In addition to the aforementioned failure modes, Rosa-Sainz et al. [7] observed a different failure mode for PC sheets in SPIF known as crazing. The study conducted experiments using PC sheets with a thickness of 2 mm and a truncated conical geometry, varying the spindle speeds at 20, 500, and 1000 rpm. The findings showed that the crazing failure mode specifically occurred at a spindle speed of 1000 rpm. This led to the development of multiple micro-cracks on the deformed specimen. These modes of failure highlight the potential challenges and failure mechanisms associated with the SPIF process in polymer sheets.

As a result of the growth in the development of SPIF process analysis, several research studies have focused on medical applications involving the manufacturing of customized prostheses using incremental forming techniques with polymeric biomaterials. These biomaterials must possess biocompatibility; that is, they should be able to interact long-term with human body tissues without causing any harm [11]. For example, Bagudanch et al. [12], Centeno et al. [13], Clavijo-Chaparro et al. [14], and Chen et al. [15] have conducted research on the implementation of SPIF using polymeric sheets for cranial prostheses in the medical field. Fiorentino et al. [16] present another noteworthy study focusing on the production of biocompatible biomedical devices using ISF with biocompatible polymers. Their research specifically explores the ISF manufacturing process of a palate prosthesis using a combination of titanium alloy and PCL (polycaprolactone), yielding promising results.

In the medical field, the biocompatible polymer Ultra-High Molecular Weight Polyethylene (UHMWPE) is used in various orthopedic applications such as hips, knees, shoulders, elbows, wrists, ankles, and spinal disks due to its exceptional properties [17]. UHMWPE is known for its excellent wear resistance, low friction coefficient, biocompatibility, and high impact strength [18]. These properties make it an ideal material for joint replacement implants and orthopedic devices. Charnley [19] introduced UHMWPE as a bearing component for total joint replacements in 1962, and since then it has become the gold standard for manufacturing articulating surfaces in total hip, total knee, and total shoulder prostheses.

Advancements in polymeric materials led to the use of high-density polyethylene (HDPE) in the 1970s, which exhibited reduced wear rates (around 0.10 mm/year) [20]. In the early 2000s, UHMWPE emerged as a significant development, demonstrating even lower wear rates (<0.02 mm/year) [21]. UHMWPE excels in wear resistance, toughness, durability, and biocompatibility, making it a preferred choice as a bearing material in combination with ceramic or metallic counter surfaces in joint arthroplasty [22]. It is worth noting that over 90% of hip implants in the past decade have utilized UHMWPE [21,23].

Within this context, the present work encompasses three distinct objectives. Firstly, it aims to obtain the forming limit diagram of UHMWPE by employing established methodologies for determining the limits of forming through necking and fracture, as documented in the authors' previous research [24]. Secondly, to provide a comprehensive investigation on this material behavior subjected to single-point incremental forming. The purpose of this investigation is to assess the process parameters that impact the material in terms of its formability and failure modes. Lastly, to incorporate a practical case within the domain of medical prosthesis manufacturing. Specifically, it focuses on the manufacture of a plastic liner utilized in total hip replacement procedures using the SPIF process with UHMWPE.

2. Methods

This section focuses on the methodology employed and is categorized into the following subsections: (i) determination of forming limits, (ii) incremental sheet forming tests, and (iii) medical application of the plastic liner with UHMWPE.

2.1. *Formability Limits*

The ultra-high molecular weight polyethylene (UHMWPE) sheets utilized in this research were procured as 2 mm thickness sheets measuring 500 × 500 mm^2 from an industrial supplier. Based on the manufacturer's data sheet provided by Goodfellow in Pittsburgh, PA, USA, these sheets have a density of 0.940 g/cm^3, a yield stress of 40 MPa, a Rockwell hardness of R70, and demonstrate a 500% elongation at fracture [25].

To assess the material's formability limits in terms of necking and fracture, Nakajima tests were conducted. These tests followed the ISO 12004-2 standard [26] and were carried out at room temperature (15–20 °C) using an universal deep drawing machine model 142-20 from ERICHSEN GmbH Co. (Hemer, Germany) (Figure 1a), depicting a tensile strain specimen being deformed by a 100 mm diameter hemispherical punch with a deformation speed of 60 mm/min.

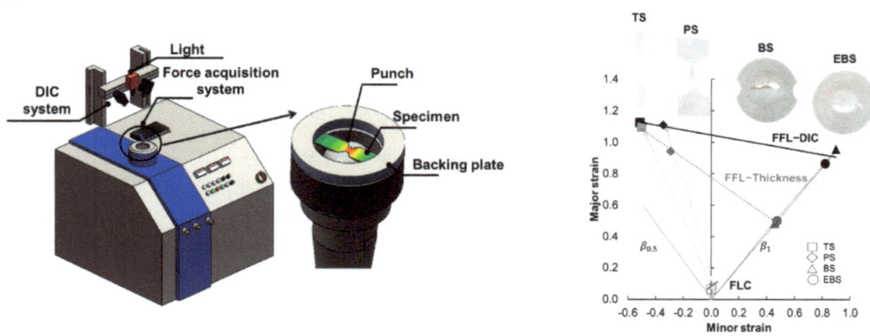

(**a**) (**b**)

Figure 1. (**a**) Schematic representation of the universal deep drawing machine along with the Nakajima test apparatus. (**b**) Formability limits for 2 mm thickness UHMWPE represented in principal strain space.

To measure the principal strains within the deforming area near the initial crack zones, an ARAMIS v6.2.0 digital image correlation (DIC) system from GOM (Braunschweig, Germany) was employed. The surface is coated with a black and white stochastic pattern which results in a fine spray of matte black speckles on a white background. The ARAMIS v6.2.0 system consists of two cameras positioned at a relative angle of 23.6° and is equipped with a lens that has a focal length of 50 mm.

The Nakajima specimen geometries were obtained by machining the sheets provided, and these specimens were aligned longitudinally with the extrusion direction. The polymer

Nakajima specimens were machined to achieve different strain loading paths from near to tensile strain conditions (approximately $\beta = \frac{\varepsilon_2}{\varepsilon_2} = -0.3$) towards equibiaxial stretching ($\beta = 1$). This resulted in various loading paths, including the tensile strain (TS), plane strain (PS), biaxial strain (BS), and equibiaxial strain (EBS).

In this sense, the dimensions of the TS specimen were as follows: the length was 200 mm, the width was 30 mm, the gauge length was 5 mm, the gauge width was 15 mm, and the notch radius was 25 mm. The PS specimen was machined with a length of 200 mm, a width of 114 mm, a gauge length of 4 mm, a gauge width of 45 mm, and a notch radius of 25 mm. As for the BS specimen, it featured a gauge length of 5 mm, a gauge width of 150 mm, a notch radius of 25 mm, and a total diameter of 182 mm. Lastly, the EBS specimen had a total diameter of 182 mm. The dimensions mentioned of the geometries used and the operating conditions can be found in a previous publication by the authors [16]. To minimize friction, two layers of polytetrafluoroethylene (PTFE) were placed between the punch and the polymeric specimen, along with three layers of Vaseline. To obtain statistically meaningful results, each Nakajima test specimen geometry was repeated three times.

The formability limit by necking (FLC) associated with necking was determined in the principal strain space using a time-dependent methodology commonly used for metals [27]. This methodology examines the temporal evolution of the major strain and its first-time derivative (the major strain rate) along a section perpendicular to the crack to identify neck initiation and its progression. On the other hand, the formability limit by fracture (FFL) was determined by measuring the thickness of specimens in the fractured surface using optical microscopy [28]. The minor strain at fracture, ε_{2f}^{*}, and major strain at fracture, ε_{1f}^{*}, were calculated by assuming a constant local loading path slope of the principal strain ratio, β^{*}, which was determined from the DIC system data ($\beta^{*} = d\varepsilon_2^{DIC}/d\varepsilon_1^{DIC}$). Notably, UHMWPE exhibits significant elastic recovery after fracture due to its relatively low Young's modulus of elasticity (~1.2 GPa according to the manufacturer). To establish the fracture forming limit line (FFL), the last pair of principal strain values provided by the DIC system before fracture was considered. Further information and details regarding the determination of forming limits for the material can be found in previous works by the authors [6,7].

Figure 1b illustrates the forming limits within the principal strain space for the analyzed 2 mm thickness UHMWPE sheets. The forming limit curve (FLC) in Figure 1b, represented by a grey straight line, was constructed using the strain pairs obtained from the Nakajima tensile strain (TS) and plane strain (PS) Nakajima specimens using the time-dependent approach. The TS strain pairs are depicted as squares, while the PS strain pairs are depicted as rhomboids in Figure 1b. The results obtained from the time-dependent methodology indicated that necking was not the failure mode for the biaxial strain (BS) and equibiaxial strain (EBS) Nakajima specimens.

In terms of the Fracture Forming Limit (FFL) based on thickness, the strain pairs at fracture obtained from the measurement of the fractured specimens were lower than the final points obtained from the Digital Image Correlation (DIC) system. This behavior is similar to what was observed for other materials studied by the authors [6–8]. Therefore, it was deemed more appropriate to characterize the FFL for the 2 mm thickness UHMWPE using the FFL−DIC method, represented by a straight line following $\varepsilon_1^{DIC} = -0.1573\varepsilon_2^{DIC} + 1.047$.

2.2. SPIF Tests

The SPIF experiments were conducted using a spherical end forming tool driven by a CNC milling machine model VMC 100 with 2.5 axes from EMCO GmbH (Hallein, Salzburger) equipped with an EMCOTRONIC TM02 controller. The experiments were performed at room temperature (15–25 °C). The dimensions of the truncated pyramidal geometry are indicated in Table 1. The temperature during the tests was recorded using a FLIR T430sc thermographic camera from Teledyne FLIR LLC (Wilsonville, OR, USA). The thermal camera is attached to a tripod positioned in front of the CNC milling machine, recording the whole surface of the SPIF setup. Temperature data was collected through the following process: multiple points on the surface of the geometries were continuously observed over a period of time. This approach accounted for potential temperature variations caused by factors such as lubricant distribution and tool motion. The selected points were positioned on two separate walls and two distinct corners (specifically for the truncated pyramidal geometry). For each of these areas, the points were positioned at different elevations. The average temperature across all the points was subsequently calculated.

Table 1. Truncated pyramidal geometry dimensions.

Geometry	Dimensions						Wall Angle
	L_A (mm)	L_B (mm)	R_1 (mm)	H (mm)	ψ	t_0 (mm)	
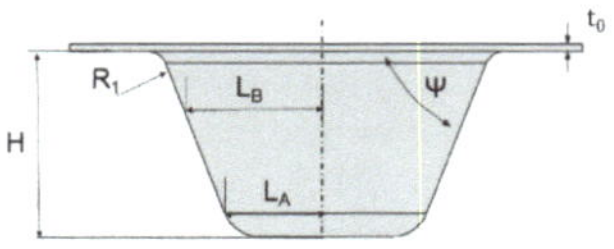	25	40	5	50	20	2	Fixed

Figure 2a illustrates a schematic view of the SPIF setup assembly. The tool followed a predetermined incremental tool path to deform the clamped UHMWPE sheet blank, gradually transforming it into the final truncated pyramid shape. To reduce friction during the forming process, water was applied as a lubricant between the tool and the surface of the sheet. To conduct the SPIF tests, two different toolpaths were utilized: the unidirectional trajectory, denoted as "U", and the bidirectional trajectory, denoted as "B". In the case of the unidirectional trajectory, as shown schematically in Figure 2b, the starting point was positioned at the midpoint of the pyramid's wall instead of a corner to avoid excessive stress concentration at the vertices. The sequence of the unidirectional trajectory was as follows: starting from the designated point, a clockwise trajectory was followed until the starting point again. Then, the tool was incrementally reduced by a certain step down and the clockwise trajectory was repeated. This process was repeated until the final desired shape was achieved. To prevent the occurrence of twisting failure mode [29,30], the tool was rotated in the opposite direction to the trajectory.

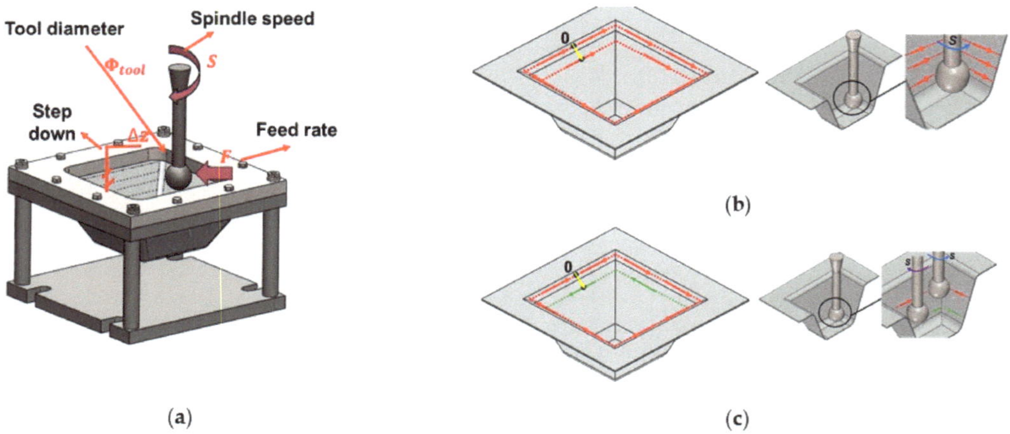

Figure 2. (**a**) Schematic of the SPIF setup and the selected trajectories named: (**b**) unidirectional and (**c**) bidirectional.

The bidirectional trajectory incorporates both clockwise and anticlockwise movements (Figure 2c). The sequence entails initiating a clockwise trajectory from the starting point until returning to the starting point, followed by the descent of the tool. Subsequently, a counter clockwise trajectory is executed. This process is repeated until the desired shape is attained.

The specific operating parameters for UHMWPE used in each experiment are summarized in Table 2. The main focus was to investigate how temperature affects failure modes, particularly twisting. Consequently, the variables analyzed included the influence of the spindle speed and the step down on temperature, as well as the impact of the imposed trajectory on the failure mode in SPIF. To conduct these investigations, the tool diameter was fixed at 10 mm and the feed rate was set at 1000 mm/min. Based on previous experience with polymeric sheet [7,8,31], it was known that spindle speed and step down significantly influence the failure modes. The authors' previous work [7,31] on PC at 20, 500, and 1000 rpm revealed that very high spindle speeds can lead to a failure mode known as crazing, specifically at 1000 rpm. However, increasing the spindle speed contributed to enhanced formability while reducing geometric precision. In this regard, the intention is to compare two significantly different spindle speed values (20 and 500 rpm) to observe the behavior of UHMWPE in terms of temperature, formability, and elastic recovery.

Table 2. Experimental SPIF work plan for UHMWPE sheets.

Test Condition	Φ_{tool} (mm)	S (rpm)	F (mm/min)	z (mm)
U TP1	10	500	1000	0.5
U TP2	10	500	1000	0.3
U TP3	10	500	1000	0.1
U TP4	10	20	1000	0.5
U TP5	10	20	1000	0.3
U TP6	10	20	1000	0.1
B TP1	10	500	1000	0.5
B TP2	10	500	1000	0.3
B TP3	10	500	1000	0.1

Additionally, three different step down values were examined: 0.1 mm, 0.2 mm, and 0.5 mm. Two trajectories were studied: the unidirectional trajectory (referred to as "U" in Table 2) and the bidirectional trajectory (referred to as "B" in Table 2). In Table 2, the test condition includes the type of trajectory and the number of the test. To ensure repeatability, three repetitions of each test were performed.

The SPIF tests were conducted until fracture. The resulting outputs considered were the final depth and the specific mode of failure observed during the tests. After the SPIF tests, a qualitative analysis of the specimens' geometries was performed using a scanner from 3D Systems model Geomagic Capture (Rock Hill, SC, USA).

In addition to the acquired failure modes, a thorough analysis of the strains in SPIF was performed. In previous studies conducted by the authors, a ARGUS system from GOM (Braunschweig, Germany) was used to measure strains after the SPIF deformation process [7,8]. The ARGUS system is an offline 3D deformation digital measurement system that utilizes circle grid analysis (CGA) methodology. To the measurement process, a dot pattern is typically applied to the surface of the specimen prior to the deformation. In the case of UHMWPE, attempts were made to print the dot pattern directly onto the surface of the sheet prior to deformation. However, it was found to be challenging to adhere the dot pattern to the UHMWPE surface during deformation. Various alternatives were explored, including the use of an ink pad and surface fixation treatments, but these attempts proved unsuccessful. The difficulty in applying the dot pattern to UHMWPE is attributed to the high viscosity [32] which restricts its flow and spread, making it unsuitable for conventional ink printing methods. Liquid ink formulations typically require specific viscosity and flow properties to be effectively deposited onto a substrate, but the high viscosity hinders its flowability and prevents successful ink printing.

In this respect, an alternative approach was employed to obtain the strain distribution in the material. This involved measuring the thickness of the SPIFed pyramids after fracture, taking into account the typical modes of deformation observed in a pyramid: plane strain on the walls and biaxial strain at the corners [2]. When the thickness measurements of the fractured pyramids were analyzed, valuable insights could be gained regarding the strain distribution within the UHMWPE material. This approach allowed for an indirect estimation of the strain levels experienced by the material during the SPIF process, particularly in the regions of interest where plane strain and biaxial strain are prevalent, namely the walls and corners of the pyramids, respectively.

2.3. Medical Application

Hip replacement surgery is a surgical procedure that is performed to replace deteriorated bone and cartilage within the hip joint with new prosthetic implants. A standard total hip replacement comprises four distinct components, which are shown schematically in Figure 3a: (i) the acetabular cup (shell), which fits in the pelvis and is made of metal; (ii) the acetabular liner, which fits into the acetabular cup; (iii) the femoral head (ball) fits on top of the stem, and the end of the stem is tapered to allow the ball to wedge into position and be held tightly in place with friction; and (iv) the femoral stem is the part of the replacement that fits into the thigh bone [33]. Collectively, these components form a complete hip replacement system (see Figure 3b), reinstating joint functionality and alleviating pain and discomfort associated with hip joint degeneration.

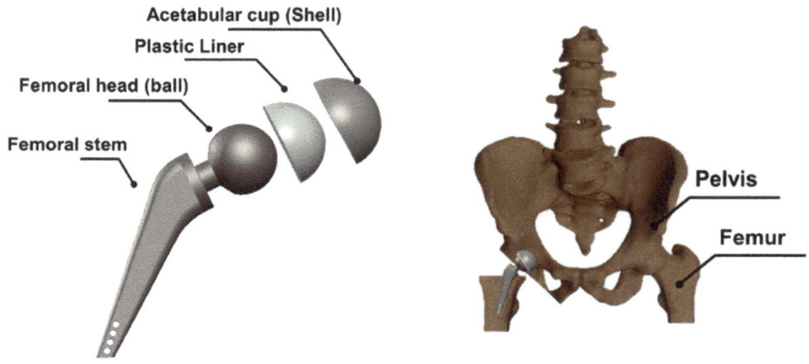

(a) (b)

Figure 3. (**a**) Schematic representation of a total hip replacement depicting the different components. (**b**) Assembly of the artificial hip joint prosthesis into the pelvis bone.

The design of the plastic liner is contingent upon the shell (acetabular cup) used in the hip replacement system. In this regard, various manufacturers provide catalogues specifying the appropriate dimensions for the liner according to the specific shell required for each patient. For the purpose of this study, the Zimmer Biomet Institute (Zählerweg, Switzerland) a manufacturer renowned for its expertise in medical applications, will be used as a reference. The manufacturer provides a catalogue for the acetabular system [34], which includes specific recommendations for the plastic liner. According to the catalogue, a 44 mm plastic liner is recommended for use with a 62 mm shell. The selection of this particular diameter is determined by both the thickness of the plastic sheet and the diameter of the shell. Typically, plastic liners designed for 62 mm shells have a thickness range between 2 and 5 mm. However, it is important to note that larger shell diameters require the use of thicker plastic sheets. This consideration ensures an optimal fit, functionality, and durability of the hip replacement system.

In Table 3, the geometry and dimensions considered for manufacturing the plastic liner device are indicated. In this regard, two different approaches were taken into consideration. According to the catalogue [34], the plastic liners have a geometry comprising a cylindrical wall and a hemispherical zone. This study will begin by examining a purely hemispherical geometry with a 44 mm diameter to observe the behavior of the plastic when subjected to SPIF. Subsequently, the geometry proposed in the catalogue will be considered, which includes a 10 mm cylindrical wall and a hemispherical zone with a 4 mm diameter.

Table 3. Geometrical parameters for the two considered manufacturing geometries of a plastic liner using UHMWPE.

Geometry	Parameters (mm)		
Hemispherical Geometry	Dimensions		
		Ø	t_0
		44	2
Cylindrical and a hemispherical zone	H	Ø	t_0
	10	44	2

The proposed medical application with UHMWPE considered two premises: (i) minimize the forming time and study the feasibility of manufacturing a plastic liner; and (ii) with the aim of reducing the forming time, improve the springback of the final part produced by SPIF. Therefore, all tests for the case of the hemispherical geometry and the case of cylinder + hemispherical geometry were carried out with a step down of 0.5 mm to reduce the forming time. All tests were performed at a spindle speed of 20 rpm to overheat the material [35,36]. Furthermore, the "cycle" parameter was taken into consideration, which is defined as a complete tool trajectory required to deform the part to its final depth. When the test comprises 3 cycles, it signifies that the forming process is repeated continuously three times. The primary objective of increasing the number of cycles was to enhance the springback characteristics of the material.

All tests were deformed using the bidirectional trajectory according to the results obtained for the pyramidal truncated geometry, and Table 4 indicates the hemispherical geometry (designated as "H" in Table 4) and for cylinder + hemispherical geometry (designated as "CH" in Table 4). To ensure repeatability, two repetitions were performed for each test.

Table 4. Process parameters for the two considered manufacturing geometries of a plastic liner using UHMWPE.

Test Condition	Φ_{tool} (mm)	S (rpm)	F (mm/min)	z (mm)	Cycles
B H1	10	20	1000	0.5	1
B H2	10	20	1000	0.5	3
B CH1	10	20	1000	0.5	3

A new backing plate was designed and manufactured to better fit the plastic liner device. The design of the backing plate was made using SolidWorks® CAD software 2020 SP 0.0 and fabrication using a 3D printing by Fusion Deposition Modelling (FDM) using PLA filament. The 3D printing process was carried out using a 3D S5 printer from

UltiMaker© (Utrecht, The Netherlands), and the UltiMaker CURA® 4.11 software was employed with the following process parameters: the printing temperature is set at 200 °C, whereas the printing speed is 60 mm/s. The chosen pattern for printing is a raster angle of 45°/−45°. The percentage of infill is 100%, indicating that the internal structure of the printed object is completely solid. Figure 4a offers a visualization of the 3D model using UltiMaker CURA® 4.11 software, whereas Figure 4b displays the SPIF setup, focusing on the 5 mm thickness of the 3D printed backing plate.

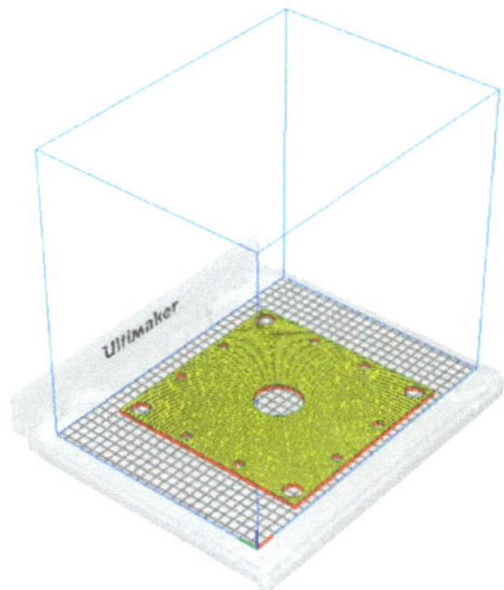

(**a**)

(**b**)

Figure 4. (**a**) 3D model printed using UltiMaker CURA® 4.11 software. (**b**) Mounting of a PLA 3D printed backing plate in the SPIF setup.

3. Results and Discussion

This section presents the results and discussions obtained, which are divided into two distinct subsections: Section 3.1 focuses on the analysis of formability and failure in the SPIF pyramid tests, while Section 3.2 delves into the results obtained for the manufacturing of the medical prostheses using SPIF.

3.1. SPIF Tests

This section provides an overview of the results obtained from the deformation of 2 mm thickness UHMWPE sheets using the SPIF process for the pyramidal geometry. The section is structured into two main parts: (i) an analysis focusing on the formability of the UHMWPE sheets; and (ii) an examination of the temperature data recorded during the SPIF experiments.

3.1.1. SPIF Analysis

The experimental work using the SPIF process is summarized in Table 5. The study involved deforming a truncated pyramidal geometry using two different trajectories: a unidirectional trajectory (trajectory U) and an alternate trajectory (trajectory B) combining clockwise and anticlockwise movements. In this study on UHMWPE, the focus was on investigating the influence of temperature on twisting occurrence and the effect of the trajectory on the observed failure modes. To achieve this objective, a total of nine tests were conducted on UHMWPE sheets, as specified in Table 5.

As explained in Section 2.2, three repetitions of each test were conducted to ensure repeatability. The tests were divided into three categories: three tests following a unidirectional trajectory with a spindle speed of 500 rpm, an additional three tests with a unidirectional trajectory but with a reduced spindle speed of 20 rpm, and finally, three tests performed using a bidirectional trajectory with a spindle speed of 500 rpm. For each test case, the resultant failure mode and the average depth of fracture, derived from the three repetitions, are indicated in Table 5.

Table 5. SPIF results for the truncated pyramidal geometry with a 2 mm thickness of UHMWPE.

Test Condition	Φ_{tool} (mm)	*S* (rpm)	*F* (mm/min)	ΔZ (mm)	Failure	Fracture Depth (mm)
U TP1	10	500	1000	0.1	T + F	29.7
U TP2	10	500	1000	0.3	T + F	36.3
U TP3	10	500	1000	0.5	T + F	39
U TP4	10	20	1000	0.1	T + F	29.4
U TP5	10	20	1000	0.3	T + F	40
U TP6	10	20	1000	0.5	T + F	32
B TP1	10	500	1000	0.1	F	28.1
B TP2	10	500	1000	0.3	F	39
B TP3	10	500	1000	0.5	F	37.5

A series of three initial tests was conducted using a unidirectional trajectory ("U") for the UHMWPE specimens with 500 rpm of spindle speed. The tests were carried out under identical spindle speed, feed rate and tool diameter conditions, with the only variation being the step down values of 0.5 mm, 0.2 mm, and 0.1 mm. Figure 5 illustrates the results obtained from these three specimens. In all three cases, the observed failure modes were a combination of twisting (indicated as "T" in Table 5) and fracture (indicated as "F" in Table 5). It can be observed that the larger step down values corresponded to a higher degree of twisting, as is evident in Figure 5a compared to Figure 5c. The research conducted by Davarpanah et al. 2015 [10] indicated that increasing the step down in SPIF of polymers (PLA and PVC) can have a positive effect in terms of formability. However, it was found that excessively large step down values could cause twisting.

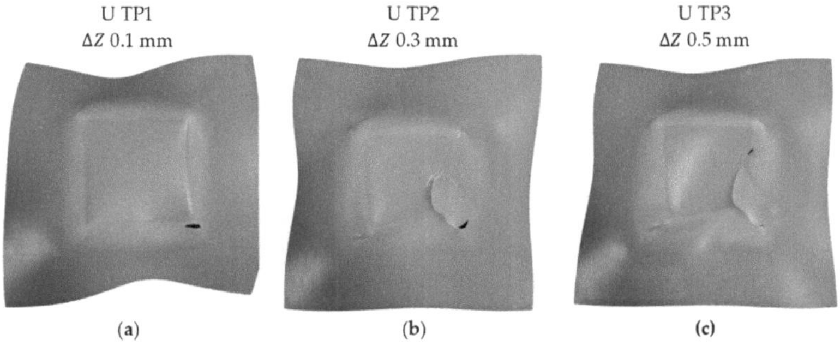

Figure 5. UHMWPE specimens formed by SPIF with the conditions of spindle speed 500 rpm, unidirectional trajectory and step downs of: (**a**) 0.1, (**b**) 0.3, and (**c**) 0.5 mm.

The phenomenon of twisting manifested primarily at the corners of the pyramid due to the inherent stress concentration at these vertices. Moreover, a noticeable trend was observed: decreasing the step-down value resulted in a reduction of the twisting occurrence, as depicted in Figure 5a. In line with this, a recent study by Formisano et al. [37] demonstrated that the twisting originated from the component of forming forces tangential to the toolpath. This specific tangential force led to in-plane shearing, ultimately inducing uncontrolled torsion in the sheets relative to the clamping frame. Additionally, the phenomenon was aggravated by heightened vertical forming forces during polymer sheet shaping. This intensification was attributed to significant indentation, further magnifying the occurrence. In this context, an increase in the step down corresponds to heightened vertical forces [3,10], thereby yielding a more pronounced twisting effect in the UHMWPE specimens.

To address the issue of twisting, which is exacerbated by the heating of the contact surface between the tool and the UHMWPE sheet due to friction [9,10], the spindle speed was reduced in an attempt to mitigate this effect. Consequently, another set of three tests was performed using a unidirectional trajectory ("U") with a spindle speed of 20 rpm. The step down values were varied, ranging from 0.1 mm to 0.5 mm. Similar to the previous tests, the failure mechanism observed in all three cases involved a combination of twisting and fracture, as depicted in Figure 6. It can be observed that the larger step down values resulted in a higher degree of twisting. The twisting phenomenon started at a corner and propagated along the edge, accompanied by the development of cracks. However, it should be noted that the level of twisting observed in these tests was not as prominent as in the cases with a spindle speed of 500 rpm, as shown in Figure 6a–c. In fact, the reduction in spindle speed, along with the smallest step down, led to a significant reduction in twisting in the "U TP4" specimen (Figure 6a), where it is nearly imperceptible.

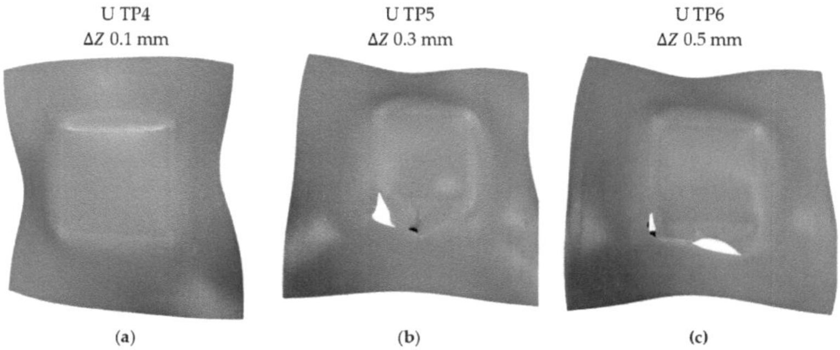

Figure 6. UHMWPE specimens formed by SPIF with the conditions of spindle speed 20 rpm, unidirectional trajectory and step downs of: (**a**) 0.1, (**b**) 0.3, and (**c**) 0.5 mm.

Subsequently, the impact of the trajectory on the failure mode was examined, despite the utilization of a high spindle speed. To investigate this, a series of three tests was conducted using a bidirectional trajectory ("B") with a spindle speed of 500 rpm. In this regard, the mode of failure associated with the bidirectional trajectory was observed to be solely fracture (Figure 7). The images presented in Figure 7 clearly demonstrate the absence of the twisting mode of failure when using the bidirectional trajectory. In this case, fracture is the sole cause of failure observed. Similar to previous tests, the fracture starts at a corner of the pyramid and proceeds to spread along its edge. In particular, the size of the fracture increases with the larger step down values. In this context, a recent investigation conducted by Yang and Chen [30] unveiled a consistent accumulation of material flow and shear strain aligned with the motion of the tool. Consequently, this phenomenon results in an uneven distribution of thickness within the formed component. In this regard, the inevitable occurrence of twisting is substantiated when utilizing the unidirectional trajectory. Yang and Chen's investigation involved experimenting with a novel alternative trajectory, employing an interpolation algorithm between three neighboring contour lines. A comparison of their findings demonstrates that this innovative alternate trajectory can significantly enhance the geometric accuracy of the formed parts.

As explained in Section 2.2, an alternative methodology was followed to obtain the strains for UHMWPE in SPIF. This procedure is based on the work conducted by [5], where measuring the thickness at the fracture is essential for determining the "gauge length" strains. In Figure 8a, a section of the truncated pyramid model is depicted, indicating Zone I (without fracture) and Zone II (with fracture). In Figure 8b, images captured by an optical microscope model SMZ800N from Nikon Corporation (Tokyo, Japan) are presented to provide a clear view of the thickness of the specimen. To accurately measure the thickness, the specimens were cut through a section perpendicular to the crack. The edges of the cut were smoothed using a polisher to ensure precise measurements. Subsequently, the thickness along the section was measured using the microscope. Measurements were taken as close to the fracture region as possible to obtain an accurate representation of the thickness of the fracture. Additionally, various sections without fracture were also measured to verify and corroborate the strain distributions within the material.

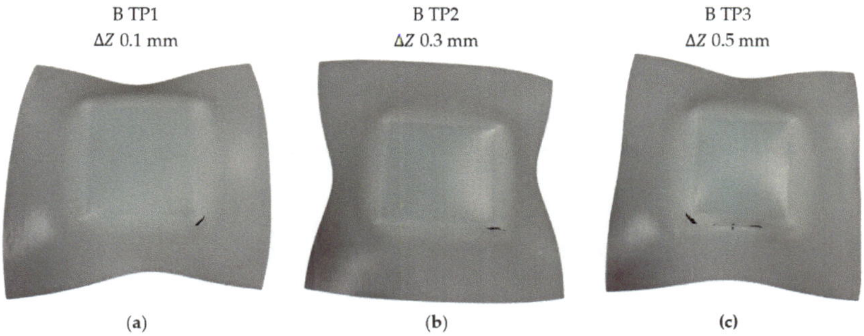

Figure 7. UHMWPE specimens formed by SPIF with the conditions of spindle speed 500 rpm, bidirectional trajectory and step downs of: (**a**) 0.1, (**b**) 0.3, and (**c**) 0.5 mm.

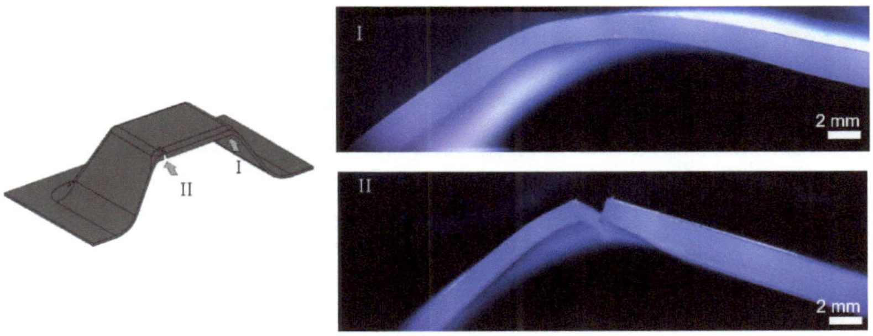

Figure 8. Image of the fracture and non-fracture zones of the UHMWPE SPIFed in the truncated pyramidal geometry. Marker "I" indicates the view of the area without the fracture, whereas "II" points to the section where the fracture is situated.

To assess the magnitude of the major strains in the walls of the truncated pyramidal geometry SPIF tests, a minor strain of zero was assumed to establish a plane strain condition [2]. Additionally, the principle of volume constancy was imposed. As explained previously, the thickness along the fracture section was measured (see Figure 9a for the case of the test B TP1).

Figure 9b illustrates the Forming Limit Diagram (FLD) for the selected SPIF tests without twisting, along with the Fracture Forming Limit (FFL) determined based on the thickness measurements (FFL−thickness) and data from the DIC system (FFL−DIC) obtained under standard Nakajima conditions. The results for the three specific tests, namely "B TP1" with a step down of 0.1 mm (represented by a square marker with grey outline), "B TP2" with a step down of 0.3 mm (represented by a triangle marker with purple outline), and "B TP3" with a step down of 0.5 mm (represented by a circular marker with yellow outline), are presented in Figure 9b. In the FLD, the filled marker represents the strains obtained by measuring the fracture thickness.

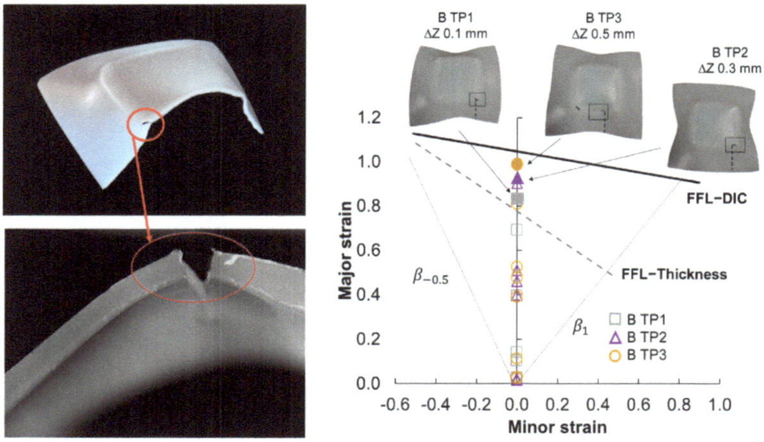

(a) (b)

Figure 9. (**a**) Procedure based on the thickness measurement to obtain the strains along a section of the SPIFed specimens. The red circle indicates the fracture that occurred in the biaxial zone (corner) and the plane strain zone (wall). (**b**) FLD for 2 mm thickness UHMWPE sheets within the principal strain space, indicating the three specimens without twisting, highlighting the fracture location, and the section where it was measured using a dashed line.

As can be observed in Figure 9b, the strains at fracture of the three tests were above the FFL thickness and below the FFL−DIC. In this regard, the FFL−DIC is calculated based on the strains in the last frame measured by the DIC system in such a way that it does not account for the immediate elastic recovery that occurs when the specimen fractures, as it is in the last moment before breaking. On the other hand, the FFL thickness is calculated from the measurement of the thickness of the specimens after the test, such that it does consider this elastic recovery of the material. In the SPIF tests, the thickness along the fracture section has been measured to obtain an order of magnitude of the strains, thus it is measuring while considering the elastic recovery. For that reason, it would be more appropriate to evaluate the strains of the SPIF tests with the obtained FFL thickness, to take into account in both cases the elastic recovery of the material. Consequently, the SPIF fracture strains are above the FFL thickness but below the FFL−DIC, demonstrating an increase in the formability of the material under SPIF.

Furthermore, the comparison between the strains obtained through SPIF and the FFL thickness, based on thickness measurements by Nakajima, reveals that the SPIF test with a step down of 0.5 mm exhibited the highest level of formability. The results of major strain values indicated 0.83 for B TP1, 0.92 for B TP2, and 0.98 for B TP3. An increase in the step down is attributed to the corresponding increase in the maximum attainable wall angle, leading to an improvement in formability. This result is consistent with the research conducted by Bagudanch et al. [3], where an increase in the step down was found to have a positive influence on improving the formability of PVC sheets in SPIF. Similarly, Davarpanah et al. [10] reported similar findings for PLA parts deformed by SPIF, indicating that higher step down values contributed to increased formability.

This investigation provides compelling evidence that the occurrence of twisting, as a mode of failure, can be effectively prevented using the bidirectional trajectory, even with 500 rpm of spindle speed, and even when using a 0.5 mm step down. This significant finding not only enhances the formability of the material but also contributes to reducing the overall duration of the forming process.

3.1.2. Temperature Analysis

This section presents the results of the temperature data analysis during the SPIF tests for the 2 mm thickness UHMWPE truncated pyramidal geometry.

Figure 10 presents the temperature evolution graphs obtained from the truncated pyramidal geometry of SPIF tests conducted on the 2 mm thickness of UHMWPE. The graphs are categorized as follows: Figure 10a represents the unidirectional strategy with a spindle speed of 500 rpm, considering three different step downs (0.1, 0.3, and 0.5 mm); Figure 10b represents the unidirectional strategy with a spindle speed of 20 rpm, also considering three different step downs (0.1, 0.3, and 0.5 mm); and Figure 10c represents the bidirectional strategy with a spindle speed of 500 rpm, again considering three different step downs (0.1, 0.3, and 0.5 mm).

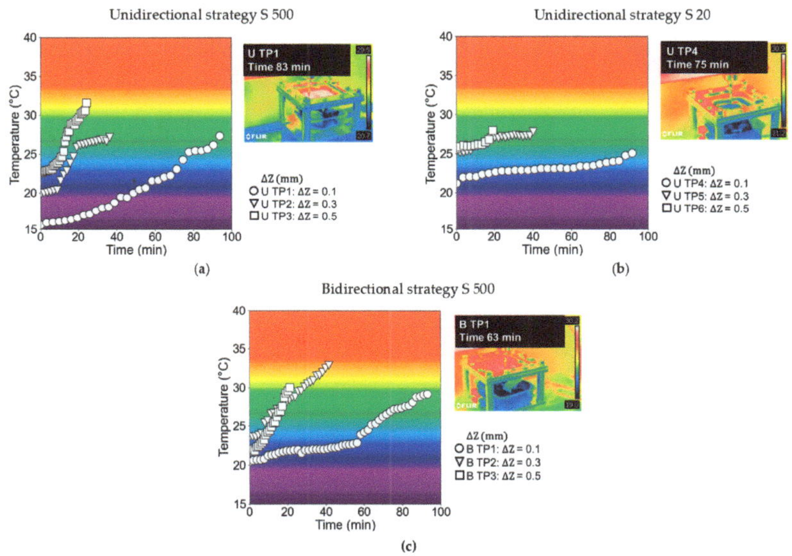

Figure 10. Temperature evolution of UHMWPE for each test in the following scenarios: (**a**) unidirectional strategy with a spindle speed of 500 rpm; (**b**) unidirectional strategy with a spindle speed of 20 rpm; and (**c**) bidirectional strategy with a spindle speed of 500 rpm.

In each graph, the temperature data for the 0.1 mm step down is depicted using circles, the 0.3 mm step down is represented by inverted triangles, and the 0.5 mm step down is represented by squares. Additionally, each graph is accompanied by a temperature contour test image captured using the thermographic camera. These images provide a visual representation of the temperature distribution during the SPIF tests.

In Figure 10a, the temperature data for the unidirectional strategy with a spindle speed of 500 rpm is presented for the tests "U TP1" (0.1 mm step down), "U TP2" (0.3 mm step down), and "U TP3" (0.5 mm step down). It can be observed that "U TP3" (0.5 mm step down) achieved a higher maximum temperature (31.6 °C) compared to the other tests and, correspondingly, it exhibited the shortest forming time (25 min). Conversely, "U TP1" (0.1 mm step down) demonstrated a stable temperature evolution throughout the test, despite using a spindle speed of 500 rpm. Bagudanch et al. [36] observed a similar outcome in their study on UHMWPE, finding that higher values of step down in SPIF led to an increase in temperature levels of the material and a reduced forming time. Similarly, in a recent study conducted by Yang et al. [38] on PEEK, a series of hot SPIF experiments were performed using truncated pyramids with constant wall angles and varying wall angle cones, revealing that as the step down value increased during SPIF for the two geometries, the proportion of the higher temperature range also increased.

Figure 10b illustrates the temperature data obtained using a spindle speed of 20 rpm for the unidirectional strategies of the tests "U TP4" (0.1 mm step down), "U TP5" (0.3 mm step down), and "U TP6" (0.5 mm step down). The temperature evolutions observed in these three tests were similar and demonstrated stability throughout all the tests. Among these tests, "U TP4" (0.1 mm step down) exhibited the lowest temperature, reaching approximately 24 °C. It should be noted that "U TP4" (step down 0.1 and spindle speed 20 rpm) had the least noticeable occurrence of failure mode by twisting on a macroscopic level. This indicates that the combination of a lower spindle speed and a smaller step down value helps to reduce the occurrence of twisting. When a spindle speed of 20 rpm was used, the temperature evolutions for the 0.3 mm and 0.5 mm step downs were similar, resulting in nearly the same maximum temperature levels at the end of the tests, and both presenting twisting as a failure mode. In this sense, as previously mentioned, as the step down value increased, the twisting effect became more pronounced.

As anticipated, the comparison between the results obtained using a spindle speed of 500 rpm and 20 rpm revealed that higher spindle speed values resulted in increased temperature data. This finding aligns with the findings of the study conducted by Lozano-Sánchez et al. [39], which involved the formation of pyramid-shaped parts using UHMWPE through SPIF. Their results demonstrated that an increase in spindle speed led to elevated heat generation due to friction between the tool and sheet. Consequently, in their study, experiments conducted with a spindle speed of 2000 rpm failed due to twisting, unlike those performed with a free rotating tool. Furthermore, Bagudanch et al. [40] also observed that an increase in spindle speed can contribute to a temperature increase during the forming process. This temperature rise is attributed to the intensified friction between the sheet blanks and tools compared to a scenario with a freely rotating tool. Additionally, in the recent work by Formisano et al. [41], the impact of spindle speed on temperature was compared for PC material at 0, 200, and 400 rpm. This study demonstrated that increasing the tool rotation speed can lead to a substantial elevation in the temperature of the contact surface. Furthermore, this study proved that raising the temperature produces both detrimental and beneficial effects. For instance, in the case of PC, processing temperatures higher than 120 °C resulted in severe deterioration. However, it was observed that an increase in temperature can also yield positive outcomes, such as reducing processing loads due to the significant reduction in the flow stress of the polymer, consequently leading to a significant reduction in springback.

On the other hand, Figure 10c provides the temperature data for the bidirectional strategies with a spindle speed of 500 rpm in the tests "B TP1" (0.1 mm step down), "B TP2" (0.3 mm step down), and "B TP3" (0.5 mm step down). Similar to the previous cases, "B TP3" (0.5 mm step down) exhibited a slight increase in temperature data and the shortest forming time compared to the other tests. Tests with 0.1 mm and 0.3 mm step downs displayed comparable temperature evolutions, but in this case, the test with a 0.3 mm step down recorded a higher maximum temperature of approximately 32.9 °C. The experimental studies conducted using the bidirectional approach revealed that fracture was the primary mode of failure, with no evidence of twisting.

Ultimately, a temperature empirical model was developed using Python language [42,43], with the specific purpose of analyzing the variations in temperature within both tool trajectories, taking into account the investigated process parameters. The following equation: $\mathrm{Tmax}(\Delta_z, S) = a \times b^{\log(\Delta_z)} + c \times S + T_0$ reproduces quite well the temperature experimental evolution, where a, b, and c are constants determined to achieve a suitable fit, and T_0 represents the initial temperature. As observed in Figure 10, the initial temperature ranged from 15 °C to 25 °C, mainly depending on the room temperature within the workspace. An empirical model was developed using various initial temperatures (15, 20, and 25 °C), and the coefficient of determination (R^2) value approached 1 as the initial temperature increased. The highest R^2 value achieved was 0.97 when using an initial temperature of 25 °C, indicating that the empirical model provided a closer fit to the experimental results.

Furthermore, a comprehensive examination of the deformed pyramids produced through SPIF is presented, employing a 3D scanner to observe and assess the extent of twisting, whether it demonstrates an increase or decrease. This analysis presents a process window based on the results obtained from deformed pyramids using SPIF with UHMWPE.

Figure 11a displays the logarithmic surface that fits the temperature results obtained for the unidirectional toolpath. Additionally, the scanned surfaces of the UHMWPE pyramid specimens are shown. It can be observed that for a spindle speed of 20 rpm (triangular markers), an increase in step down leads to an increment in temperature and, in terms of failure modes, a macroscopic increase in twisting. The same trend is observed for a spindle speed of 500 rpm (diamond markers), i.e., an increase in the step down value results in a temperature increment. Figure 11b presents the logarithmic surface obtained for the bidirectional trajectory, as per the detailed equation. Once again, an increase in step down results in elevated temperature levels. However, it was possible to inhibit twisting, leading only to a fracture failure mode.

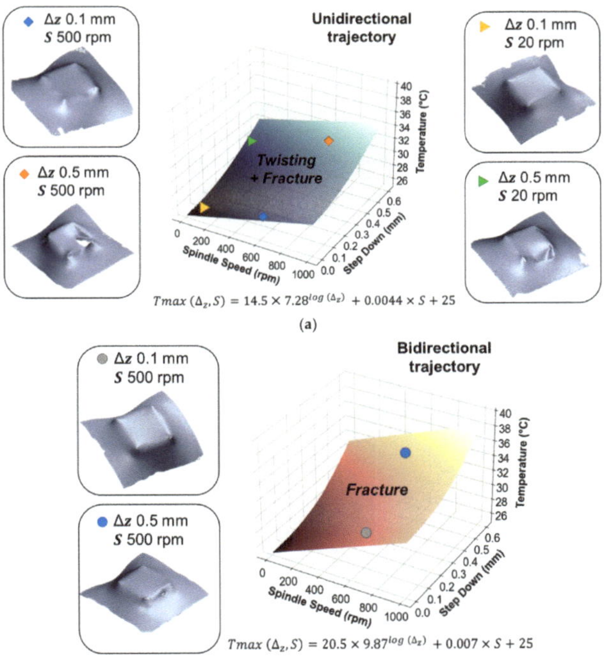

Figure 11. Developed temperature model along with scanned pyramids to illustrate the increase in twisting for the (**a**) unidirectional and (**b**) bidirectional trajectories.

This analysis highlights that, with equal process parameters and therefore similar temperature levels, utilizing the bidirectional toolpath enables achieving only fracture as the failure mode, even with moderately high spindle speeds.

3.2. Plastic Liner Device

This section presents the results of manufacturing a plastic liner device using UHMWPE for joint replacements. As explained in Section 2.3, two different geometries were considered: a hemispherical geometry ("H" in Table 6) and a geometry with a cylindrical and a hemispherical zone ("C + H" in Table 6).

Table 6. Results for hemispherical and hemispherical + cylinder plastic liner with UHMWPE in SPIF.

Test Condition	Δz (mm)	S (rpm)	Failure	Time (min)	Number of Cycles
B H1	0.5	20	no	5	1
B H2	0.5	20	no	16	3
B CH1	0.5	20	no	21	3

Based on the findings from the UHMWPE truncated pyramidal geometries produced through SPIF, it was determined that the bidirectional trajectory effectively prevents the occurrence of twisting failure, even with higher step down values (0.5 mm). Additionally, the experiments showed that using a lower spindle speed (20 rpm) resulted in a reduction of twisting failure.

With these considerations, the goal of manufacturing the plastic liner device was to establish a foundation for future improvements. Therefore, a step down value of 0.5 mm was chosen to expedite the forming process. The tests were conducted at a spindle speed of 20 rpm to minimize the temperature levels. Furthermore, a three-cycle process was used to achieve greater geometric accuracy and reduce springback, resulting in a shape that closely matches the design specifications in CATIA V5R20.

Table 6 shows the obtained results, indicating that the three plastic devices were successfully manufactured by SPIF without failure.

The alternative approach was employed to assess the formability levels. A meridional section along the specimen was selected after the plastic device and divided into distinct sections where the thickness was measured.

In the work by Jeswiet et al. [44], a hemispherical part was manufactured using SPIF. The study revealed that the value of β, representing the principal strain ratio, can range between 0 and 0.1, indicating a near plane strain condition. This information was considered in the analysis of the major strain distribution in the plastic liner device. Figure 12 presents the FLD considering a plane strain condition (β = 0) for the three specimens considered.

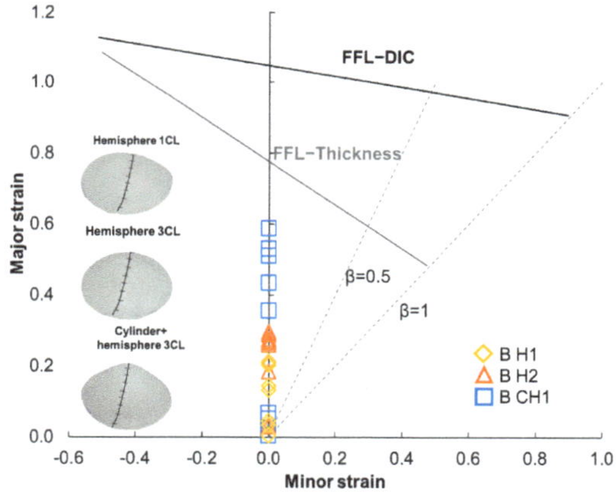

Figure 12. Strains achieved for the UHMWPE SPIF plastic liner devices by using the measurement of the final thickness in the FLD.

The cylindrical plus hemispherical geometry, with three cycles of processing (represented by a square marker with blue outline), exhibited more deformation compared to the hemispherical geometry with 3 cycles (represented by a triangle marker with orange outline), and with one cycle (represented by a rhomboid marker with yellow outline). In terms of strains, B H1 exhibited a value of 0.20, B H2 showed 0.28, and B CH1 displayed 0.58 in the major strain. In all the tests conducted, the maximum strains observed were well below the FFL thickness and/or FFL−DIC, indicating that the manufacturing of the plastic liner device using UHMWPE was feasible.

A qualitative analysis of the plastic liner geometries was performed using a 3D scanner. The process involved overlaying the CAD model with the real model, as depicted in Figure 13.

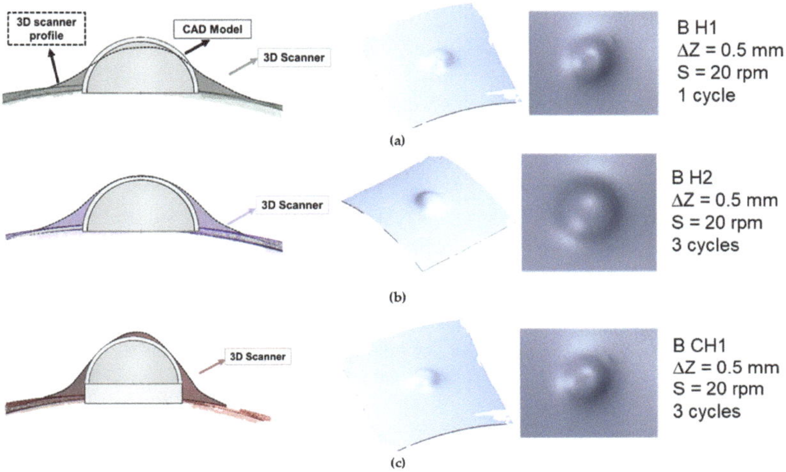

Figure 13. Differences between the 3D scanner plastic liner model and CAD model, accompanied by the image extracted from the 3D Geomatic Capture scanner software, 2016 version for the following cases: (**a**) hemispherical geometry with 1 cycle; (**b**) hemispherical geometry with 3 cycles; and (**c**) hemispherical and cylindrical geometry with 3 cycles.

As anticipated, the three conditions tested for the plastic liner geometries did not result in a perfect fit with the CAD model, primarily due to the material's elastic recovery. Elastic recovery was most prominent in the curved areas of the plastic liner, particularly along the sides. The hemispherical geometry with only one cycle exhibited the poorest geometric accuracy, while the condition with a combination of the hemispherical and cylindrical wall geometry showed better accuracy. However, it was observed that the top area of the plastic liner was not perfectly hemispherical. This discrepancy may be attributed to the influence of the tool in that particular region.

It has been established that the geometry of the hemisphere plus cylinder with three cycles achieved the highest level of geometric accuracy. However, as discussed previously, the UHMWPE material exhibits significant elastic recovery. To address this issue, further experimentation with alternative geometries and strategies is required. This study is an initial investigation to demonstrate the feasibility of producing UHMWPE-based medical devices using the SPIF process at room temperature. In this scenario, the recent research led by Cheng et al. [45] revealed that the variation in the shape of implants produced using incremental sheet forming could be controlled within the range of 1–2 mm. Nonetheless, additional efforts were required to enhance the geometric accuracy to ensure a properly fitting implant. Furthermore, optimizing the tool path emerged as the most appropriate and efficient strategy for enhancing precision, in contrast to methods such as providing additional support or utilizing hot incremental forming.

4. Conclusions

This article examines the formability and failure behavior of UHMWPE polymer through both conventional and incremental single point forming. Material characterization adapts techniques from sheet metal forming, utilizing a successful time dependent approach to detect the onset of necking. The forming limits for necking and fracture in 2 mm thickness

UHMWPE sheets were determined. The results indicated that: (i) tensile and plane strain specimens exhibited necking, whereas biaxial and equibiaxial strain specimens did not; and (ii) the latest measurement from the DIC system was utilized to determine the FFL due to the significant elastic recovery presented by the material.

On the other hand, an experimental plan of SPIF was conducted using a truncated pyramid. The research aimed to compare failure modes and formability for step down values of 0.1, 0.3, and 0.5, spindle speeds of 20 and 500 rpm, and unidirectional and bidirectional toolpaths. Under the unidirectional trajectory, the key findings were: (i) twisting and fracture were observed across all conditions; (ii) reducing step down and spindle speed decreased twisting; and (iii) regarding temperature, the highest reached was 31.6 °C in the test with 0.5 mm step down and 500 rpm spindle speed. Under the bidirectional trajectory, the conclusions were as follows: (i) it was possible to avoid twisting, with fractures occurring only under the tested conditions; and (ii) in terms of formability, an increase in step down led to higher levels of formability, with a major strain of 0.98 for the 0.5 mm step down test and 0.83 for the 0.1 mm test (an increase of 15.3%).

The article focuses on the manufacturing of a UHMWPE plastic liner using SPIF, which is a highly utilized medical component in total hip replacement. The results indicated that the part with cylindrical and semi-spherical geometry, subjected to 3 forming cycles, exhibited lower geometric deviations. In terms of strains, the manufactured part reached a major strain level of 0.58, falling below the FFL and demonstrating the potential for manufacturing medical prostheses through incremental forming. Additionally, this article critically considers that the elastic recovery of the material is a potentially improvable parameter for the manufacturing of medical prostheses using UHMWPE.

Author Contributions: Conceptualization, A.R.-S., A.M.B. and G.C.; Methodology, A.R.-S.; Formal analysis, A.R.-S., G.C. and M.B.S.; Investigation, A.R.-S.; Writing—original draft, A.R.-S.; Writing—review & editing, A.R.-S., M.B.S., A.M.B., G.C. and C.V.; Supervision, M.B.S. All authors have read and agreed to the published version of the manuscript.

Funding: The authors would like to acknowledge the funding received from the Regional Government of Andalusia under the Grant with reference P18-RT-3866 funded under the framework US/JUNTA/FEDER_UE of the "PAIDI 2020: Proyectos I+D+i", the Spanish Government for the Major Grant with reference TED2021-131153B-C22 within the call "Proyectos Estratégicos Orientados a la Transición Ecológica y a la Transición Digital".

Institutional Review Board Statement: Not applicable.

Data Availability Statement: The data presented in this study are available on request from the corresponding author.

Acknowledgments: The authors would like to acknowledge Fundação para a Ciência e da Tecnologia of Portugal, through IDMEC, under LAETA, project UIDB/50022/2020.

Conflicts of Interest: The authors declare no conflict of interest.

References

1. Popkova, E.G.; Ragulina, Y.V.; Bogovix, A.V. (Eds.) *The Industrial Revolution of the 21st Century*; Springer: Berlin/Heidelberg, Germany, 2006; Volume 169, p. 249. [CrossRef]
2. Marques, T.A.; Silva, M.B.; Martins, P.A.F. On the Potential of Single Point Incremental Forming of Sheet Polymer Parts. *Int. J. Adv. Manuf. Technol.* **2012**, *60*, 75–86. [CrossRef]
3. Bagudanch, I.; Garcia-Romeu, M.L.; Centeno, G.; Elías-Zúñiga, A.; Ciurana, J. Forming Force and Temperature Effects on Single Point Incremental Forming of Polyvinylchloride. *J. Mater. Process. Technol.* **2015**, *219*, 221–229. [CrossRef]
4. Centeno, G.; Morales-Palma, D.; Gonzalez-Perez-Somarriba, B.; Bagudanch, I.; Egea-Guerrero, J.J.; Gonzalez-Perez, L.M.; García-Romeu, M.L.; Vallellano, C. A Functional Methodology on the Manufacturing of Customized Polymeric Cranial Prostheses from CAT Using SPIF. *Rapid Prototyp. J.* **2017**, *23*, 771–780. [CrossRef]
5. Martins, P.A.F.; Kwiatkowski, L.; Franzen, V.; Tekkaya, A.E.; Kleiner, M. Single Point Incremental Forming of Polymers. *CIRP Ann. Manuf. Technol.* **2009**, *58*, 229–232. [CrossRef]
6. Rosa-Sainz, A.; Centeno, G.; Silva, M.B.; López-Fernández, J.A.; Martínez-Donaire, A.J.; Vallellano, C. On the Determination of Forming Limits in Polycarbonate Sheets. *Materials* **2020**, *13*, 928. [CrossRef] [PubMed]

7. Rosa-Sainz, A.; Centeno, G.; Silva, M.B.; Vallellano, C. Experimental Failure Analysis in Polycarbonate Sheet Deformed by Spif. *J. Manuf. Process.* **2021**, *64*, 1153–1168. [CrossRef]
8. Rosa-Sainz, A.; García-Romeu, M.L.; Ferrer, I.; Silva, M.B.; Centeno, G. On the Effective Peek Application for Customized Cranio-Maxillofacial Prostheses: An Experimental Formability Analysis. *J. Manuf. Process.* **2023**, *86*, 66–84. [CrossRef]
9. Le, V.S.; Ghiotti, A.; Lucchetta, G. Preliminary Studies on Single Point Incremental Forming for Thermoplastic Materials. *Int. J. Mater. Form.* **2008**, *1*, 1179–1182. [CrossRef]
10. Davarpanah, M.A.; Mirkouei, A.; Yu, X.; Malhotra, R.; Pilla, S. Effects of Incremental Depth and Tool Rotation on Failure Modes and Microstructural Properties in Single Point Incremental Forming of Polymers. *J. Mater. Process. Technol.* **2015**, *222*, 287–300. [CrossRef]
11. Tathe, A.; Ghodke, M.; Nikalje, A.P. A Brief Review: Biomaterials and Their Apllication. Available online: https://www.semanticscholar.org/paper/A-BRIEF-REVIEW%3A-BIOMATERIALS-AND-THEIR-APLLICATION-Tathe-Ghodke/a807e0f8fcebf4c0defc9f832a38533f9c80ac0d (accessed on 2 September 2022).
12. Bagudanch, I.; García-Romeu, M.L.; Ferrer, I.; Ciurana, J. Customized Cranial Implant Manufactured by Incremental Sheet Forming Using a Biocompatible Polymer. *Rapid Prototyp. J.* **2018**, *24*, 120–129. [CrossRef]
13. Centeno, G.; Bagudanch, I.; Morales-Palma, D.; García-Romeu, M.L.; Gonzalez-Perez-Somarriba, B.; Martinez-Donaire, A.J.; Gonzalez-Perez, L.M.; Vallellano, C. Recent Approaches for the Manufacturing of Polymeric Cranial Prostheses by Incremental Sheet Forming. *Proc. Procedia Eng.* **2017**, *183*, 180–187. [CrossRef]
14. Clavijo-Chaparro, S.L.; Iturbe-Ek, J.; Lozano-Sánchez, L.M.; Sustaita, A.O.; Elías-Zúñiga, A. Plasticized and Reinforced Poly(Methyl Methacrylate) Obtained by a Dissolution-Dispersion Process for Single Point Incremental Forming: Enhanced Formability towards the Fabrication of Cranial Implants. *Polym. Test.* **2018**, *68*, 39–45. [CrossRef]
15. Chen, L.F.; Chen, F.; Gatea, S.; Ou, H. PEEK Based Cranial Reconstruction Using Thermal Assisted Incremental Sheet Forming. *Proc. Inst. Mech. Eng. Part B J. Eng. Manuf.* **2021**, *236*, 997–1004. [CrossRef]
16. Fiorentino, A.; Marenda, G.P.; Marzi, R.; Ceretti, E.; Kemmoku, D.T.; Lopes Da Silva, J.V. Rapid Prototyping Techniques for Individualized Medical Prosthesis Manufacturing. In *Innovative Developments in Virtual and Physical Prototyping, Proceedings of the 5th International Conference on Advanced Research in Virtual and Rapid Prototyping, Leiria, Portugal, 28 September–1 October 2011*; CRC Press: Boca Raton, FL, USA, 2012; Volume 1, pp. 589–594. [CrossRef]
17. Chilukoti, G.R.; Periyasam, A.P. Ultra High Molecular Weight Polyethylene for Medical Applications. Available online: https://www.researchgate.net/publication/298626009_Ultra_high_molecular_weight_polyethylene_for_medical_applications (accessed on 31 August 2022).
18. Kurtz, S.M. From Ethylene Gas to UHMWPE Component: The Process of Producing Orthopedic Implants. In *UHMWPE Biomaterials Handbook: Ultra High Molecular Weight Polyethylene in Total Joint Replacement and Medical Devices: Third Edition*; William Andrew Publishing: Norwich, NY, USA, 2016; pp. 7–20. ISBN 9780323354011.
19. Musib, M.K. A Review of the History and Role of UHMWPE as A Component in Total Joint Replacements. *Int. J. Biol. Eng.* **2011**, *1*, 6–10. [CrossRef]
20. Ardestani, M.M.; Amenábar Edwards, P.P.; Wimmer, M.A. Prediction of Polyethylene Wear Rates from Gait Biomechanics and Implant Positioning in Total Hip Replacement. *Clin. Orthop. Relat. Res.* **2017**, *475*, 2027–2042. [CrossRef]
21. Kurtz, S.M.; Gawel, H.A.; Patel, J.D. History and Systematic Review of Wear and Osteolysis Outcomes for First-Generation Highly Crosslinked Polyethylene. *Clin. Orthop. Relat. Res.* **2011**, *469*, 2262. [CrossRef] [PubMed]
22. Hussain, M.; Naqvi, R.A.; Abbas, N.; Khan, S.M.; Nawaz, S.; Hussain, A.; Zahra, N.; Khalid, M.W. Ultra-High-Molecular-Weight-Polyethylene (UHMWPE) as a Promising Polymer Material for Biomedical Applications: A Concise Review. *Polymers* **2020**, *12*, 323. [CrossRef]
23. Wang, M.; Fu, J.; Jin, Z.-M.; Wang, J.-W. *UHMWPE Biomaterials for Joint Implants: Structures, Properties and Clinical Performance*; Springer: Berlin/Heidelberg, Germany, 2019; Volume 13. [CrossRef]
24. Rosa-Sainz, A.; Centeno, G.; Silva, M.B.; López-Fernández, J.A.; Martínez-Donaire, A.J.; Vallellano, C. On the Determination of the Forming Limits by Necking and Fracture of Polycarbonate Sheet. *Miner. Met. Mater. Ser.* **2021**, 2597–2608. [CrossRef]
25. Goodfellow UHMWPE Sheet 2 mm Thick | Properties. Available online: https://www.goodfellow.com/p/et30-sh-000100/uhmwpe-sheet (accessed on 19 July 2023).
26. *ISO12004-2*; Metallic Materials-Sheet and Strip-Determination of Forming Limit Curves Part 2: Determination of Forming Limit Curves in Laboratory. International Organization of Standards: Geneva, Switzerland, 2008.
27. Martínez-Donaire, A.J.; García-Lomas, F.J.; Vallellano, C. New Approaches to Detect the Onset of Localised Necking in Sheets under Through-Thickness Strain Gradients. *Mater. Des.* **2014**, *57*, 135–145. [CrossRef]
28. Centeno, G.; Bagudanch, I.; Martínez-Donaire, A.J.; García-Romeu, M.L.; Vallellano, C. Critical Analysis of Necking and Fracture Limit Strains and Forming Forces in Single-Point Incremental Forming. *Mater. Des.* **2014**, *63*, 20–29. [CrossRef]
29. Franzen, V.; Kwiatkowski, L.; Martins, P.A.F.; Tekkaya, A.E. Single Point Incremental Forming of PVC. *J. Mater. Process. Technol.* **2009**, *209*, 462–469. [CrossRef]
30. Yang, Z.; Chen, F. Mechanism of Twist in Incremental Sheet Forming of Thermoplastic Polymer. *Mater. Des.* **2020**, *195*, 108997. [CrossRef]
31. Rosa-Sainz, A.; Centeno, G.; Silva, M.B.; Vallellano, C. Experimental Investigation of Polycarbonate Sheets Deformed by SPIF: Formability, Micro-Mechanisms of Failure and Temperature Analysis. *J. Mater. Res. Technol.* **2023**, *25*, 7546–7565. [CrossRef]

32. Li, Y.; He, H.; Ma, Y.; Geng, Y.; Tan, J. Rheological and Mechanical Properties of Ultrahigh Molecular Weight Polyethylene/High Density Polyethylene/Polyethylene Glycol Blends. *Adv. Ind. Eng. Polym. Res.* **2019**, *2*, 51–60. [CrossRef]
33. Hallab, N.J.; Jacobs, J.J. Orthopedic Applications. In *Biomaterials Science*, 3rd ed.; Elsevier: Amsterdam, The Netherlands, 2013; pp. 841–882. [CrossRef]
34. Biomet, Z. Acetabular System. Available online: https://www.zimmerbiomet.com/en/products-and-solutions/specialties/hip/g7-acetabular-system.html (accessed on 5 September 2022).
35. Medina-Sánchez, G.; Torres-Jimenez, E.; Lopez-Garcia, R.; Dorado-Vicente, R.; Cazalla-Moral, R. Temperature Influence on Single Point Incremental Forming of PVC Parts. *Procedia Manuf.* **2017**, *13*, 335–342. [CrossRef]
36. Bagudanch, I.; Vives-Mestres, M.; Sabater, M.; Garcia-Romeu, M.L. Polymer Incremental Sheet Forming Process: Temperature Analysis Using Response Surface Methodology. *Mater. Manuf. Process.* **2017**, *32*, 44–53. [CrossRef]
37. Formisano, A.; Boccarusso, L.; Durante, M. Optimization of Single-Point Incremental Forming of Polymer Sheets through FEM. *Materials* **2023**, *16*, 451. [CrossRef]
38. Yang, Z.; Chen, F.; Gatea, S.; Ou, H. Design of the Novel Hot Incremental Sheet Forming Experimental Setup, Characterization of Formability Behavior of Polyether-Ether-Ketone (PEEK). *Int. J. Adv. Manuf. Technol.* **2020**, *106*, 5365–5381. [CrossRef]
39. Lozano-Sánchez, L.M.; Bagudanch, I.; Sustaita, A.O.; Iturbe-Ek, J.; Elizalde, L.E.; Garcia-Romeu, M.L.; Elías-Zúñiga, A. Single-Point Incremental Forming of Two Biocompatible Polymers: An Insight into Their Thermal and Structural Properties. *Polym.* **2018**, *10*, 391. [CrossRef]
40. Bagudanch, I.; Garcia-Romeu, M.L.; Sabater, M. Incremental Forming of Polymers: Process Parameters Selection from the Perspective of Electric Energy Consumption and Cost. *J. Clean. Prod.* **2016**, *112*, 1013–1024. [CrossRef]
41. Formisano, A.; Lambiase, F.; Durante, M. Polymer Self-Heating during Incremental Forming. *J. Manuf. Process.* **2020**, *58*, 1189–1199. [CrossRef]
42. Matplotlib Python v3.4.2. Available online: https://matplotlib.org/ (accessed on 25 June 2021).
43. NumPy Library Python. Available online: https://numpy.org/doc/stable/ (accessed on 22 December 2022).
44. Hagan, E.; Jeswiet, J. A Review of Conventional and Modern Single-Point Sheet Metal Forming Methods. *Proc. Inst. Mech. Eng.* **2005**, *217*, 213–225. [CrossRef]
45. Cheng, Z.; Li, Y.; Xu, C.; Liu, Y.; Ghafoor, S.; Li, F. Incremental Sheet Forming towards Biomedical Implants: A Review. *J. Mater. Res. Technol.* **2020**, *9*, 7225–7251. [CrossRef]

polymers

Article

Mechanical Properties of Weld Lines in Injection-Molded Carbon Fiber-Reinforced Nylon (PA-CF) Composites

Xianpeng Wang [1], Zuguo Bao [1,2,*], Donglin Gao [1], Shiyao Huang [1,2], Li Huang [1,2], Qiuren Chen [1,2], Hailong Zhao [2], Weijian Han [1,2] and Yahong Xu [1,*]

1 College of Materials Science and Engineering, Nanjing Tech University, Nanjing 210009, China
2 Yangtze Delta Region Institute of Advanced Materials, Suzhou 215133, China
* Correspondence: baozuguo@njtech.edu.cn (Z.B.); 13701360679@163.com (Y.X.)

Abstract: Weld lines are a common defect generated in injection molding, which apparently affects the performance of final products, but the available reports on carbon fiber-reinforced thermoplastics are still rather few. In this study, the effects of injection temperature, injection pressure, and fiber content on the mechanical properties of weld lines were studied for carbon fiber-reinforced nylon (PA-CF) composites. The weld line coefficient was also calculated by comparing specimens with and without weld lines. The tensile and flexural properties of PA-CF composites significantly increased with the rise of fiber content for specimens without weld lines, while injection temperature and pressure demonstrated slight influences on mechanical properties. However, the existence of weld lines had negative influences on the mechanical properties of PA-CF composites due to poor fiber orientation in weld line regions. The weld line coefficient of PA-CF composites decreased as fiber content increased, indicating that the damage of weld lines to mechanical properties increased. The microstructure analysis showed that there were a large number of fibers distributed vertically to flow direction in weld lines regions, which could not play a reinforcing role. In addition, increasing injection temperature and pressure facilitated fiber orientation, which improved the mechanical properties of composites with low fiber content, while weakening composites with high fiber content instead. This article provides practical information for product design containing weld lines, which helps to optimize the forming process and formula design of PA-CF composites with weld lines.

Keywords: injection molding; weld line; fiber orientation; PA-CF composites

Citation: Wang, X.; Bao, Z.; Gao, D.; Huang, S.; Huang, L.; Chen, Q.; Zhao, H.; Han, W.; Xu, Y. Mechanical Properties of Weld Lines in Injection-Molded Carbon Fiber-Reinforced Nylon (PA-CF) Composites. *Polymers* **2023**, *15*, 2476. https://doi.org/10.3390/polym15112476

Academic Editors: Yadir Torres Hernández, Ana María Beltrán Custodio and Manuel Félix Ángel

Received: 9 May 2023
Revised: 23 May 2023
Accepted: 25 May 2023
Published: 27 May 2023

1. Introduction

Weld lines have been one of the common defects puzzling the injection molding industry. When two or more strands of molten plastics meet in the mold, a weld line will be produced at the junction [1,2]. Weld lines not only affect the appearance and quality of products, but also cause negative influences on the mechanical properties [3–5]. Such influences were especially obvious for fiber-reinforced thermoplastics, which have been widely used in the automobile and home appliance industries in recent years [6]. In the weld line regions of fiber-reinforced composites, most fibers were parallel to the interface where the melt intersected, and failed to play a reinforcing role [7–9]. Thus, the weld lines are usually known as weak regions in the products. However, for the consideration of production efficiency and economic benefits, multiple gate designs were often employed in injection molding, inevitably producing weld lines on the products. Therefore, it is necessary to investigate the properties of weld lines and their contributory factors.

It has been reported that the mechanical properties of weld line regions would be enhanced with the increase in injection temperature and mold temperature [10–12]. In addition, with the increase in the injection pressure, the mechanical properties also showed an increasing trend [13,14]. However, the tensile strength and fracture strain would decrease significantly with the increase in the fiber content in weld line regions [15–17]. Andrea

Scantamburlo prepared 35wt% glass fiber reinforced polypropylene tensile specimens with weld lines using alternate dynamic fillers and rapid heat cycle injection molding (RHCM). Compared with traditional injection molded weld lines, the RHCM reduced the skin layer thickness of the specimens, and increased the strength and stiffness of the specimens by 201% and 46%, respectively [18]. Babur Ozcelik analyzed the effects of injection temperature and injection pressure on the mechanical properties of polypropylene (PP) specimens with and without weld lines. When the injection temperature and injection pressure increased, the tensile strength of the specimens with and without weld lines did not change significantly, but the impact resistance decreased significantly [19]. Azieatul Azrin Dzulkipli studied the effect of fiber addition on the formation of weld lines. Glass fiber-reinforced polypropylene had a shorter weld line length and a larger meeting angle than pure polypropylene [12]. Gyung-Hwan Oh investigated the effect of fiber content on the tensile strength of specimens with and without weld lines in glass fiber-reinforced modified polyphenylene oxide (MPPO) composites. For tensile specimens with glass fiber weight percentages of 0%, 10%, 20%, and 30%, the proportion of strength reductions due to the existence of weld lines were 12%, 34%, 52%, and 56%, respectively [20]. L.M. Martulli investigated the effect of weld lines on the mechanical properties of carbon fiber tow-reinforced vinyl ester sheet molding compound (CF-SMC) with a fiber volume fraction of 42%, and found that the tensile strength of the sample with weld lines was 48% to 88% of that without weld lines [21]. R. Selden focused on the influence of injection temperature, mold temperature, packing pressure, and injection speed on the flexural strength of five types of thermoplastics (ABS, PPO, PA6, PPS, and PP containing 40% talc) with/without weld lines, and calculated the weld line coefficient, which varied from 0.25 to 0.98 in the tested materials [22].

Currently, research on weld lines is mostly focused on pure plastics and glass fiber-reinforced plastics, and there are fewer reports on the weld lines of PA-CF composites. Compared with glass fiber, carbon fiber reinforced composites possess more excellent properties and broader potential for structural applications [23]. As the weld lines are typically weak regions in injection molding parts, the investigation of weld line properties is very critical for the design and process optimization when using this material to prepare products with weld lines. The purpose of this work is to investigate the mechanical properties of weld lines in PA-CF composites, considering the influence of injection temperature, injection pressure, and fiber content. Weld line coefficients were also calculated. Scanning electron microscope (SEM) and metallographic microscope were used to observe the microstructure of weld line samples. The software Moldex3D was employed to simulate the injection molding process and analyze the fiber orientation in the weld line regions.

2. Materials and Methods

2.1. Materials

Carbon fiber reinforced nylon 66 pellets were purchased from SABIC (No. RE0069S, SABIC Innovative Plastics Co., Ltd., Thorndale, TX, USA) with a fiber content of 30wt% (hereafter "PA-30CF"), and the nominal material properties are listed in Table 1. Due to the water absorption of nylon matrix, the raw material and specimens were dried for 4 h at 80 °C before injection molding and property testing [24,25].

Table 1. Nominal material properties of SABIC RE0069S.

Properties	Test Standards	Value	Unit
Density	ASTM D792	1.50	g/cm^3
Tensile strength	ASTM D638	228	MPa
Young's modulus	ASTM D638	26,960	MPa
Flexural strength	ASTM D790	351	MPa
Flexural modulus	ASTM D790	18,590	MPa

2.2. Injection Molding

An injection molding machine, HAITIAN-MA900III/280-A (Haitian Plastic Machinery Group Co., Ltd., Ningbo, China), was employed for the specimen preparation. The maximum clamping force was 90 tons, and it was equipped with a screw with a diameter of 32 mm and length–diameter ratio of 22.5. The setting of injection molding parameters is shown in Table 2, in which the influence of injection temperature and injection pressure on the mechanical properties of the specimens with/without weld lines were considered. The mold temperature, packing pressure, and injection speed were maintained constant with the value of 25 °C, 8 MPa, and 70 mm/s, respectively. In addition, the influence of fiber content was also considered with four loading levels, namely 0%, 10%, 20%, and 30% by weight. The specimens with lower fiber contents were prepared by mixing PA-30CF composites and pure PA proportionally in injection molding.

Table 2. Injection molding parameters.

Injection Parameters	Value	Unit
Injection temperature	295, 305, 315	°C
Injection pressure	7.5, 8.5, 9.5	MPa
Mold temperature	25	°C
Packing pressure	8	MPa
Injection velocity	70	mm/s

2.3. Specimens

The tensile specimen was designed according to ASTM D638 Type I, as shown in Figure 1a. The total length of tensile specimen was 165 mm, with the gauge length of 50 mm. The flexural specimen (Figure 1b) was designed according to ASTM D790, with a span width of 51.2 mm. The injection runners of the specimens are exhibited in Figure 2. For the specimen without weld lines, there was only a gate at one end of the cavity for the melt to flow in, as shown in Figure 2a. For the specimen with weld lines, both ends of the cavity were equipped with gates, and the two streams of melt converged in the middle area of the specimen and formed the weld line, as shown in Figure 2b.

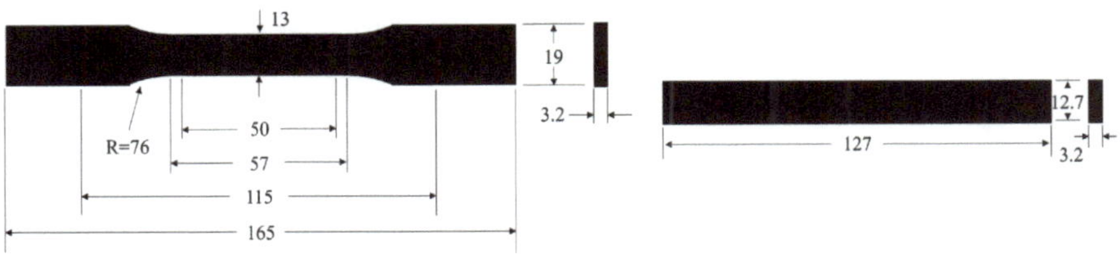

(a) (b)

Figure 1. Specimens: (**a**) tensile testing specimen and (**b**) flexural testing specimen.

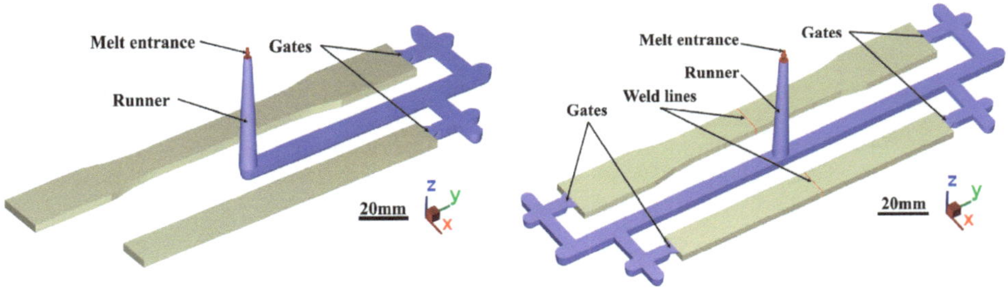

(a) (b)

Figure 2. The injection runners of the specimens (**a**) without weld lines and (**b**) with weld lines.

2.4. Mechanical Properties Testing

Mechanical properties, including tensile and flexural properties, were tested in INSTRON 5982 (Instron Co., Ltd., Boston, MA, USA). The tensile testing was conducted according to ASTM-D638, with a loading speed of 5 mm/min. The flexural testing was conducted according to ASTM-D790, with a loading speed of 1.4 mm/min. Five duplicate specimens were tested for each material, and the mean values were provided.

The weld line coefficient F_{kl} was used to describe the influence of weld lines on the mechanical properties of specimens, and the calculation formula is shown in (1) [12,26].

$$F_{kl} = \frac{\sigma_{WL}}{\sigma_{NWL}} \quad (1)$$

where σ_{WL} and σ_{NWL} are mechanical properties with and without weld lines, respectively.

2.5. Microscopic Observation

Fracture morphology was observed with the aid of scanning electron microscope (SEM) for the specimen after failure. The operation process is shown in Figure 3. The fiber orientation and distribution in the weld line regions were observed with ZEISS-Axio Imager M2m (Carl Zeiss Co., Ltd., Oberkohen, Germany) microscope, and the operation process is shown in Figure 4.

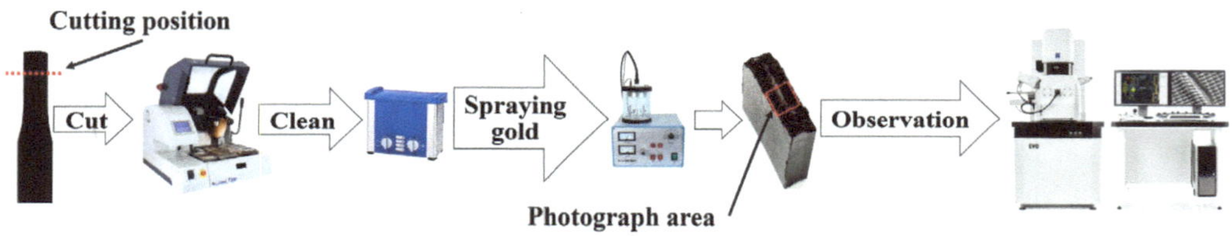

Figure 3. SEM observation process.

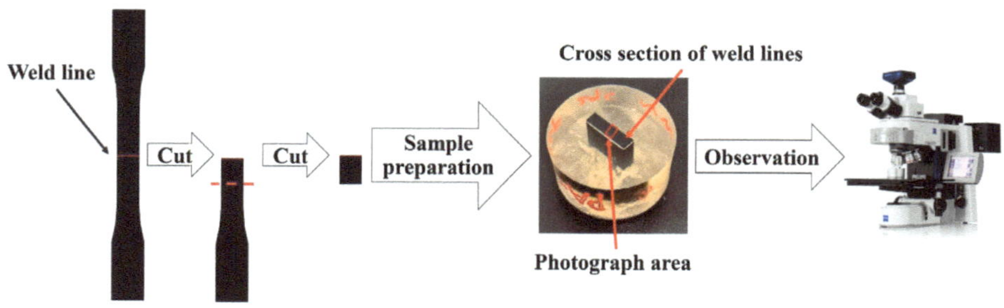

Figure 4. Metallographic microscope observation process.

2.6. DSC Measurement

The samples in the weld line regions were cut from the tensile specimens and analyzed by differential scanning calorimetry (DSC). The DSC measurement was conducted in TA-DSC25 (TA instrument Co., Ltd., New Castle, DE, USA) differential scanning calorimeter to investigate the crystallinity of nylon in samples. The samples were heated from 25 °C to 330 °C with a heating rate of 10 °C/min and kept for 5 min, and then cooled to 25 °C with a cooling rate of 5 °C/min.

2.7. Injection Molding Simulation

The mold flow analysis software Moldex3D Studio 2021 (CoreTech System Co., Ltd., Suzhou, China) was used to simulate the injection molding process, including filling, packing, cooling, and other stages, to analyze the fiber orientation tensor distribution, melt meeting angle, etc. The design of the cooling water path of the model is shown in Figure 5.

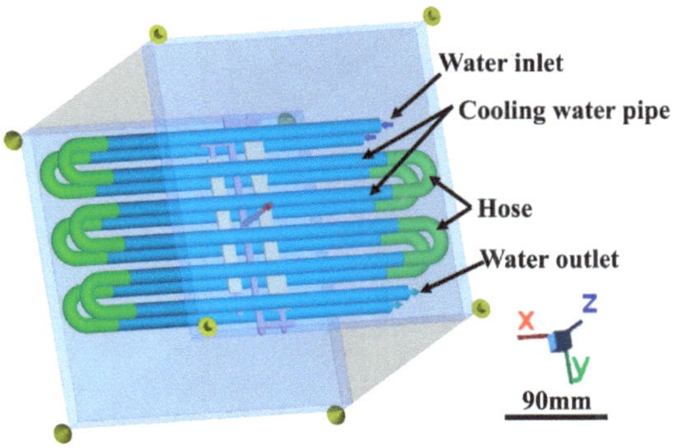

Figure 5. Moldex3D simulation model.

Melt flow is a non-isothermal process in injection molding. The melt is a typical non-Newtonian fluid, and its viscosity is affected by the temperature and shear rate changes during the filling process [27–29]. Thus, the above factors were taken into account, and Cross-WLF viscosity model was selected, as shown in Formulas (2)–(5) [30,31]. This model integrated the effects of temperature, pressure, and shear rate on the viscosity of melt, and could accurately predict the flow process accompanied by cooling effects. The values of material parameters are shown in Table 3.

$$\eta = \frac{\eta_0}{1 + \left(\frac{\eta_0 \dot{\gamma}}{\tau^*}\right)^{1-n}} \tag{2}$$

where η and η_0 are viscosity and zero shear viscosity, respectively; $\dot{\gamma}$ is shear strain rate, τ^* represents the critical stress level at the transition to shear thinning, n is the power law index in the high shear rate regime.

$$\eta_0 = D_1 exp\left[\frac{-A_1(T - T_c)}{A_2 + (T - T_c)}\right] \tag{3}$$

where D_1 and A_1 are data-fitted coefficients, T is the melting point temperature of the material, T_c is the reference temperature, and the glass transition temperature of the material is generally selected.

$$T_c = D_2 + D_3 P \tag{4}$$

$$A_2 = \tilde{A}_2 + D_3 P \tag{5}$$

where D_2, D_3, and $\tilde{A}_2$ are data-fitted coefficients; P is the pressure.

Table 3. The values of material parameters.

Parameters	PA-10CF	PA-20CF	PA-30CF
n (−)	0.449	0.474	0.536
τ^* (Pa)	4.83×10^4	1.56×10^4	1.35×10^4
D_1 (Pa·S)	8.58×10^{16}	1.68×10^{18}	1.18×10^{21}
D_2 (K)	333.15	343.15	373.15
D_3 (K/Pa)	0	0	0
A_1 (−)	42.98	48.13	52.52
$\tilde{A}_2$ (K)	51.6	51.6	51.6

3. Results and Discussion

3.1. Effect of Injection Temperature on Mechanical Properties

Mechanical properties of PA-CF composites were measured for different fiber contents and injection temperatures. As shown in Figure 6a,b, the tensile strength and Young's modulus of PA-CF composites without weld lines were significantly elevated with the increase in fiber content, indicating fibers played a reinforcing role for the nylon matrix. When carbon fiber reached 30wt%, the tensile strength and Young's modulus of PA-CF composites were 2.32 and 8.01 times higher than that of pure nylon, respectively (injection molding at 295 °C). Overall, the injection temperature had little effect on the tensile strength and modulus of PA-CF composites without weld lines, which was similar to the results of many previous studies [32,33]. Carbon fibers are the main load-bearing component in PA-CF composites, and their orientation distribution determines the mechanical properties of PA-CF composite materials. Most fibers were oriented along the flow direction in the specimens without weld lines, and the influence of injection temperature on fiber orientation was limited. Therefore, changing the injection temperature did not significantly change the mechanical properties of PA-CF composites.

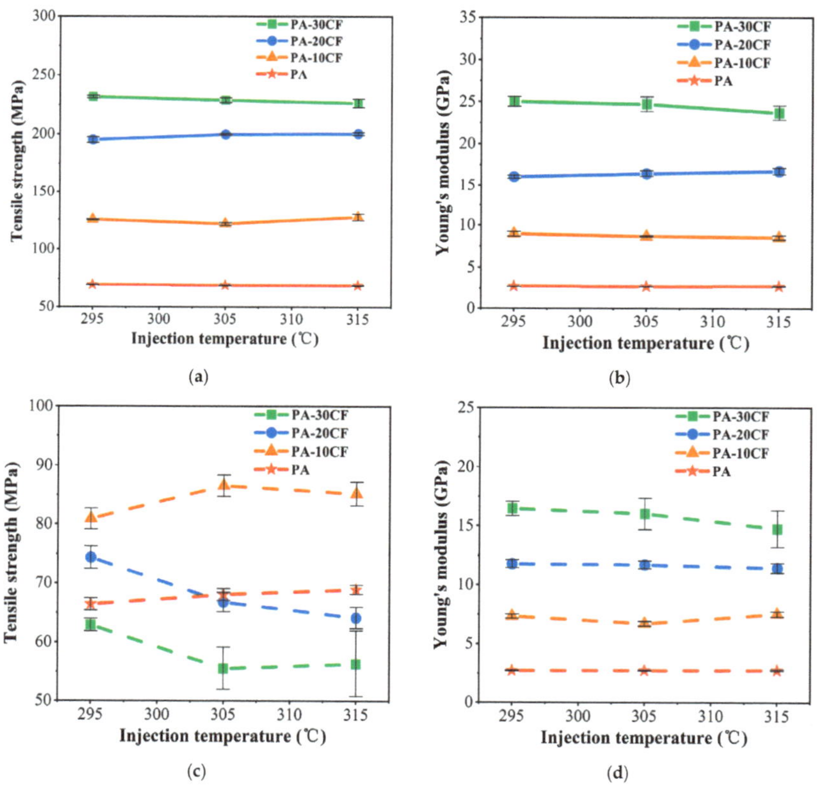

Figure 6. Tensile properties of PA-CF composites for different fiber contents and injection temperatures: (**a**) tensile strength, without weld lines; (**b**) Young's modulus, without weld lines; (**c**) tensile strength, with weld lines; and (**d**) Young's modulus, with weld lines.

However, the existence of weld lines had negative influences on the tensile properties of PA-CF composites. As shown in Figure 6c,d, the specimens with weld lines had significantly lower tensile strength and Young's modulus than those without weld lines. Meanwhile, it is worth noting that PA-20CF and PA-30CF composites had obviously lower tensile strength than PA-10CF, and PA-30CF was even lower than pure nylon, indicating that a large proportion of fibers did not play a reinforcing role, which was similar to the results of many previous studies [34,35]. Only fibers passing through the interface where the melts intersected could have a reinforcing effect, while higher fiber content hindered this process and reduced the entanglement between polymer chains. Additionally, the tensile strength of PA-20CF and PA-30CF composites with weld lines showed a downward trend when injection temperature increased (Figure 6c), while PA-10CF showed an upward trend. Compared with the tensile strength of the specimen at 295 °C, PA-20CF decreased by 10.2% and 13.7%, and PA-30CF decreased by 11.7% and 10.4%, while PA-10CF increased by 6.9% and 5.3% at 305 °C and 315 °C, respectively. The resin melt had better fluidity at higher injection temperatures, which was beneficial for fiber orientation [36]. The reinforcement effect was more obvious for those with low fiber content. A large proportion of fibers arranged parallel to the interface of the weld line for those with high fiber contents, which had greater adverse influences on the properties of composites and led to a decline in strength. In addition, samples with higher fiber content had a higher Young's modulus, which was rarely affected by injection temperature.

Flexural properties of PA-CF composites were also measured for different fiber contents and injection temperatures. Similar to tensile properties, the flexural strength and flexural modulus of PA-CF composites without weld lines significantly increased with the rise of fiber content for specimens, as shown in Figure 7a,b. The flexural strength and flexural modulus of PA-CF composites increased by 3.18 and 6.81 times, respectively, compared to pure nylon (injection molding at 295 °C) when carbon fiber reached 30wt%. The injection temperature had a negligible influence on the flexural strength and modulus of PA-CF composites without weld lines as well.

Similar to tensile properties, flexural properties of PA-CF composites were also significantly influenced by the introduction of weld lines. As shown in Figure 7c,d, the specimens with weld lines had lower flexural strength and modulus than those without weld lines. The flexural strength and modulus of the samples containing carbon fibers were higher than those of pure nylon. As the injection molding temperature increased, the flexural strength of PA-CF composites was also affected, which was similar to the effect of temperature on the tensile strength of PA-CF composites. When injection temperature reached 305 °C or 315 °C, PA-10CF composites had higher flexural strength than that of PA-20CF and PA-30CF, indicating that the reinforcement effect of fibers was limited in samples with weld lines.

To further analyze the influences of injection temperature and fiber content on the mechanical properties of PA-CF composites, weld line coefficients were calculated according to Formula (1) $F_{kl} = \frac{\sigma_{WL}}{\sigma_{NWL}}$ and are listed in Table 4. The values of pure nylon were very close to 1, indicating that the weld line had a negligible influence on its mechanical properties. With the introduction of carbon fiber, the weld line coefficient was significantly affected, and their values apparently decreased with the rise of fiber content. It demonstrated that the weld line had a greater influence on the mechanical properties of PA-CF composite when fiber content increased. With the same carbon fiber content, the tensile strength was more severely affected by weld lines than flexural strength with lower weld line coefficients, while the flexural modulus was more sensitive to weld lines than Young's modulus.

As the weld lines were the weak regions of specimens, their existence would affect the failure mode of the PA-CF composite. The fracture of specimens tended to occur close to the weld lines, as shown in Figure 8a,b. The fracture was relatively flat, and there were almost no small fragments produced by damage, indicating that weld lines were weak regions of material properties. The failure positions of specimens without weld lines are shown in Figure 8c,d. The fracture position was randomly located in the gauge length with a rough fractural surface and small fragments produced during the damage.

The fracture morphology of PA-30CF composites was observed with SEM and is illustrated in Figure 9, in which the flow direction was perpendicular to the observation surface. According to Figure 9a, the fibers of the sample without weld lines were oriented along the flow direction, and its failure modes inclufiber pulling out, fiber breakage, and resin matrix tearing. As shown in Figure 9b, a large number of carbon fibers of the sample with weld lines were oriented perpendicular to the flow direction, causing the dominant failure modes in the weld line region to be the debonding of the matrix and the fibers, and the fracture of the resin. Most carbon fibers failed to play a reinforcing role, deteriorating the mechanical properties of PA-CF composites with high fiber content.

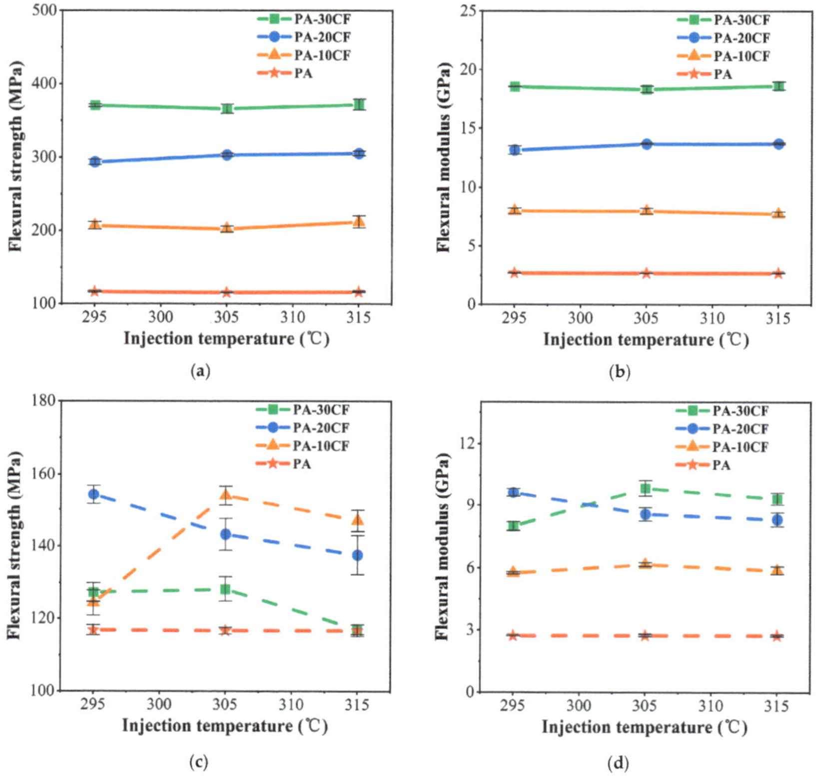

Figure 7. Flexural properties of PA-CF composites for different fiber contents and injection temperatures: (**a**) flexural strength, without weld lines; (**b**) flexural modulus, without weld lines; (**c**) flexural strength, with weld lines; and (**d**) flexural modulus, with weld lines.

Table 4. Weld line coefficients for different fiber contents and injection temperatures (Mean ± SD).

Materials	Injection Temperature (°C)	F_{kl} of Tensile Strength (−)	F_{kl} of Young's Modulus (−)	F_{kl} of Flexural Strength (−)	F_{kl} of Flexural Modulus (−)
PA	295	0.95 ± 0.04	0.99 ± 0.02	1.00 ± 0.02	1.01 ± 0.02
	305	0.98 ± 0.03	1.00 ± 0.01	1.01 ± 0.02	1.01 ± 0.02
	315	0.99 ± 0.03	1.00 ± 0.02	1.00 ± 0.02	1.01 ± 0.02
PA-10CF	295	0.65 ± 0.01	0.82 ± 0.03	0.60 ± 0.02	0.72 ± 0.03
	305	0.71 ± 0.01	0.77 ± 0.02	0.76 ± 0.03	0.77 ± 0.04
	315	0.67 ± 0.02	0.88 ± 0.02	0.69 ± 0.04	0.76 ± 0.04
PA-20CF	295	0.38 ± 0.02	0.73 ± 0.01	0.53 ± 0.01	0.73 ± 0.01
	305	0.33 ± 0.01	0.71 ± 0.01	0.47 ± 0.02	0.63 ± 0.02
	315	0.32 ± 0.02	0.68 ± 0.01	0.45 ± 0.03	0.61 ± 0.02
PA-30CF	295	0.27 ± 0.03	0.66 ± 0.02	0.34 ± 0.01	0.43 ± 0.01
	305	0.24 ± 0.02	0.65 ± 0.06	0.35 ± 0.01	0.54 ± 0.01
	315	0.25 ± 0.03	0.62 ± 0.09	0.31 ± 0.01	0.50 ± 0.02

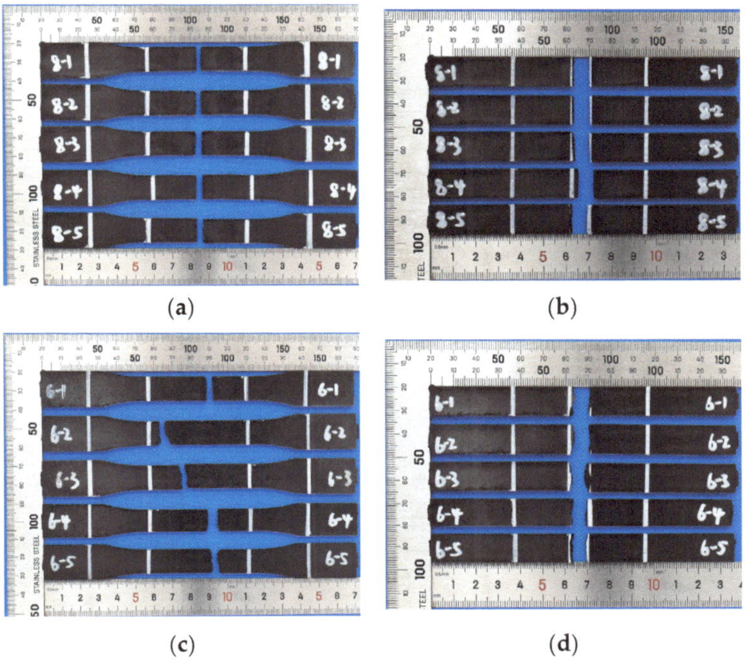

Figure 8. Fracture position of specimens with/without weld lines (injection molding at 295 °C): (**a**) tensile fracture, with weld lines; (**b**) flexural fracture, with weld lines; (**c**) tensile fracture, without weld lines; and (**d**) flexural fracture, without weld lines.

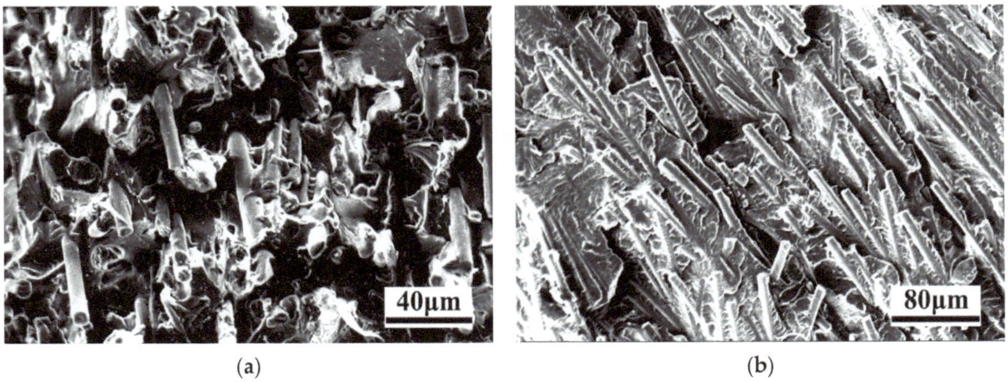

Figure 9. Fracture morphology of tensile specimens of PA-30CF composites (injection molding at 295 °C) (**a**) without weld lines and (**b**) with weld lines.

The microscopic images of PA-CF composites are shown in Figure 10, and the flow direction was perpendicular to the observation surface as well. As shown in Figure 10a, most of the fibers were oriented along the flow direction for specimens without weld lines. This structure could fully exert the reinforcement effect of fibers, and the composites obtained good mechanical properties. As shown in Figure 10b,c, in the samples with weld lines, the region presented a skin–transition–core structure, wherein the fibers of the skin layer were mainly oriented along the flow direction, while in the transition layer, a large proportion of fibers were perpendicular to the flow direction, which had negative influences on the mechanical properties of materials. The degree of fiber orientation in the core layer was intermediate between the skin layer and the transition layer. It is worth noting that the fibers in the core layer tend to orient along the flow direction when the injection temperature increases, which is beneficial for improving mechanical properties. Meanwhile, the fiber orientation in weld line regions of PA-30CF composites was observed as well, as shown in Figure 10d,e. The transition and core layers were fused together when the fiber content was high. A large number of fibers were oriented perpendicular to the flow direction, and the reinforcement effect of fibers in this structure on the matrix was very weak. At the same time, this phenomenon became more obvious with the rise of injection temperature, which did greater damage to the mechanical properties of materials. The changes in the microstructure of the weld line regions validated the previous results of mechanical properties well.

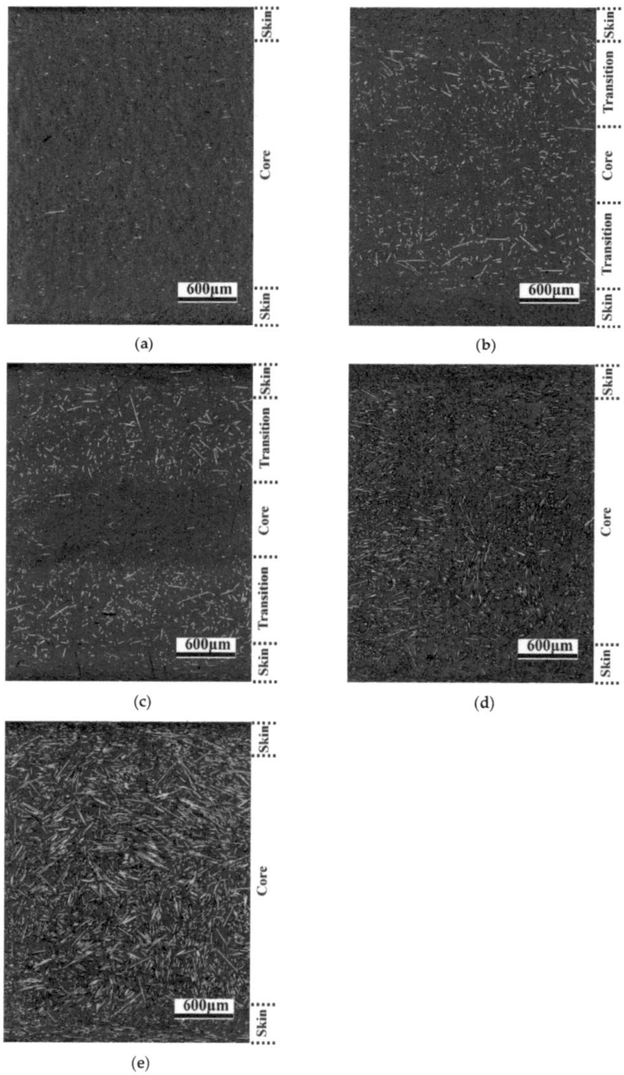

Figure 10. Metallographic micrograph (**a**) without weld lines, PA-10CF, 305 °C; (**b**) with weld lines, PA-10CF, 295 °C; (**c**) with weld lines, PA-10CF, 305 °C; (**d**) with weld lines, PA-30CF, 295 °C; and (**e**) with weld lines, PA-30CF, 305 °C.

The fiber orientation tensor could describe the probability of fiber orientation in a specified principal direction. A value close to 1 indicates a high probability of fiber orientation in the specified principal direction, while a value close to 0 indicates a low probability. The distribution of fiber orientation tensor components in the length direction of specimens obtained by simulation is shown in Figure 11, and the tensors were approximately symmetrical along the thickness direction. The orientation tensors of samples without weld lines are shown in Figure 11a–c. During the transition from the skin layer to the core layer, the orientation tensors first increased and then decreased, which was similar to the results of many previous studies [37,38]. The skin layer contacted the cavity, and the lower temperature during the filling process led to rapid condensation of the melt, which was not conducive to fiber reorientation. However, the core layer had a higher temperature, resulting in better melt fluidity, and the fibers were subjected to complex shear effects, which were not conducive to the orientation of the fibers along the flow direction. As the fiber content increased, the interaction between fibers became more pronounced, and the tensors of the skin layer decreased as the fiber content increased, while the core layer exhibited the opposite behavior. The injection temperature hardly influenced the fiber orientation tensors.

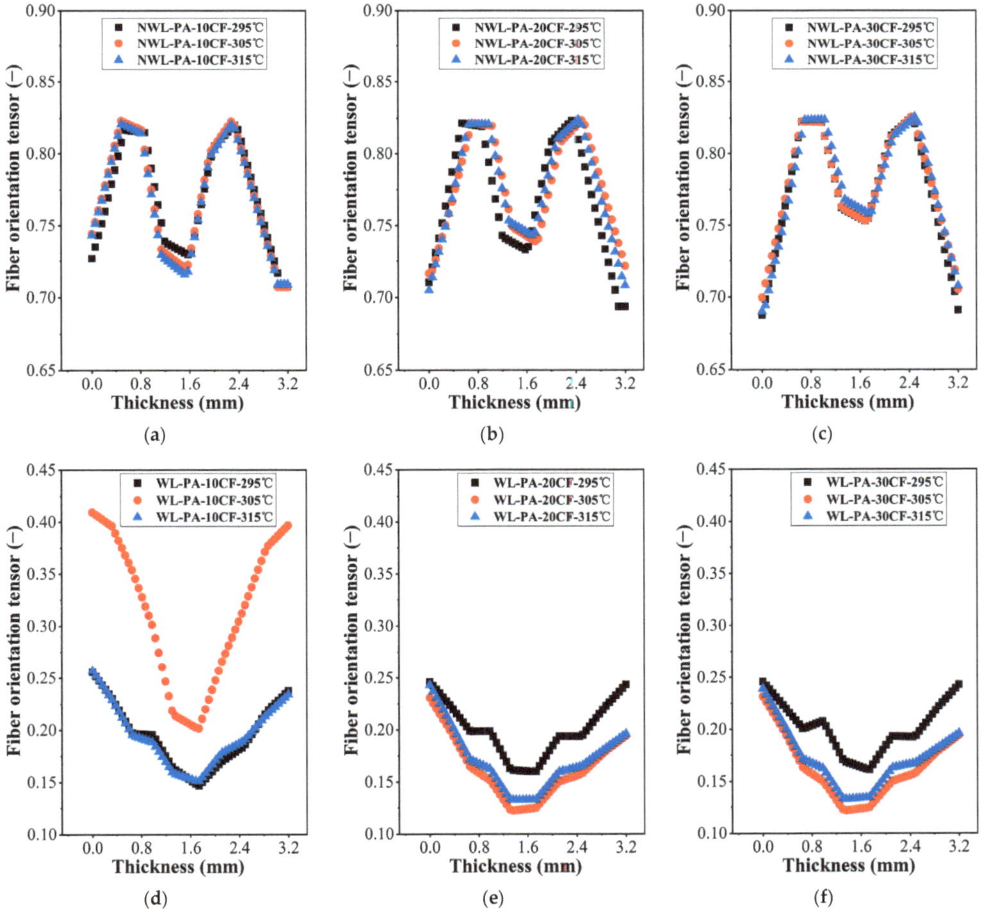

Figure 11. Fiber orientation tensors of specimens with/without weld lines for different injection temperatures: (**a**) without weld lines, PA-10CF; (**b**) without weld lines, PA-20CF; (**c**) without weld lines, PA-30CF; (**d**) with weld lines, PA-10CF; (**e**) with weld lines, PA-20CF; and (**f**) with weld lines, PA-30CF.

The fiber orientation tensors in weld line regions are shown in Figure 11d–f, in which the skin layer had larger tensors than the core layer. When two strands of melts intersected, there existed a flow process from the core layer to the skin layer, which drove the fibers in the core layer to orient perpendicularly to the initial flow direction, resulting in a lower orientation tensor of the fibers in the core layer. PA-10CF composites had the largest orientation tensor at 305 °C, while PA-20CF and PA-30CF had the largest orientation tensor at 295 °C. A larger fiber orientation tensor means better reinforcement of fibers for the matrix, corresponding to better mechanical properties. The simulation results of the fiber orientation tensor validated the change in the mechanical properties of the specimens well.

It was reported that the weld line meeting angle was related to the fiber orientation and mechanical properties of specimens [2]. Generally, a larger meeting angle causes higher mechanical properties and a more inconspicuous appearance of the weld line. The simulation results of the weld line meeting angle of PA-CF composites are exhibited in Figure 12. PA-10CF composites achieved the largest meeting angle at 305 °C, and PA-20CF and PA-30CF achieved the largest meeting angle at 295 °C. Larger weld line meeting angles corresponded to better mechanical properties as well.

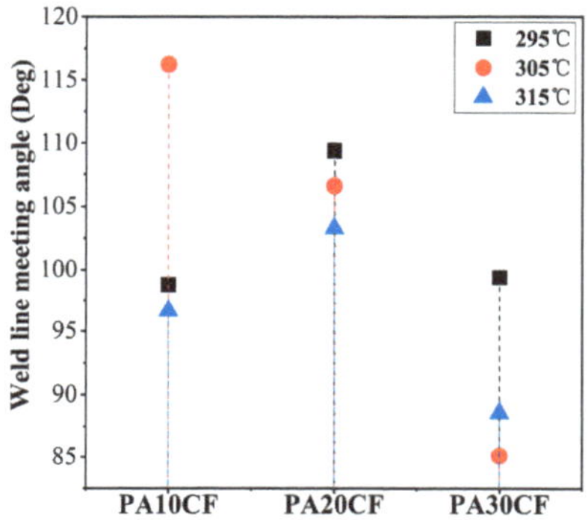

Figure 12. Weld line meeting angle for different fiber contents and temperatures.

In addition to the fiber orientation, the crystallinity of nylon in the weld line regions was also investigated. The results obtained by DSC measurement are shown in Figure 13. Generally, the crystallinity of nylon in the samples with weld lines was higher than those without weld lines, indicating that the presence of weld lines promoted the crystallization of nylon. However, the mechanical properties of the samples with weld lines were still lower than those without weld lines. Combined with the previous microstructure analysis, it was shown that fiber orientation played a decisive role in the mechanical properties of composites.

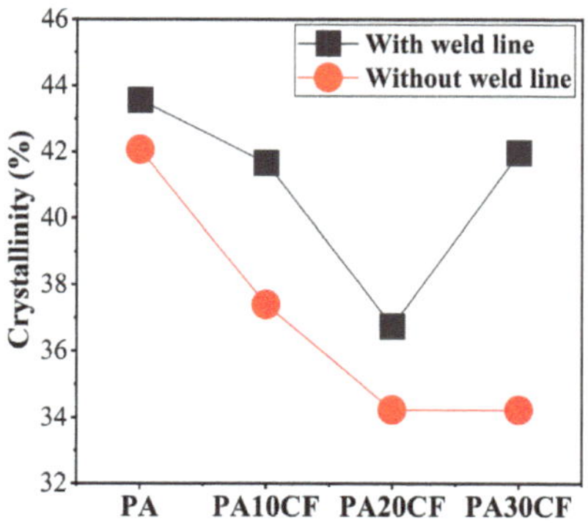

Figure 13. Crystallinity of specimens with/without weld lines for different fiber contents.

3.2. Effect of Injection Pressure on Mechanical Properties

The tensile properties of PA-CF composites were also measured for different injection pressures. As shown in Figure 14a,b, fiber content remained the most significant influencing factor for PA-CF composites without weld lines. When carbon fiber reached 30wt%, the tensile strength and Young's modulus of PA-CF composites were 2.29 and 8.43 times higher than that of pure nylon, respectively (injection molding at 7.5 MPa). In general, the injection pressure had little effect on the tensile strength and modulus of PA-CF composites without weld lines, which was similar to the results for the injection temperature. The majority of carbon fibers were oriented along the flow direction in the specimens without weld lines, and the influence of injection pressure on fiber orientation was limited as well. Therefore, changing the injection pressure did not significantly change the mechanical properties of PA-CF composites.

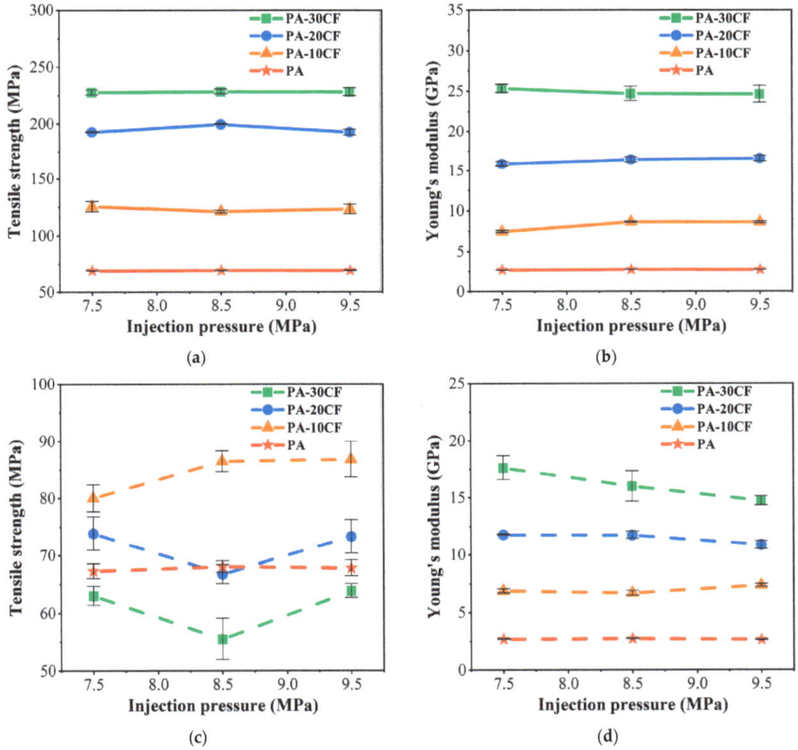

Figure 14. Tensile properties of PA-CF composites for different injection pressures: (**a**) tensile strength, without weld lines; (**b**) Young's modulus, without weld lines; (**c**) tensile strength, with weld lines; and (**d**) Young's modulus, with weld lines.

As shown in Figure 14c, PA-20CF and PA-30CF composites with weld lines had obviously lower tensile strength than PA-10CF, and PA-30CF was lower than pure nylon, which was similar to the effect of injection temperature. Only a small proportion of carbon fibers passed through the weld line region in samples with higher fiber content, and a large proportion of fibers were oriented perpendicular to the flow direction and did not play a load-bearing role. PA-20CF and PA-30CF composites obtained the lowest tensile strength at 8.5 MPa, while the values of PA-10CF increased with the rise of injection pressure. As shown in Figure 14d, Young's modulus of PA-30CF composites with weld lines decreased significantly with the increase in injection pressure. Compared with the specimen at 7.5 MPa, PA-30CF decreased by 9.2% and 16.5% at 8.5 MPa and 9.5 MPa, respectively. A large proportion of fibers were arranged parallel to the interface of the weld line for those with high fiber contents at high injection pressures, which had greater adverse influences on the properties of composites and led to a decline in modulus.

The flexural properties of PA-CF composites were measured for different injection pressures. As exhibited in Figure 15a,b, the flexural strength and modulus of PA-CF composites without weld lines were increased by 3.05 and 6.76 times, respectively, compared to pure nylon (injection molding at 7.5MPa) when carbon fiber reached 30 wt%. The injection pressure had little effect on the flexural modulus of PA-CF composites without weld lines.

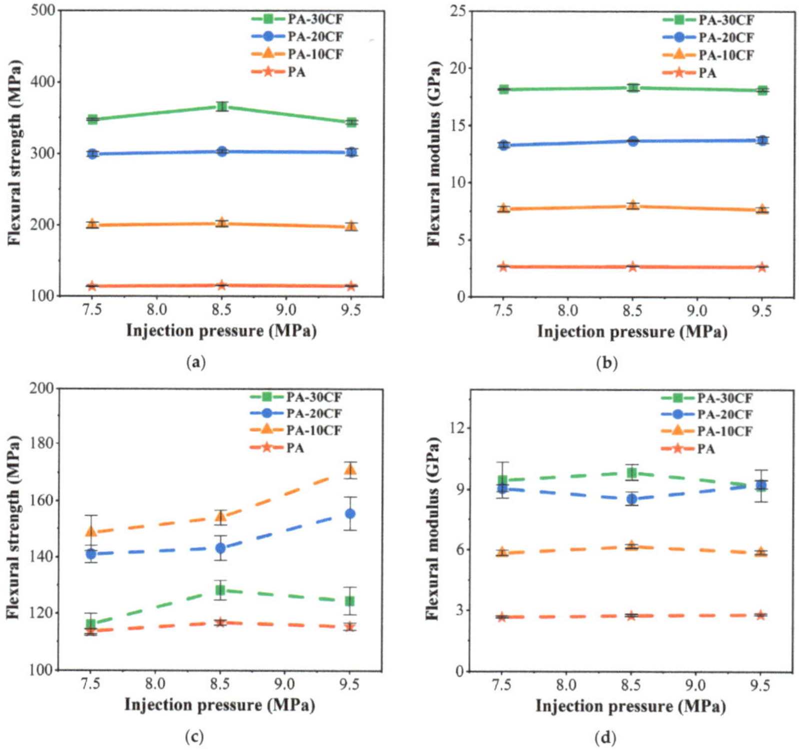

Figure 15. Flexural properties of PA-CF composites for different injection pressures: (**a**) flexural strength, without weld lines; (**b**) flexural modulus, without weld lines; (**c**) flexural strength, with weld lines; and (**d**) flexural modulus, with weld lines.

Different from the influences of injection temperature, the flexural strength of PA-CF composites with weld lines increased with the ascent of injection pressure (Figure 15c). PA-10CF composites had the largest flexural strength at each value of injection pressure. Compared with the flexural strength at 7.5 MPa, PA-10CF composites increased by 3.7% and 15.2% at 8.5 MPa and 9.5 MPa, respectively.

Weld line coefficients for different injection pressures were calculated according to Formula (1) $F_{kl} = \frac{\sigma_{WL}}{\sigma_{NWL}}$ and listed in Table 5. The weld line had a more obvious influence on the mechanical properties of PA-CF composites when fiber content increased, which was similar to the results of injection temperature. In addition, the Young's modulus weld line coefficient of PA-30CF and PA-20CF composites decreased with the ascent of injection pressure. The value of PA-10CF reached 0.92 at 7.5 MPa, indicating Young's modulus of PA-10CF with weld lines was close to that without weld lines. The flexural strength weld line coefficient of specimens containing fibers increased with the rise of injection pressure.

Table 5. Weld line coefficients for different injection pressures (Mean ± SD).

Materials	Injection Pressure (MPa)	F_{kl} of Tensile Strength (−)	F_{kl} of Young's Modulus (−)	F_{kl} of Flexural Strength (−)	F_{kl} of Flexural Modulus (−)
PA	7.5	0.97 ± 0.05	1.01 ± 0.01	1.00 ± 0.01	0.99 ± 0.02
	8.5	0.98 ± 0.03	1.00 ± 0.01	1.01 ± 0.02	1.01 ± 0.02
	9.5	0.98 ± 0.03	0.98 ± 0.02	1.00 ± 0.01	1.04 ± 0.02
PA-10CF	7.5	0.64 ± 0.03	0.92 ± 0.03	0.74 ± 0.05	0.76 ± 0.02
	8.5	0.71 ± 0.01	0.77 ± 0.02	0.76 ± 0.03	0.77 ± 0.04
	9.5	0.70 ± 0.02	0.85 ± 0.02	0.86 ± 0.03	0.77 ± 0.03
PA-20CF	7.5	0.38 ± 0.03	0.74 ± 0.02	0.47 ± 0.01	0.68 ± 0.01
	8.5	0.33 ± 0.01	0.71 ± 0.01	0.47 ± 0.02	0.63 ± 0.02
	9.5	0.38 ± 0.03	0.65 ± 0.02	0.51 ± 0.01	0.67 ± 0.01
PA-30CF	7.5	0.28 ± 0.02	0.70 ± 0.06	0.33 ± 0.02	0.52 ± 0.03
	8.5	0.24 ± 0.02	0.65 ± 0.06	0.35 ± 0.01	0.54 ± 0.01
	9.5	0.28 ± 0.01	0.60 ± 0.07	0.36 ± 0.02	0.51 ± 0.04

As shown in Figure 16, to further analyze the influences of injection pressure on microstructure, microscopic images of PA-10CF composites at 7.5 MPa and 9.5 MPa in weld lines regions were observed, showing the structure of the skin–transition–core. It is worth noting that the fibers in the transition layer tend to orient along the flow direction when injection pressure increases, which is beneficial for improving mechanical properties. The transition layer had a significantly better fiber orientation at 9.5 MPa (Figure 16b) than at 7.5 MPa (Figure 16a). The simulation also confirmed this result (Figure 17), indicating that higher injection pressure was conducive to fiber orientation in the weld line region.

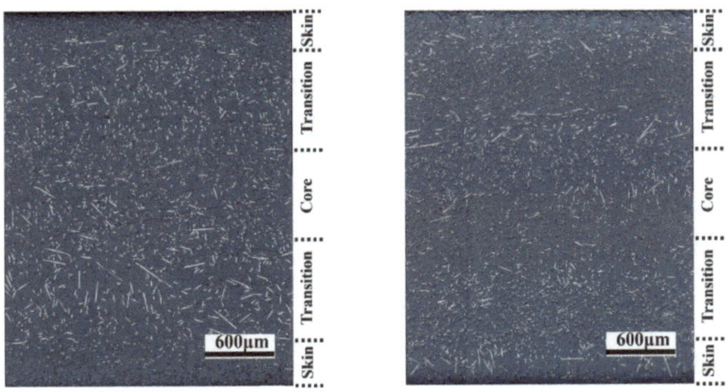

Figure 16. Metallographic micrograph of samples with weld lines: (a) PA-10CF, 7.5MPa and (b) PA-10CF, 9.5 MPa.

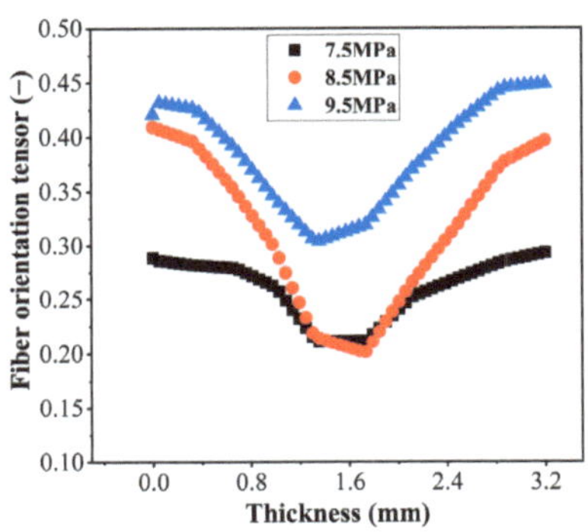

Figure 17. Fiber orientation tensor of PA-10CF with weld lines for different injection pressures.

4. Conclusions

In this work, the effects of injection temperature, injection pressure, and fiber content on the mechanical properties of weld lines were studied for PA-CF composites.

The mechanical properties of pure nylon were hardly influenced by weld lines. The existence of weld lines had negative influences on the mechanical properties of PA-CF composites. The weld line coefficient of PA-CF composites decreased as the fiber content increased, which meant that the damage of weld lines to mechanical properties increased. In weld line regions, there were a large number of fibers distributed vertically to flow direction, which could not play a reinforcing role. Thus, the damaged fractures were also concentrated in these regions. PA-10CF composites with weld lines had the largest mechanical properties due to the proportion of fibers with no reinforcement effect rising as fiber content increased. In addition, increasing the injection temperature and pressure facilitated fiber orientation, which improved the mechanical properties of composites with low fiber content. However, the mechanical properties decreased for composites with high fiber content instead.

Author Contributions: Methodology, validation, software, data curation, and writing—original draft preparation, X.W.; investigation, writing—review and editing, Z.B. and Y.X.; resources and funding acquisition, W.H.; conceptualization, supervision, and project administration, D.G., S.H., L.H., Q.C. and H.Z. All authors have read and agreed to the published version of the manuscript.

Funding: This research was funded by Key Projects of Jiangsu Industrial Technology Research Institute and Special Funding for Innovation Capacity Development in Jiangsu Province, grant number BM2021007.

Institutional Review Board Statement: Not applicable.

Data Availability Statement: Data are contained within the article.

Acknowledgments: We are grateful to the High-Performance Computing Center of Nanjing Tech University for supporting the computational resources.

Conflicts of Interest: The authors declare no conflict of interest.

References

1. Fisa, B.; Dufour, J.; Vu-Khanh, T. Weld line integrity of reinforced plastics: Effect of filler shape. *Polym. Compos.* **1987**, *8*, 408–418. [CrossRef]
2. Kovács, J.G.; Sikló, B. Experimental validation of simulated weld line formation in injection moulded parts. *Polym. Test.* **2010**, *29*, 910–914. [CrossRef]
3. Hashemi, S. Strength of single- and double-gated injection moulded short glass fibre reinforced polycarbonate. *J. Thermoplast. Compos. Mater.* **2011**, *26*, 276–295. [CrossRef]
4. Khamsehnezhad, A.; Hashemi, S. Mechanical properties of single- and double-gated injection moulded short glass fibre reinforced PBT/PC composites. *J. Mater. Sci.* **2008**, *43*, 6344–6352. [CrossRef]
5. Li, H.; Guo, Z.; Li, D. Reducing the effects of weld lines on appearance of plastic products by Taguchi experimental method. *Int. J. Adv. Manuf. Technol.* **2007**, *32*, 927–931. [CrossRef]
6. Gim, J.; Turng, L.-S. A review of current advancements in high surface quality injection molding: Measurement, influencing factors, prediction, and control. *Polym. Test.* **2022**, *115*, 107718. [CrossRef]
7. Baradi, M.B.; Cruz, C.; Riedel, T.; Régnier, G. Frontal weld lines in injection-molded short fiber-reinforced PBT: Extensive microstructure characterization for mechanical performance evaluation. *Polym. Compos.* **2019**, *40*, 4547–4558. [CrossRef]
8. Baradi, M.B.; Cruz, C.; Riedel, T.; Régnier, G. Mechanical and microstructural characterization of flowing weld lines in injection-molded short fiber-reinforced PBT. *Polym. Test.* **2019**, *74*, 152–162. [CrossRef]
9. Lim, J.K.; Shoji, T. Fiber orientation of polymer injection weld and its strength evaluation. *KSME J.* **1993**, *7*, 173–181. [CrossRef]
10. Chen, C.-S.; Chen, T.-J.; Chien, R.-D.; Chen, S.-C. Investigation on the weld line strength of thin-wall injection molded ABS parts. *Int. Commun. Heat Mass Transf.* **2007**, *34*, 448–455. [CrossRef]
11. Daiyan, H.; Andreassen, E.; Grytten, F.; Lyngstad, O.V.; Luksepp, T.; Osnes, H. Low-velocity impact response of injection-moulded polypropylene plates—Part 2: Effects of moulding conditions, striker geometry, clamping, surface texture, weld line and paint. *Polym. Test.* **2010**, *29*, 894–901. [CrossRef]
12. Dzulkipli, A.A.; Azuddin, M. Study of the Effects of Injection Molding Parameter on Weld Line Formation. *Procedia Eng.* **2017**, *184*, 663–672. [CrossRef]

13. Kagitci, Y.C.; Tarakcioglu, N. The effect of weld line on tensile strength in a polymer composite part. *Int. J. Adv. Manuf. Technol.* **2016**, *85*, 1125–1135. [CrossRef]
14. Malguarnera, S.C.; Manisali, A. The effects of processing parameters on the tensile properties of weld lines in injection molded thermoplastics. *Polym. Eng. Sci.* **1981**, *21*, 586–593. [CrossRef]
15. Chrysostomou, A.; Hashemi, S. Mechanical properties of injection moulded styrene maleic anhydride (SMA) Part II Influence of short glass fibres and weld lines. *J. Mater. Sci.* **1998**, *33*, 4491–4501. [CrossRef]
16. Fisa, B.; Rahmani, M. Weld line strength in injection molded glass fiber-reinforced polypropylene. *Polym. Eng. Sci.* **1991**, *31*, 1330–1336. [CrossRef]
17. Hashemi, S. Thermal effects on weld and unweld tensile properties of injection moulded short glass fibre reinforced ABS composites. *Express Polym. Lett.* **2007**, *1*, 688–697. [CrossRef]
18. Scantamburlo, A.; Zanini, F.; Lucchetta, G.; Sorgato, M. Improving the weld lines mechanical properties by combining alternate dynamic packing and rapid heat cycle moulding. *Compos. Part A Appl. Sci. Manuf.* **2022**, *163*, 107239. [CrossRef]
19. Ozcelik, B.; Kuram, E.; Topal, M.M. Investigation the effects of obstacle geometries and injection molding parameters on weld line strength using experimental and finite element methods in plastic injection molding. *Int. Commun. Heat Mass Transf.* **2012**, *39*, 275–281. [CrossRef]
20. Oh, G.-H.; Jeong, J.-H.; Park, S.-H.; Kim, H.-S. Terahertz time-domain spectroscopy of weld line defects formed during an injection moulding process. *Compos. Sci. Technol.* **2018**, *157*, 67–77. [CrossRef]
21. Martulli, L.M.; Kerschbaum, M.; Lomov, S.V.; Swolfs, Y. Weld lines in tow-based sheet moulding compounds tensile properties: Morphological detrimental factors. *Compos. Part A Appl. Sci. Manuf.* **2020**, *139*, 106109. [CrossRef]
22. Seldén, R. Effect of processing on weld line strength in five thermoplastics. *Polym. Eng. Sci.* **1997**, *37*, 205–218. [CrossRef]
23. Choudhari, D.S.; Kakhandki, V.J. Comprehensive study and analysis of mechanical properties of chopped carbon fibre reinforced nylon 66 composite materials. *Mater. Today Proc.* **2021**, *44*, 4596–4601. [CrossRef]
24. Guo, A.; Liu, C.; Li, S.; Zhou, X.; Wang, J.; Wang, S.; Qu, P.; Hu, Y. Water absorption rates and mechanical properties of material extrusion-printed continuous carbon fiber-reinforced nylon composites. *J. Mater. Res. Technol.* **2022**, *21*, 3098–3112. [CrossRef]
25. Ma, Y.; Jin, S.; Yokozeki, T.; Ueda, M.; Yang, Y.; Elbadry, E.A.; Hamada, H.; Sugahara, T. Effect of hot water on the mechanical performance of unidirectional carbon fiber-reinforced nylon 6 composites. *Compos. Sci. Technol.* **2020**, *200*, 108426. [CrossRef]
26. Quintana, M.C.; Frontini, P. Weld line strength factors in a reinforced injection molded part: Relationship with predicted fiber orientation. *J. Reinf. Plast. Compos.* **2020**, *39*, 219–230. [CrossRef]
27. Cao, W.; Shen, Y.; Wang, P.; Yang, H.; Zhao, S.; Shen, C. Viscoelastic modeling and simulation for polymer melt flow in injection/compression molding. *J. Non-Newton. Fluid Mech.* **2019**, *274*, 104186. [CrossRef]
28. Trotta, G.; Stampone, B.; Fassi, I.; Tricarico, L. Study of rheological behaviour of polymer melt in micro injection moulding with a miniaturized parallel plate rheometer. *Polym. Test.* **2021**, *96*, 107068. [CrossRef]
29. Xu, X.; Tian, L.; Peng, S.; Yu, P. Development of SPH for simulation of non-isothermal viscoelastic free surface flows with application to injection molding. *Appl. Math. Model.* **2022**, *104*, 782–805. [CrossRef]
30. Thakre, P.; Chauhan, A.S.; Satyanarayana, A.; Raj Kumar, E.; Pradyumna, R. Estimation of Shrinkage & Distortion in Wax Injection using Moldex3D Simulation. *Mater. Today Proc.* **2018**, *5*, 19410–19417. [CrossRef]
31. Gao, P. Three dimensional finite element computation of the non-isothermal polymer filling process by the phase field model. *Adv. Eng. Softw.* **2022**, *172*, 103207. [CrossRef]
32. Kuo, H.-C.; Jeng, M.-C. Effects of part geometry and injection molding conditions on the tensile properties of ultra-high molecular weight polyethylene polymer. *Mater. Des.* **2010**, *31*, 884–893. [CrossRef]
33. Hashemi, S.; Lepessova, Y. Temperature and weldline effects on tensile properties of injection moulded short glass fibre PC/ABS polymer composite. *J. Mater. Sci.* **2007**, *42*, 2652–2661. [CrossRef]
34. Vaxman, A.; Narkis, M.; Siegmann, A.; Kenig, S. Weld-line characteristics in short fiber reinforced thermoplastics. *Polym. Compos.* **1991**, *12*, 161–168. [CrossRef]
35. Solymossy, B.; Kovacs, J. The Examination of Weld Line Properties in Injection Molded PP Composites. *Mater. Sci. Forum* **2008**, *589*, 263–267. [CrossRef]
36. Wu, W.; Zhao, B.; Mo, F.; Li, B.; Jiang, B. In-line steady shear flow characteristics of polymer melt in rectangular slit cavities during thin-wall/micro injection molding. *Mater. Des.* **2022**, *223*, 111266. [CrossRef]
37. Wittemann, F.; Kärger, L.; Henning, F. Theoretical approximation of hydrodynamic and fiber-fiber interaction forces for macroscopic simulations of polymer flow process with fiber orientation tensors. *Compos. Part C Open Access* **2021**, *5*, 100152. [CrossRef]
38. Sasayama, T.; Sato, N.; Katagiri, Y.; Murayama, Y. Particle-level simulation for the prediction of short fiber orientation in injection molding. *Compos. Part A Appl. Sci. Manuf.* **2020**, *139*, 106115. [CrossRef]

materials

Article

Effect of pH Hydrolysis on the Recovery of Antimony from Spent Electrolytes from Copper Production

Eduardo Díaz Gutiérrez [1,*], José Antonio Maldonado Calvo [2], José María Gallardo Fuentes [1] and Antonio Paúl Escolano [1]

[1] Departamento de Ingeniería y Ciencia de los Materiales y del Transporte, Escuela Politécnica Superior, Calle Virgen de África 7, 41011 Seville, Spain; josemar@us.es (J.M.G.F.); apaul@us.es (A.P.E.)
[2] Atlantic Copper, S.L.U., Francisco Montenegro Avenue, 21001 Huelva, Spain; jmaldona@fmi.com
* Correspondence: eduardodiaz@us.es

Abstract: This study examined how pH hydrolysis affects the recovery process for antimony extracted from spent electrolytes. Various OH^- reagents were used to adjust the pH levels. The findings reveal that pH plays a crucial role in determining the optimal conditions for extracting antimony. The results show that NH_4OH and NaOH are more effective compared to water, with optimal conditions at pH 0.5 for water and pH 1 for NH_4OH and NaOH, resulting in average antimony extraction yields of 90.4%, 96.1%, and 96.7%, respectively. Furthermore, this approach helps to improve both crystallography and purity related to recovered antimony samples obtained through recycling processes. The solid precipitates obtained lack a crystalline structure, making it difficult to identify the compounds formed, but element concentrations suggest the presence of oxychloride or oxide compounds. Arsenic is incorporated into all solids, affecting the purity of the product, and water showing higher antimony content (68.38%) and lower arsenic values (8%) compared to NaOH and NH_4OH. Bismuth integration into solids is less than arsenic (less than 2%) and remains unaffected by pH levels except in tests with water, where a bismuth hydrolysis product is identified at pH 1, accounting for the observed reduction in antimony extraction yields.

Keywords: antimony metallurgy; antimony oxychloride; hydrolysis; copper electrorefining

Citation: Díaz Gutiérrez, E.; Maldonado Calvo, J.A.; Gallardo Fuentes, J.M.; Paúl Escolano, A. Effect of pH Hydrolysis on the Recovery of Antimony from Spent Electrolytes from Copper Production. *Materials* **2023**, *16*, 3918. https://doi.org/10.3390/ma16113918

Academic Editor: Vincenzo M. Sglavo

Received: 19 April 2023
Revised: 12 May 2023
Accepted: 18 May 2023
Published: 23 May 2023

1. Introduction

The concept of circular economy (CE) arises from two fundamental needs. First, to try to reduce the consumption of raw materials. Second, to reduce the production of waste which, in turn, is connected to both economic and environmental factors. This concept has traditionally been linked to the reintroduction or reprocessing of materials or elements once the useful life of the product or application they are part of is finished. However, in the field of extractive metallurgy and mineral processing, this concept can be applied to the recovery of any material that may be of interest in different phases of the production process in which streams are generated and are often treated as waste [1].

In this concern, the European Union (EU) established a specific policy on critical raw materials in 2008 [2], which focuses on ensuring sustainable and secure access to raw materials. Some of the policies and actions that the EU has implemented include: (I) identifying critical raw materials based on the assessment of factors such as supply risk, economic importance, and environmental impact; (II) promoting the circular economy and material recovery to reduce the need for new raw materials and reduce dependence on critical raw materials; (III) investing in research and innovation to develop alternatives to critical raw materials, as well as improve efficiency in their use.

Aligned with these objectives and in the search for candidate processes, during the production of high-purity copper a wide variety of materials are generated that have traditionally been treated as waste but can be converted into recyclable materials or by-products that can be marketed or used in other sectors as raw materials through certain actions. Specifically, the stream generated in the antimony and bismuth elimination plant of the electrolyte used in the final electrorefining process of copper at the Atlantic Copper metallurgical complex in Huelva turns out to be a candidate source. Some reasons support this issue. First, antimony is qualified as a critical raw material by the European Union [3,4]. Second, currently, the stream is neutralized and treated as waste. Finally, the metals are found in a liquid stream of hydrochloric acid that allows the application of hydrometallurgical methods.

The by-product stream is produced during the regeneration stage of the chelating resins used to remove antimony and bismuth from the copper electrolyte [5]. At this point, in a process known as elution, the resulting solution or discharge is commonly referred to as an eluate. It consists of an hydrochloric acid stream with chloride concentrations of approximately 170 g/L and a varying content of Sb and Bi [6]. Table 1 displays the standard composition of the eluate.

Table 1. Typical composition of the eluate (g/L) from the ion exchange facility.

As	Bi	Sb	Cl^-	Others
2–5	8–10	9–10	170	<1

Antimony is a chemical element that has been used in a variety of applications, including as a component of alloys, in flame retardants, and in medicines [7,8]. Recently, new uses of antimony have emerged, particularly in the fields of energy storage and electronics. Antimony is being investigated as a potential anode material for lithium-ion batteries [9], which are used in a wide range of devices from smartphones to electric vehicles. Antimony-based materials are also being explored for use in high-performance electronic devices, such as transistors and diodes, due to their unique electrical and optical properties [10]. Additionally, antimony could be an important material in the development of new solar cell technologies, particularly in the development of thin-film solar cells [11,12]

In this context, several authors [13–15] have published reviews about the current technological situation of antimony recovery, covering the current market and resource situation, both primary and secondary sources, and finally, the existing procedures for production and recovery depending on the medium. Within the different hydrometallurgical processes described in the reviews, hydrolysis appears as one of the procedures applied for the recovery of antimony in acidic streams.

Indeed, hydrolysis is described as one of the procedures of the hydrometallurgy method for industrial production of antimony [16], where antimony in sulfide form is transformed into chloride by leaching it in a chloride medium, reaction (1). The antimony chloride is then obtained in a metallic form by electrowinning in diaphragm cells. Antimony chloride can also be extracted by hydrolysis by adding OH^- ions, producing solid oxychlorides, reactions (2) and (3). These oxychlorides, when treated with ammonium hydroxide, produce antimony oxide, (4) and (5).

$$Sb_2S_3 + 6HCl \rightarrow 2SbCl_3 + 3H_2S \quad (1)$$

$$SbCl_3 + H_2O \rightarrow SbOCl + 2HCl \quad (2)$$

$$4SbCl_3 + 5H_2O \rightarrow Sb_4O_5Cl_2 + 10HCl \quad (3)$$

$$2SbOCl + 2NH_4OH \rightarrow Sb_2O_3 + 2NH_4Cl + H_2O \quad (4)$$

$$Sb_4O_5Cl_2 + 2NH_4OH \rightarrow 2Sb_2O_3 + 2NH_4Cl + H_2O \quad (5)$$

Hydrolysis of metal chlorides involves various processes and reactions that depend mainly on metal concentration, pH, and chloride concentration [17]. However, the presence of solid intermediate products can make the hydrolysis of antimony trichloride a particularly complex process. Therefore, researchers [18,19] have developed mathematical models and predominance diagrams to predict the behavior of the Sb (III)-Cl-H_2O system as a function of the aforementioned variables. Experimental validation of theoretical models under specific conditions and experimental studies have been published [20,21].

According to theoretical simulations carried out by Zhao [18], there is a notable correlation between the reduction of chloride concentration from 3 to 1 mol/L and the decrease in hydrolysis pH. This trend is particularly significant within the pH range of 0–4, where oxychlorides remain stable, while above pH 4, oxide becomes more stable and chloride concentration has an insignificant effect. Research conducted by Tian [19] through mathematical modeling and experimental tests on varied chloride concentrations ranging from 3–5 mol/L, had consistent observations with similar trends previously revealed in theoretical analyses. Figure 1 illustrates the precipitated species that were identified at various hydrolysis pH levels.

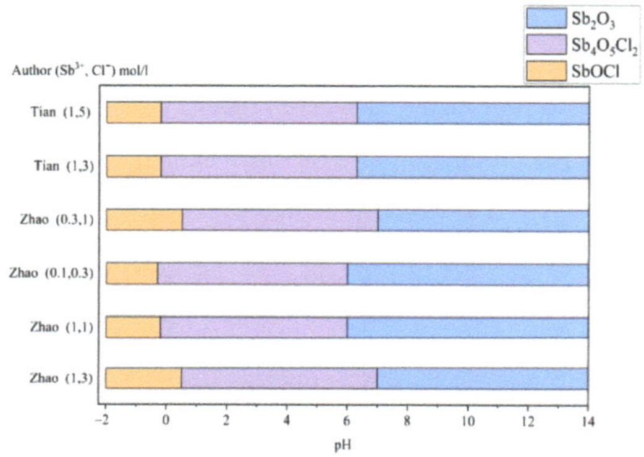

Figure 1. Distribution of species as a function of pH according to theoretical studies [18,19].

Although theoretical studies have found that SbOCl is the stable phase within the pH range of −2 to 0, none of the empirical studies have detected this compound in its basic form. The reason for this occurrence is related to the compound instability, leading to a spontaneous transformation into $Sb_4O_5Cl_2$ [19]. This compound has been obtained in most experimental studies and is typically present within a range of pH values, commonly above pH −0.1 to pH 7 [19,21] with variations depending on antimony and chloride concentrations. There are several distinct crystalline allotropic forms of this compound that have been observed. Other compounds such as $Sb_8O_{10}(OH)_2Cl_2$ have also been identified within pH range between 1.5 to 4.2 [20]. Once at a certain level, it then transforms into Sb_2O_3 when the solution reaches a certain pH value (>4.5). Another species detected during similar pH ranges was $Sb_8O_{11}C_{12}$, which appears at levels ranging from acidity to neutral states [19].

The species identified in experimental studies carried out under conditions similar to those studied in this work are shown in Figure 2.

In other studies [22], comparable phases were recognized when using an ethyleneglycol or ethanol and water mixture as the solvent instead of HCl. $Sb_4O_5Cl_2$ was detected within pH levels ranging from 1 to 2, while $Sb_8O_{11}Cl_2$ was identified at values between 4 and 5. Similarly, the presence of Sb_2O_3 became apparent within a pH range of approximately 8 to 9. Accordingly, researchers in [23] generate $Sb_4O_5Cl_2$ by adding water into an ethanol-based HCl solution with a concentration of 2 mol/L. Total precipitation is determined to occur in a volumetric ratio of 1 between the added water and original solution.

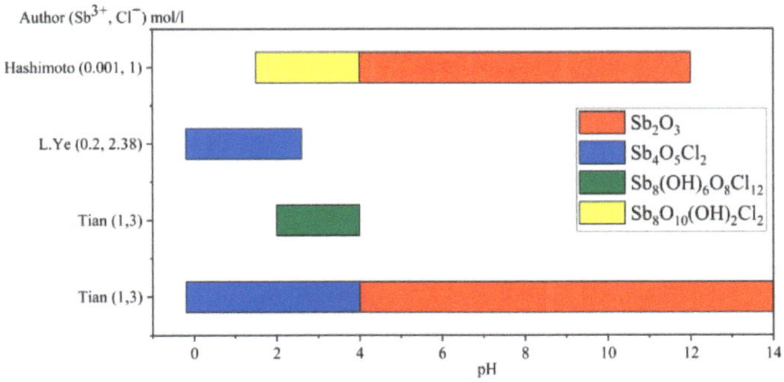

Figure 2. Distribution of species as a function of pH according to experimental studies [19–21].

It appears that the use of hydrolysis for antimony recovery is feasible in circumstances that closely resemble those present within elution-generated streams. However, more research is needed to fully understand and optimize the process parameters under specific conditions. This work presents findings from experiments carried out to extract antimony via direct hydrolysis in the eluate with water, NaOH, and NH_4OH. Additionally, product characterization was performed to confirm procedural validity and assess the impact of pH and other metals present in the eluate on extraction yield for antimony. The research aimed to identify favorable conditions that promote maximum antimony extraction efficiency.

2. Materials and Methods

2.1. Materials

Eluate samples were drawn directly from the Atlantic Copper Industrial Process. Thus, the eluate composition is subject to high dispersion, which is commonly observed in industrial operations. Through a detailed analysis of the historical data provided by Atlantic Copper, a range of values were established that can be considered representative to set the test conditions related to the composition of the eluate. Once these conditions had been established, periodic sampling was performed to obtain eluates with approximate compositions in the range selected. Table 2 shows the concentrations of metals in the eluate sample used in this work.

Table 2. Composition of the main components of eluate used in this study (g/L).

Element	As	Bi	Sb	Cl^{-}
Content (g/L)	2.3	9.5	8.9	170

The reagents used were NH_4OH and NaOH in analytical grade quality, both provided by Panreac (Barcelona,) Spain.

2.2. Methods

The hydrolysis tests were carried out in glass flasks using 50 mL of solution. The flasks were placed in a thermostatic bath to maintain the reaction temperature within ±1 °C, with continuous stirring. Distilled water with conductivity less than 5 µS/cm, NaOH (10% *w/w*), and NH_4OH (30% *w/w*) were used to modify the pH reaction. After achieving the targeted pH level, the reaction was allowed to proceed for a duration of 60 minutes.. The pH and temperature were continually monitored using a pH meter (Labprocess VioLab PH60. Barcelona, Spain) with a pH accuracy of 0.01. The validity and reliability of the results were ensured by performing a minimum of three tests for each condition.

The precipitated solid was vacuum filtered with 0.65 micron pore membrane filters and dried at 60 °C for 1 h. X-ray diffraction (XRD) patterns of solid dry hydrolyzed products were obtained in $\theta/2\theta$ geometry using an X'Pert Pro instrument (Malvern Panalytical,

Malvern, UK), with Cu-K_α radiation source (40 kV, 40 mA). Secondary K_β filter and a secondary diffracted beam monochromator were used to reduce fluorescence. XRD patterns were collected by scanning between 0° and 80° in the step-scan mode with 0.03° steps and 5 s dwell time.

Elemental analysis of hydrolysis dry products was carried out by means of X-ray fluorescence (XRF). Previously, the samples were homogenized and pressed on a boric acid tablet. XRF measurements were made on a Zetium Malvern Panalytical wavelength dispersion fluorescence spectrometer (Malvern Panalytical, Malvern, UK) using a semi-quantitative measurement method.

The content of Sb, Bi, and As in the liquid phase was measured by inductively coupled plasma optical emission spectroscopy (ICP-OES) on an Agilent 5800 spectrometer (Agilent, Santa Clara, CA, USA).

The concentration of ion chloride in the eluate samples was determined by titration with silver nitrate in an automatic Metrohm 855 tritator (Metrohm, Herisau, Switzerland).

Evaluating Speciation and Test Conditions

According to the reference [24], antimony exists in the electrolyte as Sb(III) and Sb(V), with equal distribution, while arsenic is primarily found as As(V) and bismuth exists as Bi(III). Reference [25] suggests that this resin can retain 100% of both Sb(III) and Bi(III), but its capacity for retaining Sb(V) is lower, leading to reduced durability. Thus, it can be concluded that within the eluate, antimony will be present predominantly in a 75–25% ratio between its two oxidation states (Sb III-V), while all bismuth will remain exclusively in the Bi(III) form.

Theoretical studies of the species contained in the eluate have been carried out to limit the study ranges of laboratory tests and mainly refer to arsenic, bismuth, and antimony compounds. However, the accuracy of the results depends mainly on the availability of thermodynamic data of the species involved in the process, and usually experimental tests are required to validate the simulation results. Specifically, the modeling of the Sb-Cl-H_2O and Bi-Cl-H_2O systems is complex due to the elevated number of equilibrium reactions involved and the lack of data for some of the solid species that are generated during hydrolysis. However, by using software that employs computational algorithms based on the method of minimizing free energy, it is possible to obtain an estimate of the pH values at which solid species begin to be identified.

To enhance the precision of Sb-Cl classification, the data reported in [26] are implemented. The results for the eluate's nominal conditions of the evaluation can be found in Figure 3a, which illustrates that solid phase formation takes place at a pH value of 0.3 or lower, where dominant species include $SbCl_6^{-3}$, $SbCl_5^{-2}$, $SbCl^{-4}$, and $SbCl_3$.

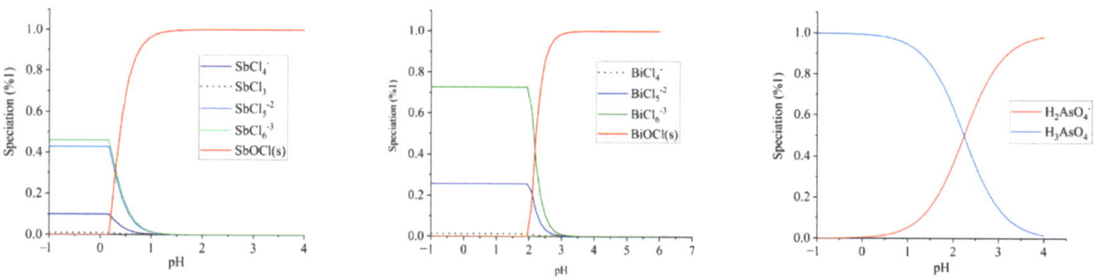

(a) (b) (c)

Figure 3. Speciation diagrams in the form of a molar fraction as a function of pH for: (**a**) Sb (III) -Cl-H_2O system, for a solution containing 4.7 mol/L HCl and 0.08 mol/L Sb. (**b**) Bi(III) -Cl-H_2O system, for a solution containing 4.7 mol/L HCl and 0.04 mol/L Bi. (c) As(V)-Cl-H_2O system, for a solution containing 4.7 mol/L of HCl and 0.04 mol/L.

A comparable examination for bismuth can be conducted, whose concentration in the eluate is similar to that of antimony and experiences hydrolysis in a chloride medium [27]. According to Figure 3b, under conditions reflecting nominal composition of the eluate, bismuth oxychloride appears to be highly stable across a wide pH range. The molar fraction versus pH diagram has been derived utilizing equilibrium constants reported by [28] for the Bi-Cl system. As indicated in Figure 3b, precipitation commences around a pH value of 2; below this level bivalent $BiCl_6^{3-}$ dominates while trivalent $BiCl_5^{2-}$ appears alongside minor amounts of tetravalent species $BiCl_4$. Similarly exhaustive results on bismuth hydrothermal chemistry have also been documented in [29]. In terms of As(V), precipitation is not expected to occur under the eluate conditions. Figure 3c, along with the literature review by [30], indicates that H_3AsO_4 and $H_2AsO_4^-$ are formed by As(V) at varying pH levels.

The pH range selected in this study was determined by reviewing pertinent literature and analyzing the results from speciation diagrams. Specifically, the investigation focused on a range of pH values from 0.25 to 1 as regards antimony extraction efficacy, which is defined here as the proportion between the Sb concentration found in liquid samples through ICP analysis and Sb content measured through the XRF assessment of solid products generated during experimentation according to Equation (6).

$$\text{Sb yield extraction } \% = (\text{Sb content in hydrolysis product (g)})/(\text{Sb content in solution (g)} \times 100) \tag{6}$$

3. Results

3.1. Effect of pH on Antimony Extraction Yield

The hydrolysis tests were carried out at 25 °C. In the tests performed with NaOH and NH_4OH, the addition speed was intentionally slow, 1 mL/min, and 0.16 mL/min, respectively, to avoid any increase in temperature resulting from the significant exothermic reaction. However, water addition does not present this constraint; thus, all the necessary volume needed to achieve the specified pH level can be directly incorporated into the eluate.

The results of antimony extraction under different pH conditions are shown in Figure 4. The data reveal that the yield of antimony extraction by hydrolysis is higher when performed with NaOH or NH_4OH compared to H_2O. Furthermore, it was observed that the degree of success in extracting antimony depended significantly on pH values; a positive correlation between pH elevation and an increase in yield efficiency can be inferred from the results presented.

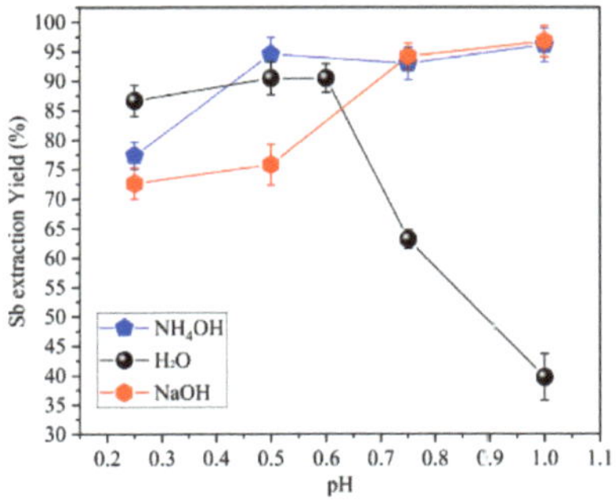

Figure 4. Antimony extraction yields results at 25 °C. (The bar represents the standard deviation.)

In terms of the reagents employed, NH_4OH exhibited higher values for antimony extraction at all the pH levels tested, with most cases producing an extraction rate surpassing 90%. Additionally, the results displayed favorable reproducibility, as evidenced by low standard deviation values.

Conversely, the use of H_2O and NaOH as hydrolysis agents produced diminished antimony extraction efficiencies, particularly under low pH conditions. Furthermore, compared to NH_4OH extractions, extractions using water and NaOH exhibited elevated standard deviation values. Overall, the results indicate that hydrolysis performed with NH_4OH is the most efficacious approach for antimony extraction in this research study.

In the analysis of results, it is crucial to consider the quantity of reagent required to attain a particular pH level. The volumetric correlations between the volume of eluate and the amount of reagent are shown in the Figure 5. A discernible distinction can be observed among NaOH, NH_4OH, and H_2O reagents. The utilization of NH_4OH results in the lowest volumetric ratio, which fluctuates from 0.48 to 0.55 and requires a 5% increase between various points. In contrast, when NaOH is used, the volumetric ratios escalate to values ranging between 1.7 and 2.3 with changes ranging from 10–15%. To achieve identical pH values, water quantities exceeding a volumetric ratio of 10 are necessary. The outcomes remain reliable despite variations in reagent concentrations and hydroxyl group availability therein.

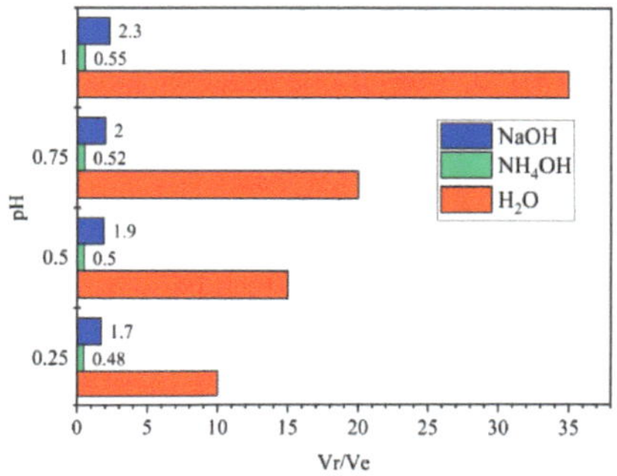

Figure 5. Ratio between the volume of reagent (Vr) used and the volume of eluate sample (Ve) required to reach the pH values indicated.

According to the analysis conducted on the quantity of the reagent, it is clear that the use of HN_4OH and NaOH provides an advantage over water. However, there are other factors to be considered when comparing all conditions. The use of highly alkaline reagents such as NH_4OH requires a slow addition rate due to the reaction's high exothermicity which extends the total reaction time. This phenomenon also occurs in the case of NaOH, though not as extensively as with NH_4OH; however, this issue does not arise with water whose addition can occur almost instantaneously. Additionally, several important factors such as gas production during reactions, handling, and storage complexities, along with the cost associated with NH_4HOH and NaOH, must influence our final choice regarding selection of a reagent, but their thorough examination requires further study.

3.2. XRD

The diffraction patterns of the solid products resulting from hydrolysis were obtained for all tested conditions. Figure 6 displays the results of X-ray diffraction experiments conducted with water. It can be noted that most of the entities are primarily amorphous, except those obtained at pH 1, where BiOCl (PDF 01-073-2060) featuring a tetragonal configuration was identified. The creation of this compound is attributed to bismuth hydrolysis, and it can occur partially or complete with antimony hydrolysis [28]. However, the results observed at pH 0.25, 0.5, and 0.75 along with those found at pH 1, show discrepancies from the findings reported in the existing literature and the thermodynamic models created based on the eluate conditions. In the case of solids obtained with NaOH and NH_4OH, as can be seen in Figure 7, they do not have a crystalline character, even at pH = 1, where no hydrolysis products of bismuth are detected as in the case of the assays carried out with water. In some of the products obtained with NaOH, low-intensity peaks of NaCl are identified which appear to be incorporated into the product. The discrepancies encountered in the comparative literature can potentially be ascribed to the existence of arsenic and bismuth, which may result in the lack of crystallinity. Studies conducted on copper electrolytes have extensively described the formation of amorphous antimony and bismuth arsenates [31]. This phenomenon may occur under these evaluation conditions, however, further investigation is required to systematically study and confirm this hypothesis.

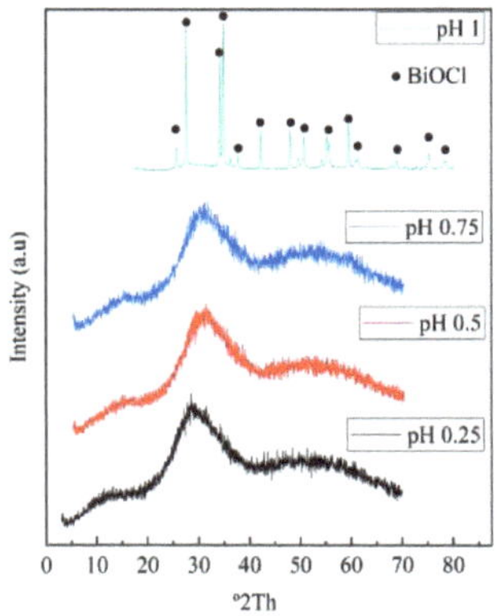

Figure 6. Representative diffractograms obtained for the hydrolysis products with water at different pH and 25 °C.

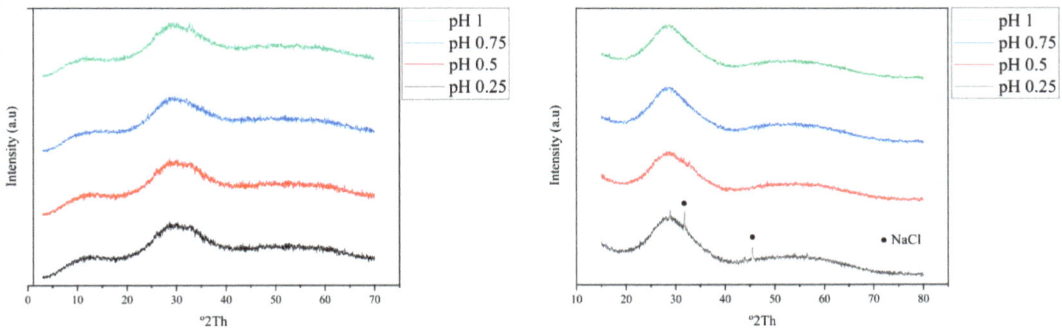

(**a**) (**b**)

Figure 7. Representative diffractograms obtained for the hydrolysis products at different pH and 25 °C. (**a**) Hydrolysis with NH_4OH. (**b**) Hydrolysis with NaOH.

3.3. Chemical Composition XRF

According to the analysis conducted through crystallography, a significant proportion of hydrolysis by-products lack crystalline structure. Consequently, investigating such samples presents substantial difficulties, especially when trying to understand the behaviour of individual components during a hydrolysis reaction.

Table 3 presents the chemical compositions of the hydrolysis products, which display stoichiometric relationships according to antimony oxychloride compounds. Nevertheless, due to notable quantities of arsenic present within these products and the lack of information of the specific phase, it is challenging to establish an unequivocal identification of each compound. Typically, water-derived products exhibit a higher concentration of antimony content compared to those obtained through the NaOH or NH_4OH methods which follow in descending order, respectively. The chlorine and oxygen exhibit comparable product compositions similar to those of water and NaOH. However, the implementation of

NH_4OH appears to generate species that could be combinations of oxides and oxychlorides. This is supported by the increase in the percentage of oxygen and the decrease in chlorine.

Table 3. Mass percent average compositions of the products obtained in the hydrolysis.

Element	Sb	Cl	O	As	Bi	Na	Others
H_2O	64.75 ± 3.59	4.01 ± 2.46	16.32 ± 1.81	10.90 ± 2.04	2.93 ± 1.73	-	0.13 ± 0.1
NaOH	62.19 ± 3.89	5.20 ± 3.89	17.02 ± 0.45	11.78 ± 0.94	1.49 ± 0.58	1.98 ± 0.79	0.34 ± 0.16
NH_4OH	56.04 ± 2.55	2.21 ± 1.43	18.32 ± 0.23	21.48 ± 1.81	1.68 ± 0.18	-	0.36 ± 0.11

The inadvertent occurrence of arsenic is of great concern, particularly as it is found in solids at levels approaching 11% when water and NaOH are involved, with values exceeding 20% in those tests involving NH_4OH. This poses a significant challenge for evaluating the effectiveness of the process and assessing product quality. In all cases, measurable amounts of bismuth are observed that range from 1.5% to approximately 3%. Experiments conducted with NaOH result in final product compositions containing around 2% NaOH.

3.4. *Effect of pH on Antimony, Bismuth, and Arsenic Content*

A study was conducted on the effect of pH on the composition of the main elements. In Figure 8, the evolution of the Sb content with pH is observed using different reagents. At pH 0.25, there is a significant increase in Sb content (68.38%) within the solid powder produced by hydrolysis using water as the medium. However, with NaOH or NH_4OH serving as alternative hydrolysis agents, there was a decrease in Sb concentration that resulted in values of 59.65% and 53.89%, respectively. Similar observations can be made at pH levels of 0.5 and 0.6, whereby the percentage of antimony in solid powder derived from water hydrolysis exceeds that obtained by NaOH or NH_4OH hydrolysis methods. However, at a pH level of 0.75, the percentage yield is comparable between Sb extracted through water and NaOH hydrolysis approaches (at 61.79%), while the use of NH_4OH results in marginally lower yields (53.95%). At pH 1, the percentage of Sb in solid powder obtained through hydrolysis with water is the lowest (36.56%). Interestingly, the percentage of Sb obtained with NaOH and NH_4OH is higher (58.86% and 59.04%, respectively), suggesting that these two hydrolysis media are more effective than water for extracting Sb at this pH.

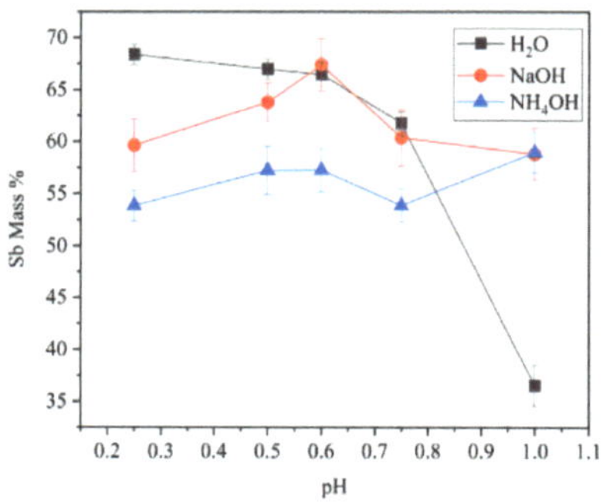

Figure 8. Evolution of the antimony content with pH (bars represent standard deviation).

The results of evolution of bismuth are shown in Figure 9. Bismuth values remain relatively constant in the solids produced using NaOH and NH_4OH with percentages below 2%. On the contrary, when H_2O is used as a reagent for solid production, values exceed this threshold across all pH levels ranging from 0.25 to 0.75, reaching 5% at their maximum value. A substantial rise in the amount of Bi is observed at a pH level of 1. It could potentially be attributed to the initiation of the bismuth hydrolysis reaction, which exists in amounts equivalent to antimony. This is substantiated by the identification of BiOCl according to X-ray diffraction outcomes.

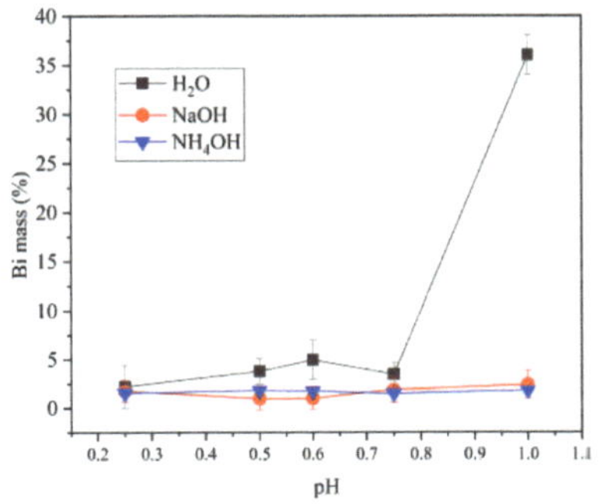

Figure 9. Evolution of the bismuth content with pH (bars represent standard deviation).

Figure 10 shows the percentage composition of arsenic in solid powders that were hydrolyzed with water, NaOH, and NH_4OH at various pH levels. The results indicate that at a pH of 0.25, the arsenic content in the powder obtained by hydrolysis with H_2O is lower than that obtained by using NaOH and NH_4OH (8.57%, 12.18%, and 23.51%, respectively). As the pH increased with time, a decrease in arsenic concentration occurred within the powders acquired by hydrolysis with water; however, there was no significant change for those produced using NaOH or NH_4OH solutions which remained constant throughout each analyzed test phase (pH = 0.75–10.36%, As-H_2O vs. 13.09%, As-NaOH vs. 22.06% As-NH_4OH). At pH 1, a decrease in solids obtained with H_2O and NH_4OH is observed.

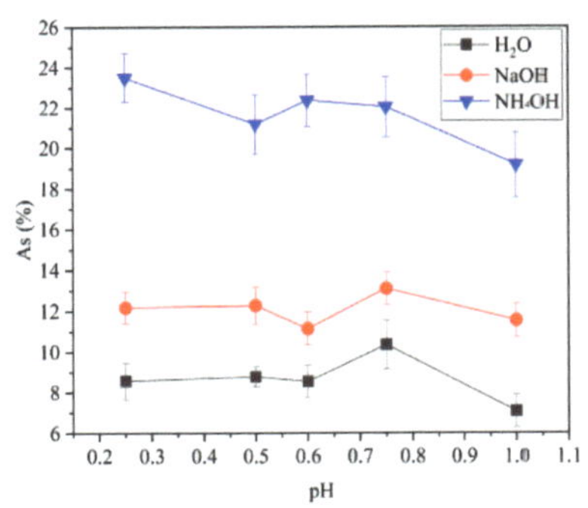

Figure 10. Evolution of arsenic content with pH (bars represent standard deviation).

4. Conclusions

1. Regarding antimony yield extraction, the results of this study demonstrate that the efficacy of antimony extraction by hydrolysis was higher when performed using NH_4OH or NaOH compared to water. Additionally, it was observed that pH values significantly impact the degree of success in the extraction of antimony. According to the results, the optimal conditions are obtained at pH 0.5 for water and pH 1 for NH_4OH and NaOH, where the average results of antimony extraction yield were 90.4, 96.1, and 96.7, respectively.
2. Most solids lack a crystalline structure, making it difficult to identify the nature of the compounds formed. Despite this limitation, element concentrations suggest that certain precipitates may contain oxychloride or oxide compounds.
3. Arsenic was generally incorporated into all solids obtained, which affects the purity of the product. In this sense, water is more selective, as the average antimony content

was higher (68.38%) and the arsenic values lower (8%) than in the case of NaOH and NH_4OH.

4. Regarding bismuth, its integration into solids was markedly inferior to arsenic (less than 2%) and remained unaffected by pH levels except in the water test where hydrolysis product of bismuth was identified at a pH of 1. This phenomenon accounts for the reduction observed in antimony extraction yields under such conditions.
5. Although the performance obtained with HN_4OH was higher than that of the other reagents, the excess arsenic that was incorporated into the solids hinders the potential commercialization of the product. Water is more selective, but the necessary amounts are much larger than those in the case of the other two reagents. NaOH offers the best compromise between performance and selectivity.

Author Contributions: Conceptualization, E.D.G. and A.P.E.; methodology, E.D.G. and J.M.G.F.; software, E.D.G.; validation, E.D.G., J.A.M.C. and A.P.E.; formal analysis, E.D.G.; investigation, E.D.G.; resources, E.D.G. and J.A.M.C.; data curation, E.D.G.; writing—original draft preparation, E.D.G.; writing—review and editing, A.P.E., J.M.G.F. and J.A.M.C.; visualization, J.A.M.C.; supervision, A.P.E., J.M.G.F. and J.A.M.C.; project administration, E.D.G. and A.P.E.; funding acquisition, A.P.E. and J.M.G.F. All authors have read and agreed to the published version of the manuscript.

Funding: This work was co-financed by the European Regional Development Fund (ERDF) through the Interreg V-A Spain-Portugal Program (POCTEP) 2014–2020.

Institutional Review Board Statement: Not applicable.

Informed Consent Statement: Not applicable.

Data Availability Statement: The data presented in this study are available upon request from the corresponding author.

Acknowledgments: The authors of this work would like to thank Atlantic Copper for providing the samples for study and allowing the dissemination of the results.

Conflicts of Interest: The authors declare no conflict of interest.

References

1. Reuter, M.A.; van Schaik, A.; Gutzmer, J.; Bartie, N.; Abadías-Llamas, A. Challenges of the Circular Economy: A Material, Metallurgical, and Product Design Perspective. *Annu. Rev. Mater. Res.* **2019**, *49*, 253–274. [CrossRef]
2. EU Commission. The Raw Materials Initiative—Meeting Our Critical Needs for Growth and Jobs. 2008. Available online: https://ec.europa.eu/growth/sectors/raw-materials/policy-strategy_en (accessed on 10 March 2023).
3. Blengini, G.A.; El Latunussa, C.; Eynard, U.; Torres De Matos, C.; Wittmer, D.; Georgitzikis, K.; Pavel, C.; Carrara, S.; Mancini, L.; Unguru, M.; et al. Study on the EU's list of Critical Raw Materials (2020) Final Report. 2020. Available online: https://op.europa.eu/en/publication-detail/-/publication/c0d5292a-ee54-11ea-991b-01aa75ed71a1/language-en (accessed on 15 March 2023).
4. Lewicka, E.; Guzik, K.; Galos, K. On the Possibilities of Critical Raw Materials Production from the EU's Primary Sources. *Resources* **2021**, *10*, 50. [CrossRef]
5. Ruiz, I.; Rios, G.; Arbizu, C.; Burke, I.; Hanschke, U. Pilot tests on Bismuth and Antimony removal from electrolyte at Atlantic Copper refinery. In *Proceedings of European Metallurgy Congress 2013*; GDMB: Clausthal-Zellerfeld, Germany, 2013; ISBN 978-3-940276-52-0.
6. Arroyo-Torralvo, F.; Rodríguez-Almansa, A.; Ruiz, I.; González, I.; Ríos, G.; Fernández-Pereira, C.; Vilches-Arenas, L. Optimizing operating conditions in an ion-exchange column treatment applied to the removal of Sb and Bi impurities from an electrolyte of a copper electro-refining plant. *Hydrometallurgy* **2017**, *171*, 285–297. [CrossRef]
7. Anderson, C. Antimony Production and Commodities. In *SME Mineral Processing and Extractive Metallurgy Handbook*; Society for Mining, Metallurgy and Exploration: Littleton, CO, USA, 2019; ISBN 9780873353854.
8. Klochko, K. Antimony Data Sheet—Mineral Commodity Summaries 2021, No. 703, pp. 2020–2021, 2021. Available online: https://pubs.usgs.gov/periodicals/mcs2021/mcs2021-antimony.pdf (accessed on 1 March 2023).
9. Xie, J.; Pei, Y.; Liu, L.; Guo, S.; Xia, J.; Li, M.; Ouyang, Y.; Zhang, X.; Wang, X. Hydrothermal synthesis of antimony oxychlorides submicron rods as anode materials for lithium-ion batteries and sodium-ion batteries. *Electrochimica Acta* **2017**, *254*, 246–254. [CrossRef]
10. Tan, C.L.; Mohseni, H. Emerging technologies for high performance infrared detectors. *Nanophotonics* **2017**, *7*, 169–197. [CrossRef]

11. Jin, Z.; Zhang, Z.; Xiu, J.; Song, H.; Gatti, T.; He, Z. A critical review on bismuth and antimony halide based perovskites and their derivatives for photovoltaic applications: Recent advances and challenges. *J. Mater. Chem. A* **2020**, *8*, 16166–16188. [CrossRef]
12. Nie, R.; Hu, M.; Risqi, A.M.; Li, Z.; Seok, S.I. Efficient and Stable Antimony Selenoiodide Solar Cells. *Adv. Sci.* **2021**, *8*, 2003172. [CrossRef]
13. Ling, H.; Malfliet, A.; Blanpain, B.; Guo, M. A review of the technologies for antimony recovery from refractory ores and metallurgical residues. *Miner. Process. Extr. Met. Rev.* **2022**, 1–25. [CrossRef]
14. Dembele, S.; Akcil, A.; Panda, S. Technological trends, emerging applications and metallurgical strategies in antimony recovery from stibnite. *Miner. Eng.* **2021**, *175*, 107304. [CrossRef]
15. Dupont, D.; Arnout, S.; Jones, P.T.; Binnemans, K. Antimony Recovery from End-of-Life Products and Industrial Process Residues: A Critical Review. *J. Sustain. Met.* **2016**, *2*, 79–103. [CrossRef]
16. Anderson, C.G. The metallurgy of antimony. *Geochemistry* **2012**, *72*, 3–8. [CrossRef]
17. Speight, J.G. Hydrolysis. *React. Mech. Environ. Eng.* **2018**, 203–229. [CrossRef]
18. Ruirong, Z.; Xichang, S.; Hanying, J. Thermodynamic equilibrium of sb-cl-h2o system®. *Trans. Nonferrous Met. Soc. China* **1997**, *7*, 127–128.
19. Tian, Q.-H.; Xin, Y.-T.; Yang, L.; Wang, X.-H.; Guo, X.-Y. Theoretical simulation and experimental study of hydrolysis separation of SbCl3 in complexation–precipitation system. *Trans. Nonferrous Met. Soc. China* **2016**, *26*, 2746–2753. [CrossRef]
20. Hashimoto, H.; Nishimura, T.; Umetsu, Y. Hydrolysis of Antimony(III)-Hydrochloric Acid Solution at 25 °C. *Mater. Trans.* **2003**, *44*, 1624–1629. [CrossRef]
21. Ye, L.; Ouyang, Z.; Chen, Y.; Wang, H.; Xiao, L.; Liu, S. Selective separation of antimony from a Sb-Fe mixed solution by hydrolysis and application in the hydrometallurgical process of antimony extraction. *Sep. Purif. Technol.* **2019**, *228*, 115753. [CrossRef]
22. Chen, X.Y.; Huh, H.S.; Lee, S.W. Hydrothermal synthesis of antimony oxychloride and oxide nanocrystals: $Sb_4O_5Cl_2$, $Sb_8O_{11}Cl_2$, and Sb_2O_3. *J. Solid State Chem.* **2008**, *181*, 2127–2132. [CrossRef]
23. Palden, T.; Machiels, L.; Regadío, M.; Binnemans, K. Antimony Recovery from Lead-Rich Dross of Lead Smelter and Conversion into Antimony Oxide Chloride ($Sb_4O_5Cl_2$). *ACS Sustain. Chem. Eng.* **2021**, *9*, 5074–5084. [CrossRef]
24. Torres, A.I.G.D.L.; Moats, M.S.; Ríos, G.; Alamansa, A.R.; Sánchez-Rodas, D. Arsenic and antimony speciation analysis in copper electrolyte by liquid chromatography coupled to hydride generation atomic fluorescence spectrometry (HPLC-HG-AFS). *Anal. Methods* **2020**, *12*, 1943–1948. [CrossRef]
25. Riveros, P.A. The removal of antimony from copper electrolytes using amino-phosphonic resins: Improving the elution of pentavalent antimony. *Hydrometallurgy* **2010**, *105*, 110–114. [CrossRef]
26. Smith, R.M.; Martell, A.E. Critical stability constant. *Inorg. Complexes* **1976**, *4*. [CrossRef]
27. Kim, D.; Wang, S. Bismuth Recovery from Hydrochloric Acid Solution. *Can. Met. Q.* **2008**, *47*, 317–326. [CrossRef]
28. Chunfa, L.; Zhenxin, X.; Jianbai, Z.; Pinguoo, J. Hydrolysis Mechanism of Bismuth in Chlorine Salt System Calculated by Density Functional Method. *Rev. de Chim.* **2020**, *71*, 178–193. [CrossRef]
29. Tooth, B. The Hydrothermal Chemistry of Bismuth and The Liquid Bismuth Collector Model. Ph.D. Thesis, University of Adelaide, Adelaide, Australia, 2013. Available online: www.lab-initio.com (accessed on 7 March 2023).
30. Kartinen, E.O.; Martin, C.J. An overview of arsenic removal processes. *Desalination* **1995**, *103*, 79–88. [CrossRef]
31. Beauchemin, S.; Chen, T.; Dutrizac, J. Behaviour of Antimony and Bismuth in Copper Electrorefining Circuits. *Can. Met. Q.* **2008**, *47*, 9–26. [CrossRef]

MDPI AG
Grosspeteranlage 5
4052 Basel
Switzerland
Tel.: +41 61 683 77 34

MDPI Books Editorial Office
E-mail: books@mdpi.com
www.mdpi.com/books

www.ingramcontent.com/pod-product-compliance
Lightning Source LLC
Chambersburg PA
CBHW041409210926
51850CB00013B/3219
9783725838394